HANDBOOK FOR

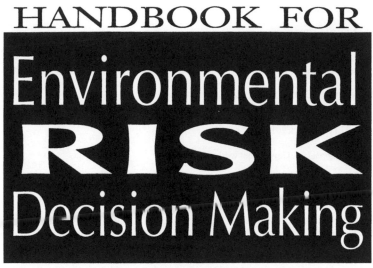

Environmental RISK Decision Making

VALUES, PERCEPTIONS, & ETHICS

C. RICHARD COTHERN, Ph.D.

Center for Environmental Statistics
Development Staff
Environmental Statistics and Information Division
U.S. Environmental Protection Agency
Washington, D.C.

CRC Press
Taylor & Francis Group
Boca Raton London New York

CRC Press is an imprint of the
Taylor & Francis Group, an **informa** business

T0175131

First published 1996 by Lewis Publishers

Published 2019 by CRC Press
Taylor & Francis Group
6000 Broken Sound Parkway NW, Suite 300
Boca Raton, FL 33487-2742

© 1996 by Taylor & Francis Group, LLC
CRC Press is an imprint of Taylor & Francis Group, an Informa business

First issued in paperback 2019

No claim to original U.S. Government works

ISBN-13: 978-0-367-45597-2 (pbk)
ISBN-13: 978-1-56670-131-0 (hbk)

Visit the Taylor & Francis Web site at
http://www.taylorandfrancis.com

and the CRC Press Web site at
http://www.crcpress.com

Library of Congress Cataloging-in-Publication Data

Cothern, C. Richard.
   Handbook for environmental risk decision making: values,
perceptions, and ethics / C. Richard Cothern.
     p.   cm.
   Includes bibliographical references and index.
   ISBN 1-56670-131-7 (permanent paper)
   1. Environmental risk assessment--Congresses. 2. Environmental
policy--Decision making--Congresses. 3. Environmental ethics--Congresses.
4. Values--Congresses. I. Title.
GE145.C68   1995
363.7′0068′4--dc20   95-16857
                                                     CIP

Library of Congress Card Number 95-16857

PREFACE

A one-day symposium on "Environmental Risk Decision Making: Values, Perceptions and Ethics" was held by the Environmental Division at the National Meeting of the American Chemical Society in Washington, D.C., August 24, 1994. The symposium consisted of 2 keynote speakers and 14 following presentations. The papers presented are combined with eight others to flesh out the topics for this volume.

WHAT DO VALUES AND ETHICS HAVE TO DO WITH ENVIRONMENTAL RISK DECISION MAKING?

Values and ethics should be included in the environmental decision-making process for three reasons: they are already a major component, although unacknowledged; ignoring them causes almost insurmountable difficulties in risk communication; and because it is the right thing to do.

Values and value judgments pervade the process of risk assessment, risk management, and risk communication as major factors in environmental risk decision making. Almost every step in any assessment involves values and value judgments. However, it is seldom acknowledged that they even play a role. The very selection of methodology for decision making involves a value judgment. The selection of which contaminants to study and analyze involve value judgments. Weighing different risks involves value judgments. We cannot, and should not, exclude values and value judgments from the environmental decision-making process as they are fundamental to understanding the political nature of regulation and decisions that involve environmental health for humans and all living things.

One of the major problems in risk communication is the failure of different groups to listen to each other. For example, many animal rights groups object to the use of animals in toxicological testing on ethical and moral grounds. The AMA and other scientific groups have mounted a response that argues that many human lives have been saved (life lengthened) by information gained from animal testing. Both sides have a point, but neither is listening to the other. These represent two different value judgments and these values are the driving force in the different groups. It is essential to understand this and include it in any analysis that hopes to contribute to understanding in this area. Any analysis must include values such as safety, equity, fairness, and justice — as well as feelings such as fear, anger, and helplessness. These values and feelings are often the major factor in effectively communicating about an environmental problem.

Lastly, including values such as justice, fairness, and equity (present and intergenerational) is the right thing to do. Any effective environmental program needs to be ethical to survive in the long term.

ENVIRONMENTAL RISK DECISION MODELS

The existing models for environmental risk assessment do not contain any explicit mention of values, value judgments, ethics, or perceptions. However, these are often the main bases used in making such decisions.

For example:

- Alar was banned to protect *children*.
- The linear, no-threshold dose response curve and the use of combined upper 95% confidence limits are based on *safety* not science.
- The Superfund program started with the idea that if I can sense it, it must be bad, while indoor radon has met with widespread apathy because it cannot be sensed, so why worry?
- The idea of zero discharge is based on the *sanctity of the individual*.
- Forests and wetlands are preserved because of *stewardship*.
- Nuclear power is avoided because of fear of *catastrophe*.

The general theme of the symposium was to examine the place of values, value judgments, ethics, and perceptions in decision models. The hypothesis is that these characteristics are directly involved in current risk decisions, but that existing models do not include them. In some decisions, attempts are made to disguise these characteristics of values and ethics with other labels such as "scientific" or "technical". Values and ethics seem like perfectly good ways to analyze, balance, and choose in the environmental risk decision-making process and since they are widely used, why not acknowledge this and formally include them in the models?

Are the current and future environmental problems and decisions more complex and of a different character that those of the past? If so, then a new decision paradigm will be needed. Some have observed that the current environmental problems are characterized by levels of complexity and uncertainty never before experienced by any society.

GOAL AND OBJECTIVES OF THE SYMPOSIUM

The goal of this volume is to examine the place values and value judgments have in the process of environmental risk decision making.

Broadly stated, there are three major objectives: viz., bring together the disparate groups that are and have been working in this area; develop a model of environmental risk decision making that includes values, perceptions, and ethics; and develop an environmental ethic.

- To bring together disparate groups to share thoughts and biases concerning the role of values in environmental risk decision making — a partial list is shown below:

  - Ethicists
  - Decision makers
  - Risk assessors
  - Economists
  - Scientists
  - Philosophers
  - Journalists
  - Theologians
  - Attorneys
  - Policy makers
  - Environmentalists
  - Regulators

- To develop a model that describes how the participants think environmental risk decision making should be conducted. This process involves several components:

  1. To explore the involvement of values and value judgments in the development of risk assessments, cost assessments, and feasibility studies
  2. To examine current environmental decisions to determine the role values and value judgments play in the process
  3. To develop approaches and methodologies that can involve the so-called objective and subjective elements into a balanced process for making environmental risk decisions
  4. Looking for what the options are, determine how to balance all the components of decision making and to be explicit about the values, perceptions and ethics

- To promote the development of an environmental ethic

One overall objective is to use the value of honesty and ask that the values, value judgments, and ethical considerations used in environmental risk decisions be expressed and discussed. To a scientist, Brownowski's comment, "Truth in science is like Everest, an ordering of the facts", is a most important value.

It is a conclusion of this line of thinking that we should unmask the use of values in environmental decisions and challenge decision makers to clearly state how they are using values.

SUMMARY

The summary presentation of the symposium consisted of three propositions and four recommendations. The strong versions of the propositions are representative of the views of many of the participants, while the weaker versions would be shared by only some of the participants.

The first proposition in strong form is that all facets of risk assessment are value laden. A weaker version of this is that risk assessment is socially constructed and thus depends on the context.

The strong version of the second proposition is that public values are relevant in standard setting. A weaker version of this proposition is that public values should trump scientific value when there is a conflict.

For the third proposition, the strong version is that risk assessment is an appropriate aid in spite of the deficiencies, while the weaker version is that we should make more use of it.

The four recommendations that emerged are

1. More attention needs to be given to the definition of values and ethics in risk assessment.
2. Given the overconfidence that we have in risk assessment, we need more humility.
3. Mistrust is one of the more serious problems that needs to be addressed.
4. Stop bashing the media and lawyers — there is enough blame to go around.

C. Richard Cothern
Chevy Chase, Maryland

COMMENTS FROM MY CO-ORGANIZER, PAUL A. REBERS

These last paragraphs in the preface are comments from the other organizer of the symposium on which this volume is based. Paul A. Rebers was not only a co-organizer of the symposium, he was the original source of the idea.

My contribution to this book is dedicated to my parents, who taught me ethics; and to Dr. Fred Smith and Dr. Michael Heidelberger who taught me the value of, and the necessity of, an ethical code in order to do good research. There can be no substitute for good mentors in and after college. After I had earned my Ph.D., Dr. Heidelberger taught me to do the "Heidelberger Control", i.e., in order to be more certain of the results, to do one more control, and to repeat the experiment. Dr. Richard Cothern helped me realize the need for looking at the broad picture in making environmental risk assessments.

This symposium was concerned with how values, ethics, and perceptions impact on the making of environmental risk assessments. Ethics were touched on in a previous symposium presented at the ACS national meeting in Boston in 1990 entitled, "Ethical Dilemmas of Chemists", which I organized, and was a basis for the present symposium and book.

If we can recognize that values, ethics, and perceptions, as well as scientific data enter into the process of environmental risk decision making, we will have made an important step forward. This should make it easier for the public to understand how difficult and indeterminate the process may be. It should also make them demand to know the biases as well as the expertise of those making decisions. By being completely honest with the media and the public, we are making an important step in gaining their confidence, and I hope this can be done more in the future than it has been done in the past.

The Editor

C. Richard Cothern, Ph.D., is presently with the U.S. Environmental Protection Agency's Center for Environmental Statistics Development Staff. He has served as the Executive Secretary of the Science Advisory Board at the U.S. EPA and as their National Expert on Radioactivity and Risk Assessment in the Office of Drinking Water. In addition, he is a Professor of Management and Technology at the University of Maryland's University College and an Associate Professorial Lecturer in the Chemistry Department of the George Washington University. Dr. Cothern has authored over 80 scientific articles including many related to public health, the environment, and risk assessment. He has written and edited 14 books, including such diverse topics as science and society, energy and the environment, trace substances in environmental health, lead bioavailability, environmental arsenic, environmental statistics and forecasting, risk assessment, and radon and radionuclides in drinking water. He received his B.A. from Miami University (Ohio), his M.S. from Yale University, and his Ph.D. from the University of Manitoba.

Contributors

Richard N.L. Andrews
Department of Environmental
  Sciences and Engineering
University of North Carolina
Chapel Hill, North Carolina

Jeffrey Arnold
Department of Environmental
  Sciences and Engineering
University of North Carolina
Chapel Hill, North Carolina

Scott R. Baker
Director, Health Sciences Group
EA Engineering, Science and
  Technology, Inc.
Silver Spring, Maryland

Lawrence G. Boyer
Department of Public
  Administration
The George Washington University
Washington, D.C.

Donald A. Brown
Director
Department of Environmental
  Resources
Bureau of Hazardous Sites
Commonwealth of Pennsylvania
Harrisburg, Pennsylvania

Thomas A. Burke
School of Hygiene and Public
  Health
Johns Hopkins University
Baltimore, Maryland

Bayard L. Catron
Department of Public
  Administration
The George Washington University
Washington, D.C.

Victor Cohn
American Statistical Association
Washington, D.C.

William Cooper
Biology Department
Michigan State University
East Lansing, Michigan

C. Richard Cothern
Center for Environmental Statistics
Development Staff
Environmental Statistics and
  Information Division
U.S. Environmental Protection
  Agency
Washington, D.C.

Douglas J. Crawford-Brown
Institute for Environmental Studies
Department of Environmental
  Sciences and Engineering
University of North Carolina
Chapel Hill, North Carolina

William R. Freudenburg
Department of Rural Sociology
University of Wisconsin
Madison, Wisconsin

Jennifer Grund
PRC Environmental
  Management, Inc.
McLean, Virginia

Rachelle D. Hollander
National Science Foundation
Ethics and Values Studies Program
Arlington, Virginia

P. J. (Bert) Hakkinen
Senior Scientist, Toxicology and
  Risk Assessment
Paper Product Development and
  Paper Technology Divisions
The Procter & Gamble Company
Cincinnati, Ohio

John Hartung
Office of Policy Development
U.S. Department of Housing and
  Urban Development
Washington, D.C.

Carolyn J. Leep
Chemical Manufacturers
  Association
Washington, D.C.

Douglas MacLean
Department of Philosophy
University of Maryland
Baltimore, Maryland

Hon. Mike McCormack
The Institute for Science and
  Society
Ellensburg, Washington

James A. Nash
The Churches' Center for Theology
  and Public Policy
Washington, D.C.

Bryan G. Norton
School of Public Policy
Georgia Institute of Technology
Atlanta, Georgia

Christopher J. Paterson
Northeast Center of Comparative
  Risk
Vermont Law School
South Royalton, Vermont

Van R. Potter
Department of Oncology
University of Wisconsin
Madison, Wisconsin

Resha M. Putzrath
Georgetown Risk Group
Washington, D.C.

David W. Schnare
Office of Enforcement and
  Compliance Assurance
U.S. Environmental Protection
  Agency
Washington, D.C.

Virginia A. Sharpe
Departments of Medicine and
  Philosophy
Georgetown University
Washington, D.C.

Kristin Shrader-Frechette
Environmental Sciences
  and Policy Program and
  Department of Philosophy
University of South Florida
Tampa, Florida

Table of Contents

Section I: INTRODUCTION

1. Values and Value Judgments
   in Ecological Health Assessments   3
   William Cooper

2. Strange Chemistry: Environmental Risk Conflicts
   in a World of Science, Values, and Blind Spots   11
   William R. Freudenburg

Section II: ISSUES IN ENVIRONMENTAL
   RISK DECISION MAKING   37

3. An Overview of Environmental Risk
   Decision Making: Values, Perceptions, and Ethics   39
   C. Richard Cothern

4. Introduction to Issues in
   Environmental Risk Decision Making   71
   Scott R. Baker

5. Industry's Use of Risk, Values,
   Perceptions, and Ethics in Decision Making   73
   P. J. (Bert) Hakkinen and Carolyn J. Leep

6. Regulating and Managing Risk:
   Impact of Subjectivity on Objectivity   83
   Scott R. Baker

7. Back to the Future: Rediscovering the Role of
   Public Health in Environmental Decision Making   93
   Thomas A. Burke

8. Telling the Public the Facts —
   or the Probable Facts — About Risks   103
   Victor Cohn

9. The Urgent Need to Integrate Ethical
   Considerations into Risk Assessment Procedures   115
   Donald A. Brown

10. The Problem of Intergenerational Equity: Balancing
    Risks, Costs, and Benefits Fairly Across Generations   131
    Bayard L. Catron, Lawrence G. Boyer, Jennifer Grund,
    and John Hartung

Section III: VALUES AND VALUE JUDGMENTS   149

11. Introduction to Quantitative Issues   151
    David W. Schnare

12. Ecological Risk Assessment:
    Toward a Broader Analytic Framework   155
    Bryan G. Norton

13. Environmental Ethics and Human Values   177
    Douglas MacLean

14. Moral Values in Risk Decisions   195
    James A. Nash

15. Values and Comparative Risk Assessment   213
    Christopher J. Paterson and Richard N.L. Andrews

16. Risk and Rationality in Decision Making:
    Exposing the Underlying Values Used
    When Confronted by Analytical Uncertainties   227
    David W. Schnare

17. Comparing Apples and Oranges:
    Combining Data on Value Judgments   245
    Resha M. Putzrath

18. The Ethical Basis of Environmental Risk Analysis   255
    Douglas J. Crawford-Brown

19. Ethical Theory and the Demands of Sustainability   267
    Virginia A. Sharpe

20. The Cardinal Virtues of Risk Analysis: Science
    at the Intersection of Ethics, Rationality, and Culture   279
    Douglas J. Crawford-Brown and Jeffrey Arnold

21. Value Judgments Involved in Verifying
    and Validating Risk Assessment Models   291
    Kristin Shrader-Frechette

22. The Stewardship Ethic —
    Resolving the Environmental Dilemma   311
    David W. Schnare

Section IV: COMMENTARY   333

23. Introduction to the Commentary Section   335
    C. Richard Cothern

24. Awakenings to Risk in the Federal
    Research and Development Establishment   337
    Rachelle D. Hollander

25. The Citizenship Responsibilities of Chemists   349
    Hon. Mike McCormack

26. Global Bioethics: Origin and Development   359
    Van R. Potter

Section V: SUMMARY   375

27. Ethics and Values in Environmental
    Risk Assessment — A Synthesis   377
    Bayard L. Catron

The Contributors   383

Index   393

Dedication

To Ellen Grace, Hannah Elizabeth, and all future generations we pass on the torch of attention to the impact of values, perceptions, and ethics in life's decision making.

SECTION I
Introduction

1 VALUES AND VALUE JUDGMENTS IN ECOLOGICAL HEALTH ASSESSMENTS

William Cooper

CONTENTS

Introduction 3
Values in Ecological Risk Assessments 4
Values in Other Risk Assessments 7
Value Trade-Offs 9
Conclusion 10
References 10

INTRODUCTION

It is unusual to start a symposium and this volume[1] on the subject of values and ethics with the views of an ecologist. Doing so is a conscious value judgment. The importance of values and ethics is seldom discussed during ecological debates. This discussion is an empirical observation about the way judgments are made and on how data are weighted when we make decisions involving ecological health.

My background involves, among other things, serving for 14 years on a state environmental review board that ran public hearings for federal and state governments concerning environmental impact statements. You really get an education in a meeting with 200 people who are madder than hell. I also chaired the Ecology and Welfare panel on the Reducing Risk project[2] for William Riley, the Administrator of the U.S. Environmental Protection Agency. The panel had to rank ecological and welfare effects (dollars and "dickey birds"[3]) to show where the priorities are and where one can get the biggest bang for the buck in environmental problem setting.

1-56670-131-7/96/$0.00+$.50

In these experiences, I had to deal with values. Any time you confront issues like risk assessment, land use, and wetlands, everyone has their own perceptions as to what is important or unimportant. They bring different backgrounds, experiences and biases to this pragmatic and empirical surrounding and provide a realistic forum to examine the importance of values.

VALUES IN ECOLOGICAL RISK ASSESSMENTS

Ecologists think differently from everyone else. We have our own mind set of how one makes optimal trade-offs of ecological criteria based largely on evolution. The only criterion for good or bad is whether the gene pools survive to have more offspring. Everything else as a value is secondary. A quick empirical example of how natural selection determines "ecological correctness" is the following. A couple of years ago there were some whales caught in the ice off Alaska and every night on TV for about 6 nights there were stories of the whales and the millions of dollars being spent trying to free them. This approach set marine biology back about 100 years in its lack of common sense. About halfway through that week, I got a phone call from our local newspaper reporter requesting a quote. I replied, "Any whales caught up in the ice are so dumb that you don't want them to have kids. That is how population models work. You do not support the continuation of that kind of maladapted behavior and the best thing to do is to stop that line."

There is a value judgment in comparing ecological and human health — which is more important? This relationship is not competitive, it is complimentary. In real life you do not have a choice — if you want to maintain a high level of human health you must invest in the environment. For example, look at Eastern Europe. It is a false argument to play ecological health and human health against each other. The first thing you must deal with in addressing environmental issues is social perception — the values involved. Often folks do not want to hear what the scientist has to say. Too many have a romantic idea of how nature should work — they envision Bambi, Flipper, Smokey the Bear, and a warm and fuzzy Walt Disney love-in. In nature just the opposite is the case. Ecological systems are very harsh, abrupt, and chaotic. There are high mortality rates. This is a completely different perception than that in the average public mind. Take the case of Smokey the Bear, a value in the Forest Service for three generations. Any school kid will tell you that Smokey is a good guy. However, with the forest fires in the West this season you need to re-educate a whole population that to keep the forest landscape safe includes natural events like fires. Fire is a natural and central need in landscape management. There is a difference in perceptions. I am not sure I can translate these values and perceptions into quantitative parameters, but I understand the distinction.

Table 1 Ecological Rankings

High
Habitat alteration/destruction
Species extinction
Stratospheric ozone depletion
Global climate change

Medium
Herbicides and pesticides
Toxics, nutrients, BOD in water
Acid deposition
Airborne toxics

Low
Oil spills
Groundwater pollution
Radionuclides
Acid runoff
Thermal pollution

From Reducing Risk: Setting Priorities and
Strategies for Environmental Protection,
U.S. EPA, Science Advisory Board, SAB-
EC-90-021, September, 1990.

Perceptions become critical when we try to combine social values, land use, economic issues, property rights, expectations, and all the human values to determine trade-offs between ecological balance and social opportunity. It is easier for the ecologist to rationally balance ecological rankings among the four most important problem areas in ecology: global warming, stratospheric ozone, biological diversity, and loss of habitat (see Table 1). The panel on ecology and welfare for the Reducing Risk project concluded that these were more important than groundwater and Superfund problems. This conclusion is a value judgment itself.

The human health panel was asked to rank contaminants, carcinogens, and noncarcinogens, but finally this group of eminent scientists decided that they could not agree concerning the relative rankings. It is the ethical structure of medicine that does not allow them to say what is more important to society — how to compare a 75-year-old dying of cancer or an urban child of 6 months who is mentally retarded because of ingested or inhaled lead. There was no way they could find to rank these — and so they did not. We ranked our ecology problems with no ethical problems whatsoever. In estimating the value of an eagle I did not have to worry whether it was happy or someone had violated its individual rights. In population assessments, the fate of an individual is irrelevant. You might have noted recently in the *New York Times* the article in the science section on hormonal copycats — the estrogenic compounds. Incidences of alligators with small penises, men with low sperm counts, and diseases in populations of

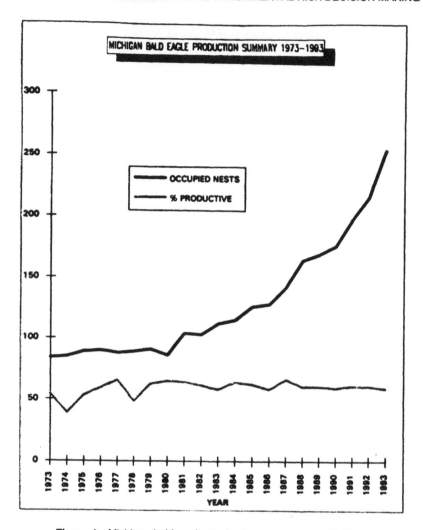

Figure 1   Michigan bald eagle production summary, 1973–1993.

wildlife in the Great Lakes were attributed to exposures to toxic substances. The compounds involved are PCBs, dioxins, furans, chlorinated pesticides, and a few others. If you actually look at the data of eagle populations in the Great Lakes it shows that the populations have been growing exponentially since the 1970s (see Figure 1). There is no indication of ecological impact whatsoever at the level of the population. It is absolutely true that you can go out and find hot spots such as Saginaw Bay and some areas in Wisconsin where there are very large amounts of chlorine-containing compounds in the sediments, and you can find eagle populations where there are deformed chicks — one or two here and there. Thus, there is evidence that individual eagles are impacted. Again, one does not have to protect and/or value individual survivorship in order to protect and preserve the ecological resource.

VALUES IN OTHER RISK ASSESSMENTS

There is a big debate going on right now proposing the virtual elimination of all chlorine-containing compounds. The idea is to eliminate most chlorine-containing compounds in feedstock and products in all the Great Lakes states and the entire U.S. This proposal is based on the idea that these compounds are affecting eagles, Beluga whales, mink, and everything else that has decreased in population in the last 20 years. It is guilt by temporal association with these chlorine-containing compounds. Can you imagine banning an element in the periodic table based on that kind of data? This is a value judgment and involves trade-offs. Even if there has been such an impact, so what? Do you know what the impact would be of banning the majority of the chlorine-containing compounds in pharmaceuticals, insecticides, and the chlorination of water? Even if you can show effects, one of the problems is, at what level? Is this not an extreme example of overemphasizing pristine ecological conditions? Any analysis of chlorine compounds and their ecological health effects involves risk assessment. When we first got into the area of ecological risk assessment, the thought was to take the National Academy of Sciences's "red book"[4] and paint it green. For example, generate dose-response curves for eagles and other wildlife animals for expected individual exposures. That is exactly what the Fish and Wildlife Service is doing. Many conservation groups argue that the goal of risk assessment is to get the level of permissible concentrations as low as you can get them, as close to zero as possible, no matter what the costs are. If you are a conservationist and your major value in life is to protect "dickey birds" you might take the same approach. However, from a scientific view this is wrong. It is wrong because you cannot protect an individual in the wildlife population. The human health risk assessment model puts a value on an individual life; but in nature, the individual is expendable. The only thing of value in an ecological population model is the perpetuation of the gene pool. The major mechanism is called natural selection. The Endangered Species Act does not protect individuals — it protects habitat. If you protect the habitat, the populations will take care of themselves. You cannot legislate that an individual baby eagle must survive. You can argue that the Endangered Species Act protects the individual and use this as an excuse to crank down the ambient concentrations as low as you can get them, but ecological science does not justify that kind of draconian measure. This argument represents the value judgment of whether or not one makes the decision based only on science.

Back when life was simple, and chemists weighed things in milligrams, everything was black and white, dirty or clean. We passed the Delaney amendment stating that if a contaminant is a known or suspect carcinogen there cannot be one molecule of it in a food additive. That was fine when we were measuring things in milligrams, because a value below the detection limit was assumed to be zero. The problem is that we spent all kinds of money to push detection limits orders of magnitude lower, and in the case of 2,3,7,8-dioxin (dichloro-p-dibenzodioxin) the detection level is down to ten to the minus sixteenth (10^{-16}). Now, what does below detection limit mean? The question is one of values and what is an acceptable risk has become more complicated. The normal kind of command-and-control EPA regulations involve setting a number for the discharge to the air or water or whatever so that there is a certain level of protection. This level is of the order of 10^{-5} to 10^{-6} in a lifetime of exposure. As an example, PCBs have an action level of 2 parts per million in fish (if you eat 6.5 grams per day, 365 days per year for 72 years, the probability of dying from exposure to this compound at this concentration is about 1 in 100,000). You can look at that number as a level of protection that EPA and FDA have guaranteed the public. Or you can view this number as a license to kill — the permitting of a discharge knowing that it will be accumulated in the fish population and that people will die at the level of 1 in 100,000. Is the cup half empty or half full? This level is considered a safe number and it can be interpreted either way. The choice is a value judgment.

The idea that we are safe when exposed to levels below a standard is one option. Another is to mandate zero discharge or nondetectable discharge, no matter what the cost is. If you accept this as a level of protection, then when do you stop spending money on the mitigation and spend it on alternate social problems like schools, housing, and the poor? You are out in the real world and if you value human life, then perhaps it is better to focus on problems like alcohol, tobacco, diet, and automobiles. Here the mortality rates are in the range of 10^{-2} and 10^{-3} where the environmental regulation is in the range of 10^{-5} and 10^{-6}. Where do your values and priorities lie?

Involved here are trade-offs in comparative risk assessment and balancing values. Some difficulties include different endpoints and different conclusions on the same set of data, all depending on our values. What you will find out when you do risk assessment for human health is that it is halfway between black magic and a ouija board, and for ecological risk the problem is much worse. It basically comes down to common sense and best professional judgment.

For many of the multitude of environmental issues you cannot put probabilities on the events. I cannot draw a dose–response curve for global warming, stratospheric ozone, loss of habitat, or species loss, all the kind of things that ecologists worry about. These major issues are not probabilistic or intensity dose–response curves. They are scenarios or events; they are occurrences. Almost all of these kinds of scenarios begin with a little bit of data, a few simulation models, maybe some regression lines, extrapolations in time and space for anything you can measure, and then you have to fill in the gaps with good old professional best judgment.

Sometimes lawyers and juries have problems with too much professional judgment and not enough hard data. They say that there is too much uncertainty and they cannot make a decision. However, decisions will be made whether you like it or not. Either scientists belly up to the bar and contribute their own professional judgment, or others will make the decisions without the constraints of thermodynamics, evolution, or calculus. The fallback position is that economists and lawyers will make the decisions, and they are not constrained by the laws of science.

You cannot print money fast enough to solve all the environmental problems to the level of zero risk simultaneously. The bottom line is that our society is like a bunch of spoiled brats who want an affluent lifestyle based on a throwaway society, supplied by synthetic chemistry and risk free at the same time. How are you going to do that?

VALUE TRADE-OFFS

Having a certain level of risk, a nonzero risk, involves cost. That is the kind of balancing act that we are involved in right now, and if you expect either extreme you will have trouble. A couple of quick scenario examples of what I mean by trade-offs follow.

Remember back in the late 1950s Lake Erie was supposed to be dying? We look back with hindsight, and realize that our freshwater lakes in the U.S. are usually phosphate limited. In a phosphate-limited lake, green algae dominate, not blue-green algae. The zooplankton will eat the green algae and, therefore, Lake Erie had a grazing food chain in the 1930s and 1940s. Lake Erie had one of the highest social values of any lake in North America. It was one of the more popular sport fishing lakes in the country — very productive, very shallow, and very warm, with walleyed pike, yellow perch, northern pike, white perch, blue pike — a tremendous resource. What happened over a 20-year period is that we loaded the lake with more phosphorus from cities, farms, and industry than nitrogen. Before phosphates were the limiting factor, and now nitrates were the limiting factor. This shifted the competitive balance of the algae and led to the production of blue-green algae. They can fix their own atmospheric nitrogen, so they have an independent reservoir. This shifted the whole base of the food chain over to blue-green algae which are generally inedible. The lakes looked yellow-green and scummy, but this was a perfectly healthy blue-green algae population. Some of the algae died and settled to the bottom. The bacteria consumed this dead algae and took up the dissolved oxygen and the lake became anaerobic. The lake was then declared dead. All we did was to shift from a grazing food chain to a detritus food chain. We had, however, more fish in Lake Erie when it was dead than before. But they were carp, suckers, catfish, draum: fish that loved to eat garbage, not fish that liked to eat the green food chain. Now, if you stop and think about it, two thirds of the world eats carp that feed in sewage ponds. They do not eat beef and potatoes, because they are too poor. We have a value choice — walleyed pike or carp? If you want a walleyed pike lake and you get one, you are happy. If you want a carp pond and get one, you are happy. If you want a walleyed pike lake and get a carp pond, you call it polluted. You can operate an ecologically stable carp pond as easily as you can a walleyed pike pond. If you use the criteria of self-reproducing health stocks of fish and sufficient biodiversity, then Lake Erie would be considered good under either condition. The problem is that as ecologists we can operate landscape at many different shapes, sizes and biological forms, all of which are ecologically stable. The trade-off is not ecology, it is economics and social utility. In the same way, with most of the pollution events that have happened over the years, the impacts have not been ecological; they have been economic. Even with Kepone in the James River, all the basic stocks of fish, crabs, and oysters continue to exist. So there is a lot more flexibility in designing landscapes than people give ecology credit for. The problem is, who is going to make those decisions like making Lake Erie a carp pond or a walleye pike pond? That is a value judgment. Ecologically, Lake Erie can be operated either way.

Another example: back in the 1950s the sea lamprey virtually wiped out the lake trout in the Great Lakes. The fish stocks have been restored by decisions of the states and Canadian provinces. Michigan not only put back lake trout, but put back Coho, Atlantic, and other varieties of salmon. It is now a great big fish farm with a different array of species than were there before — a value judgment. It is now a $3.2-billion sport-fishing industry. In order to protect sport fishing, Michigan shut down commercial gill netting — another value judgment. This was done by political fiat. Here we are managing the largest lake system in the world using values and value judgments, not ecological reasons. Basically, we value sport fishing more than commercial fishing.

CONCLUSION

As we discuss values and value judgments, keep in mind that there is more ecological flexibility and redundancy than most people recognize. This does not mean that you can do anything you want; there are constraints. Be careful not to shoot yourself in the foot. The question is who determines the scenarios and chooses from the different options and on what basis.

REFERENCES

1. This chapter is adapted from the first keynote address at the symposium on "Environmental Risk Decision Making: Values, Perceptions and Ethics" held by the Environmental Division of the American Chemical Society at the National Meeting in Washington, D.C., August 24, 1994.
2. Reducing Risk: Setting Priorities and Strategies for Environmental Protection, U.S. Environmental Protection Agency, Science Advisory Board, SAB-EC-90-021, September 1990. This project involved three panels; viz., human health, ecology and welfare and strategic options.
3. The term "dickey bird" is used here to describe a generic wildlife creature that has aesthetic value and in that sense requires protection.
4. National Research Council, *Risk Assessment in the Federal Government, Managing the Process,* Washington, D.C., National Academy Press, 1983.

STRANGE CHEMISTRY: ENVIRONMENTAL RISK CONFLICTS IN A WORLD OF SCIENCE, VALUES, AND BLIND SPOTS

2

William R. Freudenburg

CONTENTS

Introduction 11
What Is the Sound of One Fact Speaking? 12
How Do We See What It Is We Do Not See? 14
What Is Going On Here? 19
    Can We Blame the Mass Media? 19
    Can We Blame the Public? 21
    Do We Need to Blame Ourselves? 21
Do Things Have to Work Out This Way? 24
What Can One Chemist Do? 27
Endnotes 31

INTRODUCTION

Science and technology have achieved many remarkable successes, but it would be difficult to argue that dealing well with the public should be counted among them. Increasingly, whether scientists and engineers are searching for oil offshore or attempting to dispose of nuclear or other wastes onshore, the efforts have become less likely to be welcomed with open arms than to open the public policy equivalent of armed warfare.

The underlying reasons involve a different kind of chemistry than is normally studied in a laboratory setting — a strange kind of interpersonal chemistry that often seems as exotic to scientists and engineers as the real chemical compounds can seem to members of the general public. While my presentation is intended to follow some of the best scientific traditions of the

1-56670-131-7/96/$0.00+$.50
© 1996 by CRC Press, Inc.

American Chemical Society, in which a brave explorer of hitherto unknown chemical reactions comes back with a report that is intended to clarify the mysteries of a scientific frontier, the main difference is that I will focus on a set of reactions so exotic and often so toxic, that most ACS members would not want to touch them with a 10-foot test tube.

Strictly speaking, this "new" kind of interpersonal chemistry is not exactly new. In the past, however, it could often be ignored. In practice, when objections arose, it would often be sufficient to denounce the critics, to commission an additional study, and then to push forward more or less as originally planned. In a growing number of cases today, however, to follow that same formula is to move "forward" into a buzz saw of opposition.

While the public reactions are not exactly new, what *is* new — and what I take to be hopeful — is the growing recognition that the problem needs to be addressed with the same kinds of systematic, scientific approaches that have long been sought for the underlying technical questions. One sign of the degree of change is that philosophers and sociologists have been invited to offer keynote addresses to the annual meetings of the American Chemical Society. Another sign, which is at least equally telling, is that most of you who have shown up have actually done so of your own free will, and in spite of knowing what kinds of speakers were going to be on the program.

It seems only fair, accordingly, that I should also give you some advance warning of what my message is likely to be. My overall conclusion will be that, while science and technology have important contributions to offer to the making of decisions about environmental risks, those are often not the contributions that are being sought; instead, scientists are often put into a position of trying to help in a way that can often *seem* reasonable, at least in the sense of a short-term tactic, but that can actually create longer-term consequences that are bad both for science and society. Rather than spending most of my time stating those conclusions, however, what I want to do instead is to summarize for you the evidence that causes me to reach them, doing so in terms of four questions that need to be answered before we can really claim to understand the role of science, values, and blind spots in making decisions about environmental risks. The questions, which I shall consider in order, are as follows. What is the sound of one fact speaking? How do we see what it is we do not see? What is going on here? Do things have to work out this way?

WHAT IS THE SOUND OF ONE FACT SPEAKING?

As philosophers are fond of pointing out, but as most of the people who have scientific training should also know by now, science is at its best where the questions involved are factual ones, while often being at its worst where the need is to make choices from among competing values. Knowing that in the abstract, however, can sometimes be quite different from remembering it in practice. Particularly given the amount of time that scientists and engineers spend in dealing with questions that are indeed factual ones — what happens when compound X is added to compound Y at temperature T and pressure P? — members of the technical community are often heard to say something along the lines of "the facts speak for themselves."

Mark Twain, of course, had a different assessment. At one point, he wrote, "figures don't lie, but liars figure." At another point, he elaborated his philosophy, providing what remains one of history's most elegant epistemological typologies: "There are three kinds of lies: lies, damned lies, and statistics."

In more recent years, those who study what scientists actually do — sociologists and historians of science, among others — have also shown increasing skepticism toward the notion that facts "speak for themselves." Most of their assessments are not nearly so succinct nor memorable as those produced by Mark Twain, but these researchers have produced a large and ultimately compelling literature, documenting the degree to which scientific "facts" are not so much "discovered" (as if revealed by some divine hand) as they are "socially constructed" by mere mortals — by fallible individuals and groups who not only encounter research results, but also then go about trying to make sense of those results.

Because of the degree to which this literature is a reaction against the notion that facts "speak for themselves," the "social constructivists," as they are called, have sometimes seemed, at least to their critics, to be arguing that the facts do not matter at all. They, of course, have then accused their critics of claiming that the social construction processes — or even the human weaknesses of scientists in general — do not matter at all. To someone like me, who has read a fair amount of literature on both sides but is not a direct combatant, these claims and counter-claims can sometimes seem reminiscent of an argument over whether a glass is half empty or half full. For the purposes of the present presentation, accordingly, I will spare you the more detailed versions of the arguments involved, and instead boil the matter down to what at least some of my distinguished colleagues[1] have called "critical realism." The title is intended to show a degree of deference both to the "critics" and to those who say that "reality" matters; the perspective itself, to put it in the simplest possible English, essentially holds that the glass is indeed half empty as well as half full. The central conclusion is that what is out there in the empirical world does matter, but at the same time, the *way* in which it matters is so heavily dependent on human interpretation that, for all practical purposes, there is no such thing as a "fact" that is meaningful in the absence of human interpretation.

So what does that mean? One answer is that even the need to ask this question is an example of the importance of "human interpretation" — of context — but it is also possible to be a bit more specific, even while still staying close to the English language. If I were making this talk to a group of die-hard constructivists, of course, what I might need to defend would be the argument that "what is out there" really does matter, although most of the constructivists are really not that unreasonable. Given that this presentation is being made to an audience of scientists and scientific practitioners, however, let me instead illustrate the second half of the argument, namely, that there is no such thing as a "fact" that is meaningful in the absence of human interpretation.

Since some of you will not believe that, let me give you a pair of examples. Here they are: 92.5% and 94.0%. Do those facts "speak" to you? If so, what do they say?

Human beings, of course, can say, that 94.0% is larger than 92.5%, but it is not even possible to say whether the difference is large, small, or irrelevant, without knowing a bit more about these otherwise-isolated facts — without having some kind of interpretative context; the context, and the interpretations, can often be every bit as important as the figures themselves in deciding what, in practice, these facts "mean."

Social scientists, of course, are often accused of not making falsifiable predictions, so I should perhaps point out that I am effectively making a pair of implicit, but in fact falsifiable, predictions at this very moment. One of them is that, if my pair of facts do indeed "speak" to you, they will not be saying much that conveys a meaningful message. The other is that many if not most of you will already have been trying out a few possibilities, in the backs of your minds, about what these facts might "mean" — and that for all of you, the facts will become more meaningful (and may even start to *seem* to speak for themselves) when I let you know that these happen to be two of my favorite statistics from a recent study. More specifically, they are the proportions of variance in total local employment in two parishes (counties) of southern Louisiana, over a 20-year period, that could be statistically explained, ignoring national employment trends and even the previous year's employment in the same parish, in a time-series regression analysis that instead used the independent variables of the world price of oil, the worldwide count of active oil-drilling rigs, U.S. oil consumption, and whether the year in question came before or after the 1973/74 oil embargo.[2,3] By the standards of social science research, these two findings are reasonably "clean" ones, "facts" that are relatively easy to interpret. Even at that, however, they do not provide the kinds of information from the offshore oil industry that are most relevant to our deliberations; the more important lessons are enough broader that they will need to wait for the closing section of this presentation.

HOW DO WE SEE WHAT IT IS WE DO NOT SEE?

If there is no such thing as a self-interpreting fact, what does that say about the ways in which scientific facts — and the interpretations of scientific facts — can influence the making of decisions about environmental risks? A simple answer might be, "Proceed with extreme caution." Another might be, "That's not our job."

As scientists, we are experts about what our answers are, and even about what our answers mean for scientific theory, but we have no particular dispensation for decreeing what our answers "mean" for real-world risk decisions. Most real-world problems, of course, are not sufficiently polite to stay within the boundaries of any of the disciplines we have established, but the bigger challenge is the fact that they do not stay within the boundaries of "science," even as broadly defined. A *strictly* scientific decision, by its nature, is likely to be something along the lines of "92.5%," not something along the lines of, "Here's what the 92.5% answer means for resolving this controversy."

Instead, even if we are lucky, the best we can hope for in almost any case of technological controversy is to boil the matter down to three questions. The first, which can at least in principle be answered scientifically — although the answers are often a good deal more difficult to come up with in practice[4] — would be, "How safe is it?" The second question is one that, by its nature, cannot be answered scientifically: "is that safe enough?" The third one, which often contributes in unseen ways to the difficulty of answering both of the first two, is, "Are we overlooking something?" To put it simply, in other words, technological controversies almost invariably combine three very different types of questions — questions about facts, about values, and about blind spots.

When it comes to questions about values — questions along the line of "how safe is safe enough?" — the limitations of scientific expertise become evident quickly enough. Scientists are as capable of offering answers to this question as are any other citizens, but our answers are neither more valid nor "more scientific" by virtue of having been offered by scientists. There simply *is* no such thing as a "scientific" way to compare apples against oranges against orangutans. Instead, at least in a democracy, when it comes to matters of values — and to questions such as "how safe is safe enough?" — another word for "scientist" is "voter."

Even among scientists, moreover, values play a powerful role in decisions. To take the example of the question of how safe is "safe enough" — or to paraphrase Rayner and Cantor, "How safe is fair enough?"[5] — our answers may very well depend on obvious if nonscientific factors, such as whether we are neighborhood parents or far-away investors. Recent findings by Flynn and his colleagues[6] show that the answers are also likely to depend on less obvious factors, such as whether or not we are white males. Across a wide range of hazards, white males consistently rated the risks as being more acceptable than did white females, but also as more acceptable than did either the females or the males who were members of minority groups. That finding is one that would be worth pondering simply in view of the fact that white males are a distinct minority in society as a whole, but it is especially important given that we still make up a high proportion of the scientists in the U.S., even today — particularly when there is now disturbingly consistent evidence that race may be even more important than poverty in shaping citizens' exposure to real environmental risk.[7-10] As if to underscore the point that scientists are not immune to the broader trends of society, incidentally, Barke and his colleagues Jenkins-Smith and Slovic[11] have recently found physical scientists were significantly more likely to view risks of nuclear technologies as acceptable if those scientists happened also to be men instead of women.

The underlying reasons, of course, include differential distributions of values, not just across but also within the many segments of society. The study by Flynn and his colleagues also found, for example, that environmental risk decisions depended on whether the specific white males in question were ones who thought that we did not need to worry about environmental risks being passed on to future generations or imposed on people today without their consent; beyond that, decisions that things were "safe enough" were more likely to come from white males who think we have gone too far in pushing equal rights in this country, while not going far enough in using capital punishment. Even when the focus is limited to scientists, studies have found that answers about the acceptability of environmental risks can depend on factors that would to be irrelevant if the decisions were merely factual, such as the specific discipline of training, or whether the scientists within a given discipline are employed by industry, government, or academia.[12,13]

These are among the many reasons why, while it can make a great deal of sense for narrow, technical questions to be delegated to narrow, technical experts, we also need to recognize that many of the most important questions about technology are neither narrow nor technical. The strength of the expert is in the technical details, not the broader philosophy; the job of the technical specialist is to implement societal choices about values, not to make them.

It may not be without reason that a popular definition describes an expert as someone who knows more and more about less and less; a more formal definition, offered by Yale University Professor Charles Perrow, is that an expert is "a person who can solve a problem better or faster than others, but who runs a higher risk than others of posing the wrong problem."[14] For deciding on value-based questions, including the question of what is the right problem, perhaps the fundamental principle of a democracy has less to do with the technical expertise than with whether the answer reflects "the will of the governed."

It is important to recognize, however, that distortions of vision are created not just by values, but also by blind spots — by blind spots that include not just "unknowns," but "unknown unknowns." All of us, it turns out, have a significant risk of failing to understand how powerfully our view of the world can be shaped by the spot from which we do our viewing — the risk of being prisoners of our own perspectives. Sometimes the limitations on our vision do come from our values, which lead us to focus more intently on some parts of the picture than on others, but often, the problem is almost literally a matter of "blind spots" — of parts of the picture that are obscured from our view or that we simply fail to see. What makes the blind-spot problem so vexing is that, not only do we often fail to see something, but that we fail to see *that* we fail to see.

Let me be more specific, starting with the easy stuff. Probably the simplest examples involve what most of us do not know about other people's areas of expertise. Most of us who have scientific training tend to be reasonably cautious about not overstepping the bounds of our own expertise, so long as the questions are ones that involve our own disciplines. Somewhat surprisingly, however, the same caution tends not to be in evidence when discussion turns to matters that lie entirely outside of our training and expertise — particularly when the questions at issue are ones that involve human behavior.[15,16] Part of the reason may be that we often know just enough about the various specialties within our disciplines that we can also have some sense of how much we do not know about them, while we often do not know enough about other disciplines to have the same kind of awareness of what we do not know. Another part of the reason may have to do with a strange statistical anomaly: almost all of us like to think we are significantly better than most other people when it comes to being experts on "people," and we tend to hold to this belief whether we are familiar with the relevant research or not.

The blind spots, unfortunately, can be found even when we are limiting estimates to our own areas of expertise. Most scientists, it turns out, are susceptible to the same malady that afflicts most other mortals, namely, a pervasive overconfidence, or to put it differently, a pervasive tendency to *under*estimate how many unknowns still remain to be discovered within the field we think we know something about.

While this is a strong claim, it can be illustrated in fields that are as well-developed as physics and when dealing with a quantity as fundamental and as carefully measured as the speed of light. A compilation of the 27 published surveys of the speed of light between 1875 and 1958 that included formal estimates of uncertainty found that the measurements differed from the official 1984 value by magnitudes that would be expected to occur less than 0.0005 of the time, by chance alone, according to the original estimators' *own* calculations of the uncertainties in their estimates.[17] The absolute magnitude of the errors declined significantly over time, with improved measurement techniques, but there was no real improvement in the accuracy with which the remaining uncertainty was estimated. The 1984 estimate of the speed of light (which has since been used to calibrate the length of a meter, rather than vice versa) falls entirely outside the range of standard error (1.48 × "probable error") for *all* estimates of the "true" speeds of light that were reported between 1930 and 1970.[17]

Other examples can be reported for scientists who range from engineers to physicians. One study asked a group of internationally known geotechnical engineers for their 50% confidence bands on the height of an embankment that would cause a clay foundation to fail; when an actual embankment was built, not one of the experts' bands was broad enough to enclose the true failure height.[18] Another study followed a group of patients who were diagnosed on the basis of an examination of coughs to have pneumonia; of the group listed by physicians as having an 85% chance of having pneumonia, less than 20% actually did.[19] Other studies of the ability to assess probabilities accurately[20,21] — the problem of calibration — have found that calibration is unaffected by differences in intelligence or expertise,[22] while there is some evidence that errors will be increased by the importance of a task.[23] Overall, one would expect that only about 2% of the estimates having a confidence level of 98% would prove to be surprises, but in empirical studies, it is more common to find a "surprise index" on the order of 20 to 40%.[21,24,25]

In general, scientists can be expected to do a reasonably good job of "predicting" something that is already familiar, but we are much less impressive when it comes to specifying the likelihood of surprises. For example, Henshel[26] notes that, while demographers have often done a reasonably good job of "predicting" populations during times of relative stability, they have generally failed to anticipate "surprises" such as the baby boom of the 1940s–1950s or the birth dearth of the 1960s–1970s.

Other studies, however, suggest a more troubling implication: it may be that people who have the most experience with a given kind of risk will not be the ones with the most accurate assessments of that risk. Instead, there is evidence that the underestimation of risks may be particularly likely when people are reflecting on the risks of the activities in which they normally engage. Weinstein[27-29] reports a pervasive "it can't happen to me" attitude that shows up when people are estimating the risks of a wide range of everyday risk concerns, ranging from automobile accidents to arteriosclerosis. Newcomb[30] found a similar "macho or omnipotent response" toward the possibility of accidents among nuclear power plant workers, and Rayner[31] found comparable reactions in technically trained hospital personnel, particularly among higher-status workers. Indeed, the underestimation of risks appears to be a pervasive characteristic of "risky" occupations, ranging from commercial fishing,[32] to high steel work,[33,34] to coal mining,[35,36] to police work,[37] to offshore oil drilling;[38] it extends even to recreational parachute jumping[39] and professional gambling.[40]

While it is possible to hypothesize that experience will breed accuracy, in short, a growing number of studies suggest just the opposite: in all too many fields, familiarity appears to generate at least complacency, if not a kind of contempt. The "cognitive dissonance" perspective[41] suggests a stronger conclusion: it may be that persons in "risky" occupations will generally tend to ignore, minimize, or otherwise underestimate the risks to which they are exposed, thus reducing the "dissonance" that might be created by focusing on the risks that are implicit in an occupational choice that has already been made.

I have not yet seen any studies that have followed up on chemists' own estimates of the risks that are presented by chemicals, but I have seen enough studies of enough other kinds of specialists to be able to tell you that if chemists *do not* suffer from overconfidence — the form of overconfidence, more specifically, that results from a failure to be aware of blind spots — the chemistry profession would be unusual indeed. I am aware of only two such groups of "technical experts" — weather forecasters[42] and the very different group of experts who publish forecast prices for horse betting.[43] In both of these exceptional groups, the experts receive enough feedback on their predictions to be able to calibrate the accuracy of their probability estimates — and intriguingly, they are unlike most other kinds of technical specialists, including chemists, in that they have enough "lay experts" examining their predictions that they would have a relatively high likelihood of hearing about any errors in calibration if they failed to recognize their errors themselves.

WHAT IS GOING ON HERE?

If facts do not speak for themselves, and if the problems we do not see are often just as important as the ones we do see, what comes next? Nothing much, really — just fundamental errors in the usual views about the ways in which science relates to the rest of society. The errors fall into three main categories — misunderstandings about the ways in which the mass media convey information to the public, about the ways in which the public thinks about environmental risks, and about the nature of the societal roles that are played by those of us who are technical specialists.

Can We Blame the Mass Media?

In all likelihood, this audience needs little convincing that public views of science and technology are affected less by what goes on in the laboratories than by what gets reported in the media. Even "hard scientists," after all, often learn about other scientific fields from the "news pages" of journals such as *Science*, and it is hard to imagine how else most members of the public would get their information.

Considerably more effort, unfortunately, is likely to be required to overcome the usual assumptions about what gets reported, and how. One of the easiest ways to win the hearts of a technical audience such as this one is to move into vigorous denunciations, railing at the irresponsibility of the reporters and the ignorance of the readers. Unfortunately, while such an approach can win the hearts of a scientific audience, it has very little to do with the minds. Instead, as is often the case in the physical or biological sciences, empirical findings in the social sciences often show a stubborn refusal to agree with what you may expect.

When most scientists express their spontaneous *impressions* about media reports, however, we stress something far different. The reason may involve something akin to "availability bias:"[44] Like members of the general public, we often tend to remember not the daily diet of good news, but the much more dramatic news associated with specific technological failures — ranging from the *Challenger* to Chernobyl, from Bhopal to Times Beach, and from the Hubble space telescope to the *Exxon Valdez*. As is the case with studies of the general public, however, the examples we remember are often not representative of the coverage as a whole.

The relevant findings have been summarized elsewhere, but in essence, there are a couple of key problems with the argument, so popular at scientific conferences, that the mass media are putting out consistently antiscientific reporting, which is being swallowed whole by a gullible public. The first is that, in contrast to the usual complaints about a fixation on scientific disasters, the reality is that the mass media carry a steady diet of good news and reports on the impressive accomplishments of the scientific community, ranging from superconductivity to biochemical breakthroughs in medicine to improved understanding about how organizations function. Rather than supporting the common belief, systematic studies have managed to find little in the way of a true antiscience bias in any of the media outlets this side of the lunatic fringe.[45-48] The second problem is that, much to the chagrin of generations of mass communication researchers who had assumed they would earn tenure by demonstrating the effectiveness of one type of media message over another, perhaps the most common finding in the mass communication literature is in essence a nonfinding: In general,[49,50] studies find mass media reports have precious little influence on people's actual views.[45,46,51]

There is by now a reasonably substantial body of literature on this topic, but if you prefer not to read it, you may wish instead to reflect on two pieces of information. The first is that if you think back on the era that has seen the greatest growth of public criticism toward science and technology, namely, the past three or four decades, you will realize that much of the growth in criticism has taken place during precisely the era that has also seen unprecedented expenditures, both by government bodies and by private entities, trying to tell people just how safe and wonderful our technology has become. Remember the vast sums spent on "risk communication" messages about "My Friend, the Atom," during the 1950s and 1960s? Remember the vast mobilization that arose against nuclear power during the 1970s and 1980s? There are some lessons there for those of you who, like me, are old enough to remember "Better Living Through Chemistry," but those lessons will need to wait for the end of this presentation.

The second piece of information is that, more recently, systematic studies have found that opposition toward nuclear waste facilities, for example, has actually increased during the very time periods when project proponents were spending large amounts of money on "public education campaigns" that were intended to have the opposite effect, namely, to convince the citizens of intended host regions that the facilities would be safe, as well as being good for the local economy. These finding have proved to be surprisingly consistent, incidentally — not just in the U.S.,[52] but also in nations that are commonly seen as being more deferential toward technology, including Japan,[53] Taiwan,[54] and in a less systematic case, in Korea.[55] Just as would be the case in the physical or biological sciences, in other words, you can choose to ignore the social science findings, but you do so at your own peril; ignoring the findings from the past does little to guarantee that you will have immunity to the same kinds of findings in the future.

Can We Blame the Public?

So what about the broader public? You have all heard the complaint: if only people had the facts, they would support our technology. Perhaps you have even attended conferences where slick, expensive presentations have put forth a similar point: The people who know the facts — people like us — tend to think that we are great. Opposition comes either from ignorance or irrationality, since, after all, to know us is to love us.

A few of the earlier and better known pieces in the risk literature featured the same kind of speculation, but on the basis of the empirical studies that have now been done, the safest conclusion is that those early speculations had everything going for them but the facts. A growing body of literature has found that the opponents of technology tend to be just as well-informed as the supporters.[56-64] In case studies of specific facilities, researchers have even found that opponents were characterized by an active searching for information, while the *supporters* of the controversial facilities were the ones who, "by their own accounts, were noticeably and — in many cases — intentionally uninformed."[65]

While it would be premature to describe the differential information-seeking as universal, it has been encountered repeatedly in studies done to date.[66-69] One study even found that a group of citizens became amateur but reasonably skilled epidemiologists in an effort to obtain the kinds of answers that relevant health authorities were unable or unwilling to provide.[70] Particularly in studies of technological disasters, in fact, members of citizen groups often describe one of their greatest frustrations as being the difficulty of obtaining the credible, scientific information they actively seek.[62,65,66,68,71-73] In short, to the extent to which many of us have long assumed that "public ignorance" has been the factor at fault, the evidence shows we have been barking up the wrong fault tree.

Do We Need to Blame Ourselves?

If the problem cannot be blamed on the usual scapegoats — irresponsible journalists and ignorant citizens — then who do we have left to blame? The obvious possibility was stated long ago by the Walt Kelley cartoon character, Pogo, "We have met the enemy, and he is us" — but that, too, would be a bit simplistic. Let me suggest a slightly different approach — one that involves not the search for scapegoats but the effort to understand the changing structure of society.

This suggestion starts with the need to deal with a widespread misunderstanding about what it means to say we live in an "advanced, technological" society. One of the usual assumptions about the scientific and technological advances of the past century is that those of us who live today "know more" than did our great-great-grandparents. Collectively, of course, that is true. Individually, however, to note a point made initially by one of the earliest and most articulate proponents of "intellectualized rationality," Max Weber, roughly three-quarters of a century ago,[74] we know far *less* today about the tools and technologies on which we depend.

In the early 1800s, roughly 80% of the American population lived on farms, and for the most part, those farm residents were capable of repairing, or even of building from scratch, virtually all of the tools and technologies upon which they depended. By contrast, today's world is so specialized that even a Nobel laureate is likely to have little more than a rudimentary understanding of the tools and technologies that surround us all, from airliners to ignition systems, and from computers to corporate structures.

Far more than was the case for our great-great-grandparents, we tend to be not so much in control of as dependent upon our technology; in most cases, in other words, we have little choice but to "depend on" the technology to work properly. That means, in turn, that we also need to depend on whole armies of the fallible human beings who are specialists, most of whom we will never meet, let alone be able to control. In this sense, too, we are very much unlike our great-grandparents: in the relatively few cases where they needed to buy an item of technology from someone else, it was often from a "someone" that they knew quite well, or that they would know how to find if something went wrong.

For most of us, most of the time, we find we *can* depend on the technologies, and the people who are responsible for them, but the exceptions can be genuinely troubling; one of the reasons is that our increases in *technical* control have come about, in part, at the cost of decreases in *social* control. The achievements of science and technology have been important and even impressive, and they have helped to bring about a level of physical safety, as well as of material wealth, that is simply unprecedented. But they have done so at a cost — a cost of substantially increased vulnerability to the risks of interdependence.[75]

This point is summarized in Figure 1, which illustrates "The Technological Risk Crossover." Using a very simple index of interdependence, based on the proportion of the citizenry *not* involved in growing their own food, this figure shows that, during the very era during which society has been enjoying a substantial decline in the risks that have been the traditional focus of the scientific community — namely, the risks of death — there has been a substantial increase in the significance of interdependence. The most serious of the resultant new risks, if you will pardon the one technical term I hope you'll remember at the end of this presentation, involves *recreancy* — in essence, the likelihood that an expert or specialist will fail to do the job that is required. The word comes from the Latin roots *re-* (back) and *credere* (to entrust), and the technical use of the term is analogous to one of its two dictionary meanings, involving a retrogression or failure to follow through on a duty or a trust. The term is unfamiliar to most, but there is a reason for that. We need a specialized word if the intention is to refer to behaviors of institutions or organizations as well as of individuals, and importantly, if the focus of attention is to be on the

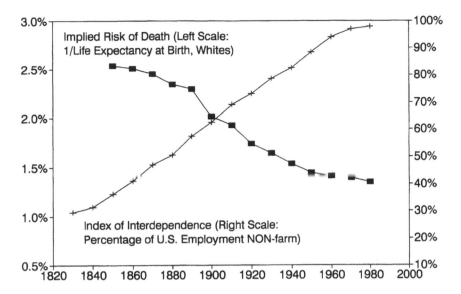

Figure 1   The technological risk crossover. [Adapted from William R. Freudenburg, "Risk and Recreancy: Weber, the Division of Labor, and the Rationality of Risk Perceptions," *Social Forces* 71 (#4, June 1993): 909–932. Data are drawn from *Historical Statistics of the United States: Colonial Times to 1970* (Washington, D.C.: U.S. Census Bureau, 1975), and *Statistical Abstract[s] of the United States* (Washington, D.C.: U.S. Census Bureau, various years). National data are available only as far back as 1900; data from 1850–1900 are drawn from Massachusetts, where 99% of the enumerated population at the time was white. Information for 1870 represents an interpolation.]

facts instead of the emotions. In a fact that may tell us something about the societal importance of trustworthiness, virtually all of the common words having comparable meanings have come over time to take on a heavily negative set of value connotations. To say that a technical specialist is responsible, competent, or trustworthy, for example, is to offer at least a mild compliment, but to accuse that same person of being *ir*responsible, *in*competent, or of having shown a "betrayal" of trust, is to make a very serious charge, indeed. While "recreancy" may not be an everyday term, the need for such a term grows quite directly out of the need to avoid the emotional and/or legal connotations of the available alternatives.

Unlike media coverage and public knowledge levels, trustworthiness and recreancy *have* been shown by systematic research to catalyze the strange kinds of interpersonal chemistry that have greeted an ever-increasing range of technologies. An analysis of attitudes towards a proposed low-level nuclear waste facility, for example, found that sociodemographic variables were only weak predictors of attitudes, being associated with differences of only 7 to 15% in levels of public support; even values-based and/or ideological items were associated with differences of only 10 to 25%. Three measures of recreancy, by contrast, were associated with differences of 40 to 60%+. In regression analyses, the recreancy variables alone more than tripled the amount of variance that could be explained by the sociodemographic and the ideological variables combined.[75]

The growth in interdependence, and in the risks of recreancy, appear to be among the reasons why trust and trustworthiness have also been found to be key variables in a growing range of other studies.[76-86] The underlying causes appear to include not just the statistically dramatic increase in the extent to which we have become dependent on the actions of unknown others, but also the pragmatically dramatic increase in societal awareness of cases in which those "others" have proved not to be deserving of that trust. In a growing number of cases, moreover, members of the general public have started to question the trustworthiness even of scientists — a point that leads to the final question of this chapter.

DO THINGS HAVE TO WORK OUT THIS WAY?

The loss of trust needs to be kept in perspective: while public support for science and technology does appear to have fallen in recent years, the public continues, at least so far, to have more faith in science than in most other institutions of society.[87] For the most part, that is good news, but it comes as a kind of double-edged sword. Partly, that is because of the increasing tendency for politicians (and sometimes, members of the general public) to ignore the kinds of evidence summarized thus far in this presentation, and to call for technological controversies to be resolved by "scientific decisions."

In some ways, these calls represent a longing that is not only quite understandable, but quite old. As Frank[88] pointed out in a classic law article, societies have long shown distrust for the "human" element in decision making. Under what might be called the first models for dealing with this problem, at least in the U.S., namely, the early modes of trial, the emphasis was on ordeals, judicial duels, "floating" tests for witches, and other ways of making decisions that were "considered to involve no human element. The judgment [was] the judgment of the supernatural, or 'the judgment of God'."[88] The 19th century equivalent of the earlier distrust for human judgment, according to Frank, was reliance on a body of impersonal legal rules. Under this second model of decision making, "rationality" was thought to emerge through a dependence on rules that were derived from self-evident principles, thus reducing the human element in decision making to a minimum.[89] In the 20th century, the emphasis on such abstract "universal principles" declined, and a third model emerged, one that placed increased emphasis on empirical evidence. One could even argue that with society's increasing replacement of the sacred by the secular, this model attempted to replace the presumably fair or even sacred decisions of the supernatural with those of the scientists, where in the extreme case, the call for "scientific" decisions is an expression of the hope that the scientist would replace the judge or the elected official as the actual decision maker.

As might have been expected on the basis of the evidence summarized in this paper, unfortunately, the reality has proved almost never to be that simple; judges proved to have many of the same weaknesses as scientists — or nearly anyone else — when it came to being swayed by human fallibilities. One of Frank's major conclusions was that "the human element in the administration of justice by judges is irrepressible... [T]he more you try to conceal the fact that judges are swayed by human prejudices, passions and weaknesses, the more likely you are to augment those prejudices, passions and weaknesses... For judges behave substantially like the human beings who are not judges."[88]

In the case of scientists, unfortunately, the problem may be even worse, because at least a few of the key characteristics of "good scientists and engineers" are often different from the characteristics of "good judges," particularly when it comes to making decisions about values and blind spots. There are good reasons why most of us, if forced to choose, would rather have technicians who are specialized and efficient, but judges who are broad and fair. While scientists, too, tend to behave substantially like the human beings who are not scientists, there do tend to be a number of consistent differences, some of which can show up precisely in making values-based decisions about technology. Scientists and engineers, for example, tend to place more emphasis on cost-containment and efficiency than do most citizens, while placing less emphasis on long-term safety.[90-93] These choices, to note the obvious, do not exactly provide the kind of extra protection of public health and safety that is often being sought by the citizens who call for a "more scientific" approach to environmental risk decisions.

Most citizens' calls for "scientific" decisions, in reality, are a request for something a bit broader — in most cases, a call for ways of assuring that "the human element" of societal decision making will be not just technically competent, but equitable, fair, and responsive to deeply felt concerns. In most cases, what is important is not just technical expertise, but the broader ability to expect that the "responsible" authorities will indeed behave responsibly — making decisions that will not work to the detriment of some groups for the benefit of others, for example, and taking care not to ignore the values that affected citizens hold to be most dear. In a growing number of cases, unfortunately, this is precisely the expectation that fails to be upheld.

Scientists and engineers are probably as well-equipped as anyone in society for providing technical competence, and most scientists and engineers, as individuals, tend to place a high level of importance on responsibility. Despite these individual-level characteristics, however, a growing body of work[65,75,76,94,95] suggests that the organizational-level characteristics of technological institutions may be parts of the problem, contributing to the growing societal perception that decisions are being made in ways that are *not* equitable, fair, or sad to say, even responsible.

Instead, what is increasingly happening in practice is that societal institutions that do *not* enjoy the high credibility of science and technology — government and industry, to name two — will look to science and technology for help. The help is sought not just in the area where it is most appropriate — involving questions of fact — but also in legitimating what are actually value decisions and in glossing over what often remain blind spots. For many years, the technique worked reasonably effectively, and thus a number of institutions are tempted to keep on trying the old techniques today. As shown by the growing number of cases in which the interpersonal chemistry turns toxic, however, increasing segments of the public are no longer buying the old sales pitch. To make matters worse, the effort to disguise value decisions as "technical" ones, while it sometimes still seems to "work" in the case of individual facilities or disputes, often does so at the cost of greatly increased public distrust toward the institutions of science and technology — a distrust that often explodes when the next proposal comes along. In an ever-increasing number of cases, as a result, the net effect of the technique is not so much an increased credibility of decisions, but a reduced credibility for science and technology more broadly.

So what is the proper way to respond when the institutions of society, even powerful institutions that have long been good to science and technology, ask for help not just in answering factual questions, but in disguising value questions, or in diverting attention away from blind spots? A lot of the "obvious" answers to this question have to do with the eternal vigilance of scientists and engineers toward evil, sinister people who will try to get us to do something that is ethically wrong, and most scientists and engineers have a very strong sense of ethics for coping with just such pressures. As should be obvious by now, however, the need is not just to be vigilant in resisting the pressures that are visible, but the ones that are so subtle we often fail to see them at all.

Michael Davis[96] gives an example from the decisions that led up to the disaster on the space shuttle *Challenger*. In essence, the final responsibility for allowing the shuttle launch was vested in an engineer, not a politician, for reasons you can easily imagine. That engineer's technical people had expressed concerns to him about the integrity of O-ring seals at low temperatures, and accordingly, in the big meeting the night before the planned launch, the engineer made his decision: "No." A few minutes later, he changed his mind, allowing the launch to take place — and a few hours later, the entire shuttle crew had been killed.

What happened during those few key minutes? Was it the application of "political" pressure? The threat that he would be fired if he refused to change his mind? Apparently not — and in fact, my guess is that if this engineer had been subjected to these or any other kinds of obvious pressures, he would have resisted, probably even at the cost of losing his job. In reality, however, he did not face that kind of a clear, black or white choice.

What did happen was that the engineer's boss called a break, brought a small group to one side of the room, and effectively expressed his admiration for the engineer's ethics. What he did next, however, was to remind that same engineer that he was now in the position of wearing two hats — one as an engineer, and one as a manager — and he urged the engineer to "take off [your] engineering hat and put on [your] management hat."[97] A manager's hat, apparently, comes complete with a different set of blinders than does an engineer's hat. Thanks in part to the new blind spots created by his new hat, the engineer reconsidered, the launch went forward, and the rest was disaster.

WHAT CAN ONE CHEMIST DO?

If my informal conversations with a number of you can be taken as providing any indication, chemistry stands poised today in an awkward balance between comfort and terror. On the one hand, there is something quite comforting about the world of a chemistry laboratory, where H_2O is always H_2O, with known properties, one of which is that the H_2O will not suddenly decide to transform itself into SO_2, to change its boiling point, or to behave in other ways that would shake the very foundations of chemical science. While the job of a scientist is to probe the unknown, most of the probing takes place at the edge of the known, adding comfortably and incrementally to what we can feel safe to say we know, contributing further to a sense of collective mastery over the physical world.

In the rest of the world, however — the part that begins just outside the door of the chemistry lab — things are no longer what they used to be. I am part of a generation that grew up hearing about, and believing in, "better living through chemistry," and appreciating what it was that chemistry had done for me lately. The coming generation, however, is one that is more likely to be wondering what it is that chemistry has done *to* them lately.

It is possible to speak nostalgically and even passionately about the era that has passed, but it is not possible to bring it back. Still, we do have both the option and the responsibility to select the path by which we decide to move forward. In recent years, the direction that has often been chosen, particularly by those who claim and may genuinely believe that they have the best interests of the scientific community at heart, has been to embrace the age-old if not always accurate precept of political power that the best defense is a good offense. The details vary from case to case, but the basic principle is that, if you can successfully characterize your opponents as lacking in legitimacy — by accusing them, for example, of being ignorant, irrational, selfish, or opposed to all that is scientific, prosperous or patriotic — then you can often dismiss rather than dealing with their concerns.

Because this technique could be expected to work reasonably effectively in a world where citizens genuinely believed in "better living through chemistry," there are many people who believe it ought to continue to work today. Some of those people are in the business of selling things for a living, and some of them will even sell you the advice that their approaches will still work, particularly if you use the approaches on people who are relatively compliant, poor, or powerless.[98]

It is possible that some of you will soon be faced with making a decision about whether or not to follow such advice. Before you make such a decision, however, I urge you to consider three questions. The first and probably most important is to ask yourself whether, even in what you may see as a defense of a scientific project, you are comfortable making claims that you know to be contradicted by most of the relevant — which is to say empirical — scientific literature.

The second question is whether, even when public relations specialists offer you confident assurances that you can "win" by characterizing your opponents as "antiscience," for example, you might be dealing with "experts" who suffer from the same kinds of overconfidence — and the same kinds of conviction that "it can't happen to me" — that have been found by empirical studies to afflict so many other disciplines. More specifically, before we move to the third important question, I would urge you to consider what has already happened to two of the nation's most powerful industries, which happen also to be among the ones where the efforts to attack the credibility of critics have been pushed most forcefully — those involving nuclear power and the search for offshore oil.

In the case of nuclear power, people who presumably thought they were acting in the best interest of the industry not only backed one of the most extensive "public education" campaigns in history to inform people about the potential uses of "the peaceful atom" — the U.S. Atomic Energy Commission's 1968 *Annual Report to Congress*[99] noted that nearly 5 million people had seen its exhibits and demonstrations in 1967 alone — but for a time also backed a campaign to "study" opposition to nuclear power as a "phobia."[100] Did it work? I will let you be the judges, but I, at least, would not want to buy a used public relations strategy from this industry. Not a single new nuclear power plant has been ordered since 1977, and the projections for nuclear electricity have dropped by over 80% in the last two decades.[100] The reasons seem to center around the high costs of the completed plants, rather than around public opposition per se, but given the number of ways in which opposition can translate into increased costs[101] — and given the amounts of money and energy that have continued to be invested into the effort to turn those attitudes around[102] — an objective observer would be hard pressed to characterize the opposition as irrelevant.

In the case of the offshore oil industry, the evidence is more mixed, in that there are still any number of communities, particularly along the coasts of Louisiana and Texas, where the offshore industry remains quite popular. We have long known, however, that we have been nearing the end of new discoveries in the Gulf of Mexico; for decades, the eyes of oil industry geologists, and the efforts of oil industry public relations firms, have been directed toward other areas, such as the coast of California. Particularly after the election of

President Reagan in 1980, followed by the 1988 election of George Bush, the former Texas oil executive, at least the national-level policy context became as favorable toward the offshore oil industry as could ever be hoped for in the latter part of the 20th century, but many Californians had a much less favorable view of the industry. Given what many saw as the best way to deal with local opposition, increasing effort went into some of the most sophisticated public relations campaigns that money could buy. In combination with a coordinated effort to present the offshore industry in a more favorable light, and to "educate the public" about the industry's benefits, the opponents to offshore drilling were called a variety of names, most of which will sound familiar to anyone who has followed public debates about environmental risks and the chemical industry — ignorant, superstitious, ill-informed, selfish, parochial, Luddite, even Communist. What happened? By the time President Bush left office near the end of 1992, he had been told by the National Academy of Sciences that the government did not have enough information to make scientifically informed decisions about offshore drilling in some of the areas where it was most contentious, including Florida as well as California; faced with an unbroken string of congressional moratoria that effectively precluded any further drilling not just in California, but along most of the coastline for the rest of the country as well, the former Texas oilman himself declared a moratorium to the year 2000.[103]

Now, back to that third question. If you do find yourself in the position of trying to decide whether to use the tactic sometimes known as "diversionary reframing" — diverting attention away from questions about a technological proposal by accusing the opponents of the proposal as being opposed instead to science itself — I urge you to ask yourself whether you want to run the risk of creating what Robert Merton[104] once called "a self-fulfilling prophesy."

Even today, in other words, there is a chance that the tactic of diversionary reframing will work, but the odds continue to go down, and there is also another possibility — one that tends to be profoundly troubling to anyone who is convinced of the value of the scientific approach. To put the matter plainly, every time an opponent of a given facility is accused of being opposed to science in general, there is a significant risk of contributing to exactly that outcome. The risk is particularly high in cases where the accusation is endorsed by the kinds of scientists who would normally be expected by the public to play a more neutral role. It is also important to remember that, while those of us *within* the scientific community often differentiate carefully between what we think about a specific scientist and what we think about science in general, members of the broader public often obtain their most vivid evidence about the credibility of "science," as a whole, from just this kind of contact with specific, individual scientists.

The world is full of people who fail to live up to their responsibilities — of stereotypical used car salesmen, fast-buck operators, con artists, and others who have something to gain from promising more than they can deliver. Given the realities of an increasingly complex and interdependent world, it is not the least bit unreasonable for a citizen to become suspicious if there is evidence that yet another specialist might be inclined to be a little less than upstanding. What *is* unreasonable, in my view, is when any member of the scientific community provides evidence, even unintentionally, that "scientists" should be tarred with the same stereotypes that afflict hucksters and hired guns, or that science, as an institution, is more interested in profit than in truth and the broader public good.

While scientists have suffered far less than have most groups in society from the erosion of public confidence, at least to date, past experience gives us no reason to conclude that scientists are immune to the broader malaise. My own belief is that science has continued to enjoy a reasonably high level of public support not because of a happy accident, but because of a real behavioral regularity: in the vast majority of cases, scientists have shown by actual behaviors that we *can* be trusted.

That credibility, it seems to me, is far too precious to be put up for sale — not even for the institutions that employ us and support us, and perhaps especially not for them. As Paul Slovic has noted, following up on observations made by earlier observers of democracy who have included Abraham Lincoln,[105] any examination of trust needs to take note of a powerful "asymmetry principle" — the fact that trust is hard to gain, but easy to lose.[106] As Slovic so aptly illustrates the point, if a specialist such as an accountant takes even one opportunity to steal, that single act creates a level of distrust that is not counterbalanced if the accountant does not steal anything the next time around, or even the next several dozen times around.

It would probably be melodramatic, although it might nevertheless be true, that as scientists, all of us are in effect trustees for something more important than money — for the credibility of science and technology more broadly. Unlike the accountant, moreover, we need to be alert not just for outright embezzlement, but for far more subtle kinds of blind spots, such as the one that afflicted and may still continue to haunt the chief engineer for the *Challenger*. Indeed, the problem may even be more subtle than that; in the case of the *Challenger*, the cost of the error became obvious within a matter of hours. In areas of science that are more probabilistic, as in the case of environmental risks, the consequences of a willingness to fudge a technical judgment — or a political one, as in characterizing an opponent to a given facility as being opposed to science in general — may not become obvious for years. In the long run, however, it may prove to be no less serious. The public trust is valuable, but it is also fragile, and it is highly susceptible to the corrosive effects of scientific behaviors that fall short of the highest standards of responsibility. At least for those of us who care about the continued viability of the scientific enterprise, in short, the risks of recreancy may ultimately prove to be the greatest risks of all.

ENDNOTES

1. Dietz, T., R.S. Frey and E.A. Rosa. "Risk, Technology and Society," in R.E. Dunlap and W. Michelson, Eds., *Handbook of Environmental Sociology* (Westport, CT: Greenwood, in press).
2. For more systematic analyses of these two facts, see, e.g., R. Gramling and W.R. Freudenburg. "A Closer Look at 'Local Control': Communities, Commodities, and the Collapse of the Coast," *Rural Sociology,* 55:541–558 (1990).
3. Freudenburg, W.R. and R. Gramling. *Oil in Troubled Waters: Perceptions, Politics, and the Battle over Offshore Oil* (Albany: State University of New York Press, 1994).
4. Weinberg, A. "Science and Trans-Science," *Minerva* 10: 209–222 (1972).
5. Rayner, S. and R. Cantor. "How Fair Is Safe Enough? The Cultural Approach to Technology Choice," *Risk Analysis* 7:3–9 (1987).
6. Flynn, J., P. Slovic and C.K. Mertz. "Gender, Race, and Perception of Environmental Health Risks", *Risk Analysis* 14:1101–1108.
7. Commission for Racial Justice. *Toxic Wastes and Race in the United States. A National Report on the Racial and Socio-Economic Characteristics of Communities with Hazardous Waste Sites* (Washington, D.C.: Public Data Access, Inc., 1987).
8. Bullard, R.D. "Anatomy of Environmental Racism and the Environmental Justice Movement," in R.D. Bullard, Ed., *Confronting Environmental Racism: Voices from the Grassroots* (Boston: South End Press, 1993), pp. 15–39.
9. Mohai, P. and B. Bryant. "Race, Poverty, and the Environment," *EPA J.* March/April:6–10 (1992).
10. *National Law Journal.* "Unequal Protection: The Racial Divide in Environmental Law," *Natl. Law J.* Sept. 21:S1–S12 (1992).
11. Barke, R., H. Jenkins-Smith and P. Slovic. "Risk Perceptions of Men and Women Scientists" (unpublished manuscript, 1994).
12. Dietz, T. and R.W. Rycroft. *The Risk Professionals* (New York: Russell Sage Foundation, 1987).
13. Lynn, F. "The Interplay of Science and Values in Assessing and Regulating Environmental Risks," *Sci. Technol. Human Values* 11:40–50 (1986).
14. Perrow, C. *Normal Accidents: Living with High-Risk Technologies* (New York: Basic, 1984).
15. Fischhoff, B., P. Slovic and S. Lichtenstein. "Lay Foibles and Expert Fables in Judgments about Risks," in T. O'Riordan and R.K. Turner, Eds., *Progress in Resource Management and Environmental Planning,* Vol. 3. (New York: Wiley, 1981), pp. 161–202.
16. Freudenburg, W.R. "Social Scientists' Contributions to Environmental Management," *J. Soc. Issues* 45:133–152 (1989).
17. Henrion, M. and B. Fischhoff. "Assessing Uncertainties in Physical Constants," *Am. J. Physics* 54:791–798 (1986).
18. Hynes, M.E. and E.H. Vanmarche. "Reliability of Embankment Performance Predictions," in R.N. Dubey and N.C. Lind, Eds., *Mechanics in Engineering,* Presented at Specialty Conference on Mechanics in Engineering (Toronto: University of Waterloo Press, 1977), pp. 367–384.

19. Christenson-Szalanski, J.J.J. and J.B. Bushyhead. "Physicians' Use of Probabilistic Information in a Real Clinical Setting," *J. Exp. Psychol.* 7:928–935 (1982).

20. DeSmet, A.A., D.G. Fryback and J.R. Thornbury. "A Second Look at the Utility of Radiographics School Examination for Trauma," *Am. J. Radiol.* 43:139–150 (1979).

21. Lichtenstein, S., B. Fischhoff and L.D. Phillips. "Calibration of Probabilities: The State of the Art to 1980," in D. Kahneman, P. Slovic and A. Tversky, Eds., *Judgment under Uncertainty: Heuristics and Biases* (New York: Cambridge University Press, 1982), pp. 306–333.

22. Lichtenstein, S. and B. Fischhoff. "Do Those Who Know More Also Know More About How Much They Know?" *Organ. Behav. Human Performance* 20:159–183 (1977).

23. Sieber, J.E. "Effects of Decision Importance on Ability to Generate Warranted Subjective Uncertainty," *J. Pers. Soc. Psychol.* 30:688–694 (1974).

24. Fischhoff, B., P. Slovic and S. Lichtenstein. "Knowing with Certainty: The Appropriateness of Extreme Confidence," *J. Exp. Psychol.* 3:552–64 (1977).

25. Freudenburg, W.R. "Nothing Recedes Like Success? Risk Analysis and the Organizational Amplification of Risks," *Risk* 3:1–35 (1992).

26. Henshel, R.L. "Sociology and Social Forecasting," *Ann. Rev. Sociol.* 8:57–79 (1982).

27. Weinstein, N.D. "The Precaution Adoption Process," *Health Psychol.* 7:355–86 (1988).

28. Weinstein, N.D. "Why It Won't Happen to Me," *Health Psychol.* 3:431–57 (1984).

29. Weinstein, N.D., M.L. Klotz and P.M. Sandman. "Optimistic Biases in Public Perceptions of Risk from Radon," *Am. J. Pub. Health* 78:796–800 (1988).

30. Newcomb, M.D. "Nuclear Attitudes and Reactions: Associations with Depression, Drug Use and Quality of Life," *J. Pers. Soc. Psychol.* 50:906–20 (1986).

31. Rayner, S. "Risk and Relativism in Science for Policy," in B.B. Johnson and V.T. Covello, Eds., *The Social and Cultural Construction of Risk* (Dordrecht, Holland: D. Reidel, 1987), pp. 5–23.

32. Tunstall, J. *The Fishermen* (London: MacGibbon & Lee, 1962).

33. Haas, J. "Binging: Educational Control among High Steel Ironworkers," *Am. Behav. Sci.* 16:27–34 (1972).

34. Haas, J. "Learning Real Feelings: A Study of High Steel Ironworkers' Reactions to Fear and Danger," *Soc. Work Occupations* 4:147–72 (1977).

35. Fitzpatrick, J.S. "Adapting to Danger: A Participant Observation Study of an Underground Mine," *Soc. Work Occupations* 7:131–80 (1980).

36. Lucas, Rex A. *Men in Crisis: A Study of a Mine Disaster* (New York: Basic, 1969).

37. Skolnick, J. "Why Cops Behave as They Do," in S. Dinitz, R.R. Dynes and A.C. Clarke, Eds., *Deviance: Studies in the Process of Stigmatization and Societal Reaction* (New York: Oxford, 1969), pp. 40–47.

38. Heimer, C. "Social Structure, Psychology and the Estimation of Risk," *Ann. Rev. Sociol.* 14:491–519 (1988).

39. Epstein, S. and W.D. Fenz. "The Detection of Areas of Emotional Stress through Variations in Perceptual Threshold and Physiological Arousal," *J. Exp. Res. Personality* 2:191–99 (1967).

40. Downes, D.M., B.P. Davies, M.E. David, and P. Stone. *Gambling, Work and Leisure: A Study Across Three Areas* (London: Routledge and Kegan Paul, 1987).

41. Festinger, L. *A Theory of Cognitive Dissonance* (Evanston, IL: Row, Peterson, 1957).

42. Murphy, A.H. and R.L. Winkler. "Can Weather Forecasters Formulate Reliable Probability Forecasts of Precipitation and Temperature?" *Nat. Weather Dig.* 2:2–9 (1977).

43. Dowie, J. "On the Efficiency and Equity of Betting Markets," *Economica* 43:139–150 (1976).

44. Cognitive "availability" involves the tendency to judge the probability of an event by the ease with which examples can be recalled — a pattern that causes an overestimation of the probability of events that are more vivid or memorable. For a useful overview of literature on this and other "heuristics," or judgmental rules of thumb, see Slovic, P., B. Fischhoff, and S. Lichtenstein. "Perception and Acceptability of Risk from Energy Systems," in *Public Reactions to Nuclear Power: Are There Critical Masses?*, W.R. Freudenburg and E.A. Rosa, Eds. (Boulder, CO: American Association for the Advancement of Science/ Westview, 1984), pp. 115–135.

45. Dunwoody, S., M. Friestad and M. Shapiro. "Conveying Risk Information in the Mass Media," paper presented at convention of International Communication Association, Montreal (1987).

46. Raymond, C.A. "Risk in the Press: Conflicting Journalistic Ideologies," in *The Language of Risk*, D. Nelkin, Ed. (Beverly Hills: Sage, 1985), pp. 97–133.

47. Stallings, R. "Media Discourse and the Social Construction of Risk," *Social Prob.* 37:80–95 (1990).

48. Molotch, H. "Media and Movements," in *The Dynamics of Social Movements*, M.N. Zald and J.D. McCarthy, Eds. (Cambridge, MA: Winthrop, 1979), pp 71–93.

49. For the most forcible statement of an alternative perspective that most observers would find credible, see Mazur, A., "Communicating Risk in the Mass Media," in *Psychosocial Effects of Hazardous Toxic Waste Disposal on Communities*, D.L. Peck, Ed. (Springfield, IL: Charles C. Thomas, 1989), pp. 119–137.

50. For a more subtle reading of the underlying dynamics, see A. Szasz, *EcoPopulism: Toxic Waste and the Movement for Environmental Justice* (Minneapolis: University of Minnesota Press, 1994).

51. Freudenburg, W.R., C.L. Coleman, C. Helgeland and J. Gonzales. "Media Coverage of Hazard Events," paper presented at Annual Meeting, Society for Risk Analysis, Baltimore (1991).

52. Slovic, P., M. Layman and J.H. Flynn, *What Comes to Mind When You Hear the Words, 'Nuclear Waste Repository'? A Study of 10,000 Images* (Carson City, NV: Nevada Nuclear Waste Projects Office, 1990).

53. Budd, W., R. Fort, R. Rosenman. "Risk Externalities, Compensation and Nuclear Siting in Japan," *Environ. Prof.* 12:208–13 (1990).

54. Liu, J.T. and V.K. Smith. "Risk Communication and Attitude Change: Taiwan's National Debate over Nuclear Power," *J. Risk and Uncertainty* 3:327–45 (1990).

55. United Press International. "Islanders Protest Construction of Nuclear Waste Storage Facility." New York: United Press International News Wire Story, Nov. 8 (1990).

56. Dunlap, R.E. and M.E. Olsen. "Hard-Path Versus Soft-Path Advocates: A Study of Energy Activists," *Policy Studies J.* 13:413–428 (1984).
57. Gould, L.C., G.T. Gardner, D.R. DeLuca, A.R. Tiemann, L.W. Doob and J.A.J. Stolwijk. *Perceptions of Technological Risks and Benefits* (New York: Russell Sage, 1988).
58. Mitchell, R.C. "Rationality and Irrationality in the Public's Perception of Nuclear Power," in *Public Reactions to Nuclear Power: Are There Critical Masses?*, W.R. Freudenburg and E.A. Rosa, Eds. (Boulder, CO: American Association for the Advancement of Science/Westview, 1984), pp. 137–179.
59. Rosa, E.A. and W.R. Freudenburg. "Nuclear Power at the Crossroads," in *Public Reactions to Nuclear Power: Are There Critical Masses?*, W.R. Freudenburg and E.A. Rosa, Eds. (Boulder, CO: American Association for the Advancement of Science/Westview, 1984), pp. 3–37.
60. Dietz, T., P.C. Stern and R.W. Rycroft. "Definitions of Conflict and the Legitimation of Resources: The Case of Environmental Risk," *Sociol. Forum* 4:47–70 (1989).
61. Johnson, B.B. and V.T. Covello, Eds. *The Social and Cultural Construction of Risk: Essays on Risk Selection and Perception.* (Dordrecht, Holland: D. Reidel, 1987).
62. Kraft, M. and B.B. Clary. "Citizen Participation and the NIMBY Syndrome: Public Response to Radioactive Waste Management," *West. Polit. Q.* 44:299–328.
63. Kasperson, R.E. "Six Propositions on Public Participation and Their Relevance for Risk Perception," *Risk Anal.* 6:275–281 (1986).
64. Fiorino, D.J. "Technical and Democratic Values in Risk Analysis," *Risk Anal.* 9:293–99 (1989).
65. Fowlkes, M.R. and P.Y. Miller. "Chemicals and Community at Love Canal," in *The Social and Cultural Construction of Risk: Essays on Risk Selection and Perception*, B.B. Johnson and V.T. Covello, Eds. (Dordrecht, Holland: D. Reidel, 1987), pp. 55–78.
66. Levine, A.G. *Love Canal: Science, Politics, and People* (Lexington, MA: Lexington, 1982).
67. Krauss, C. "Community Struggles and the State: From the Grassroots — A Practical Critique of Power," paper presented at Annual Meeting, American Sociological Association, Chicago (1987).
68. Freudenberg, N. "Citizen Action for Environmental Health: Report on a Survey of Community Organizations," *Am. J. Pub. Health* 74:444–448 (1984).
69. Edelstein, M.R. *Contaminated Communities: The Social and Psychological Impacts of Residential Toxic Exposure* (Boulder, CO: Westview, 1988).
70. Brown, P. "Popular Epidemiology: Community Response to Toxic Waste-Induced Disease in Woburn, Massachusetts," *Sci. Technol. Human Values* 12:78–85 (1987).
71. Couch, S.R. and J.S. Kroll-Smith. "Chronic Technical Disaster: Toward a Social Scientific Perspective," *Soc. Sci. Q.* 66:564–75 (1985).
72. Kroll-Smith, J.S., S.R. Couch, and A.G. Levine. "Technological Hazards and Disasters," in *Handbook of Environmental Sociology*, R.E. Dunlap and W. Michelson, Eds. (Westport, CT: Greenwood, in press).

73. Finsterbusch, K. "Community Responses to Exposures to Hazardous Wastes" in *Psychosocial Effects of Hazardous Toxic Waste Disposal on Communities*, D.L. Peck, Ed. (Springfield, IL: Charles C Thomas, 1989), pp. 57–80.

74. Weber, M. "Science as a Vocation," in *From Max Weber: Essays in Sociology* H.H. Gerth and C.W. Mills, Trans. and Ed. (New York: Oxford, [1918] 1946), pp. 129–156.

75. Freudenburg, W.R. "Risk and Recreancy: Weber, the Division of Labor, and the Rationality of Risk Perceptions," *Soc. Forces* 71:909–932 (1993).

76. Slovic, P. "Perceived Risk, Trust, and Democracy," *Risk Anal.* 13:675–682 (1993).

77. Stoffle, R.S., M.W. Traugott, C.L. Harshbarger, F.V. Jensen, M.J. Evans and P. Drury. "Risk Perception Shadows: The Superconducting Super Collider in Michigan," *Practicing Anthropol.* 10:6–7 (1988).

78. Rayner, S. and R. Cantor. "How Fair Is Safe Enough? The Cultural Approach to Technology Choice," *Risk Anal.* 7:3–9 (1987).

79. Mushkatel, A.H., K.D. Pijawka, P. Jones and N. Ibitayo. *Governmental Trust and Risk Perceptions Related to the High-Level Nuclear Waste Repository: Analyses of Survey Results and Focus Groups* (Carson City, NV: Nuclear Waste Policy Office, 1992).

80. Jacob, G. *Site Unseen: The Politics of Siting a Nuclear Waste Repository* (Pittsburgh: University of Pittsburgh Press, 1990).

81. Freudenburg, W.R. and T.I.K. Youn. "A New Perspective on Problems and Policy," *Res. Soc. Prob. Pub. Policy* 5:1–20 (1993).

82. Freudenburg, W.R. and S.K. Pastor. "NIMBYs and LULUs: Stalking the Syndromes," *J. Soc. Issues* 48:39–61 (1992).

83. Clarke, L. and J.F. Short, Jr. "Social Organization and Risk: Some Current Controversies," *Ann. Rev. Sociol.* 19:375–99 (1993).

84. Bella, D.A. "Organizations and Systematic Distortions of Information," *J. Prof. Issues Engin.* 113:117–29 (1988).

85. Bernard B. *The Logic and Limits of Trust* (New Brunswick, NJ: Rutgers University Press, 1983).

86. Slovic, P., M. Layman, N. Kraus, J. Flynn, J. Chalmers and G. Gesell. "Perceived Risk, Stigma, and Potential Economic Impacts of a High-Level Nuclear Waste Repository in Nevada," *Risk Anal.* 11: 683–96 (1991).

87. Marrett, C.B. "Public Concerns About Nuclear Power and Science" in E.A. Rosa and W.R. Freudenburg, Eds., *Public Reactions to Nuclear Power: Are There Critical Masses?* (Boulder, CO: American Association for the Association for the Advancement of Science/Westview, 1984), pp. 307–328.

88. Frank, J. "Mr. Justice Holmes and Non-Euclidean Legal Thinking," *Cornell Law Q.* 16:568–603 (1932).

89. For a further discussion of this approach, also known as "formal jurisprudence," see Monahan, J., and L. Walker. *Social Science in Law: Cases and Materials* (Mineola, NY: Foundation Press, 1985), pp. 1–31.

90. Nealey, S.M. and J.A. Hebert. "Public Attitudes toward Radioactive Wastes," in *Too Hot to Handle? Social and Policy Issues in the Management of Radioactive Wastes*, C.A. Walker, L.C. Gould and E.J. Woodhouse, Eds. (New Haven, CT: Yale University Press, 1983), pp. 94–111.

91. Johnson, R. and W.L. Petcovic. "What Are Your Chances of Communicating Effectively with Technical or Non-Technical Audiences?" paper presented at Annual Meeting, Society for Risk Analysis, Washington D.C. (1988).

92. Freudenburg, W.R. and S.K. Pastor. "Public Responses to Technological Risks: Toward a Sociological Perspective," *Sociol. Q.* 33:389–412 (1992).

93. Freudenburg, W.R. "Perceived Risk, Real Risk: Social Science and the Art of Probabilistic Risk Assessment," *Science* 242:44–49 (1988).

94. Bella, D.A. "Organizations and Systematic Distortions of Information." *J. Prof. Issues Engin.* 113: 117–29 (1987).

95. Sterling, T.D. and A. Arundel. "Are Regulations Needed to Hold Experts Accountable for Contributing 'Biased' Briefs of Reports that Affect Public Policies?," in C. Whipple and V.T. Covello, Eds., *Risk Analysis in the Private Sector* (New York: Plenum, 1985), pp. 243–256.

96. Davis, M. "Explaining Wrongdoing," *J. Soc. Philos.* 20:74–90 (1989).

97. This discussion draws from Davis, note 96; the quotation is drawn from volume 1, page 93, of *The Presidential Commission on the Space Shuttle Challenger Disaster* (Washington, D.C.: June 6, 1986).

98. See, for example, Cerrell Associates, Inc., *Political Difficulties Facing Waste-to-Energy Conversion Plant Siting.* California Waste Management Board, Technical Information Series. (Los Angeles, CA: Cerrell Associates, Inc., 1984).

99. U.S. Atomic Energy Commission. *Annual Report to Congress of the Atomic Energy Commission for 1967* (Washington, D.C.: U.S. Atomic Energy Commission, 1968).

100. For further discussion, see the contributions in W.R. Freudenburg and E.A. Rosa, Eds., *Public Reactions to Nuclear Power: Are There Critical Masses?* (Westview Press/American Association for the Advancement of Science, 1984).

101. Freudenburg, W.R. and R.K. Baxter. "Nuclear Reactions: Public Attitudes and Public Policies Toward Nuclear Power Plants," *Policy Studies Rev.* 5:96–110 (1985).

102. Flynn, J. "How Not to Sell a Nuclear Waste Dump." *Wall Street Journal* (April 15, 1992), p. A20.

103. The key report from the National Academy of Sciences/National Research Council was *The Adequacy of Environmental Information for Outer Continental Shelf Oil and Gas Decisions: Florida and California* (Washington, D.C., National Academy Press, 1989). The decision by President Bush was announced on June 26, 1990. For further discussion, see W.R. Freudenburg and R. Gramling *Oil in Troubled Waters: Perceptions, Politics, and the Battle Over Offshore Drilling* (Albany, NY: SUNY Press, 1994).

104. Merton, R.K. *On Theoretical Sociology: Five Essays, Old and New* (New York: Free Press, 1967).

105. Slovic, P. "Perception of Risk." *Science* 236:280–285 (1987).

106. For supportive evidence on this point, see Rothbart, M. and B. Park. "On the Confirmability and Disconfirmability of Trait Concepts." *J. Pers. Soc. Psychol.* 50:131–42 (1986).

AN OVERVIEW OF ENVIRONMENTAL RISK DECISION MAKING: VALUES, PERCEPTIONS, AND ETHICS*

3

C. Richard Cothern

CONTENTS

Introduction 38
Values 39
    Introduction 39
    General Characteristics 39
    Views of Values 40
    Values in Environmental Risk Decisions 41
Perceptions (Including the Idea of Values in Quantitative Risk
Assessment) 43
    General 43
    Examples 44
Ethics 47
    Introduction 47
    What Is Ethics? 48
    Risk and Ethics 49
    Theology 50
    Environmental Ethic 51
Decision Making 52
    Introduction 52

* This volume is based on the proceedings of a symposium "Environmental Risk Decision Making: Values, Perceptions and Ethics" held by the Environmental Division of the American Chemical Society at their National Meeting in Washington, D.C., August 24, 1994. The participants in the symposium provided chapters for this volume and additional chapters were added to flesh out some themes.

The thoughts and ideas expressed in this paper, symposium, and book are those of the contributors and participants and do not necessarily reflect the policies of the U.S. Environmental Protection Agency.

1-56670-131-7/96/$0.00+$.50
© 1996 by CRC Press, Inc.

Environmental Risk Decision Models: Values, Perceptions
and Ethics 53
    Ideal Model... 54
    The National Academy of Sciences "Red Book" Model 54
    Cost–Benefit Analysis 54
    A Framework Model 55
    A Channel Model 56
    An Overlay Model 57
    Continuous Model 58
    Conclusions... 60
    The Big Picture.. 60
    Risk 63
Conclusions 64
Bibliography.. 65

INTRODUCTION

Values and ethics should be included in the environmental decision-making process for three reasons: they are already a major component although unacknowledged; ignoring them causes almost insurmountable difficulties in risk communication; and it is the right thing to do.

Values and value judgments pervade the process of risk assessment, risk management, and risk communication as a major factor in environmental risk decision making. Almost every step in any assessment involves values and value judgments. However, it is seldom acknowledged that they even play a role. The very selection of methodology for decision making involves a value judgment. The selection of which contaminants to study and analyze involves value judgments. Weighing different risks involves value judgments. We cannot, and should not, exclude values and value judgments from the environmental decision-making process, as they are fundamental to understanding the political nature of regulation and decisions that involve environmental health for humans and all living things.

One of the major problems in risk communication is the failure of different groups to listen to each other. For example, many animal rights groups object to the use of animals in toxicological testing on ethical and moral grounds. The American Medical Association and other scientific groups have mounted a response that argues that many human lives have been saved (life lengthened) by information gained from animal testing. Both sides have a point, but neither is listening to the other. These represent two different value judgments and these values are the driving force in the different groups. It is essential to understand this and include it any analysis that hopes to contribute to understanding in this area. Any analysis must include values such as safety, equity, fairness, and justice, as well as feelings such as fear, anger, and helplessness. These values and feelings are often the major factor in effectively communicating about an environmental problem.

Last, including values such as justice, fairness, and equity (present and intergenerational) is the right thing to do. Any effective environmental program needs to be ethical to survive in the long term.

This chapter includes sections on values, perceptions, and ethics followed by a discussion of how and where these enter in the environmental risk decision-making process.

VALUES

Introduction

Different people looking at the same set of environmental data and information can come to different conclusions due to different value systems. Values and value judgments enter at every stage of environmental decision making and thus affect the outcome in a real, continuous, and profound way. Even the selection of which problems to study involves a value judgment. There is no value-free inquiry. Values enter the process when the information is incomplete. The choice of assumptions or default involves a value judgment. Because the world, nation, state, locality, or even two professionals can have different value systems, the place of value judgments in environmental risk decision making is central.

Values are different in different cultures. Americans say the squeaky wheel gets the grease, while the Japanese say the nail that stands out gets pounded down. The cardinal American virtues of self-reliance and individualism are at odds with those of most non-Western cultures.[1] Our current linear and Cartesian way of thinking shows "an imbalance in our thoughts and feelings, our values and attitudes, and our social and political structures" along with our ethical sensibilities.[2] An example of a value judgment in a major decision occurred in the few weeks before the University of Utah announced that a member of their chemistry department had discovered cold fusion. Someone in the group asked the question, "what if this gives a terrorist the ability to make a nuclear bomb for fifty dollars?" They decided that this was too profound to contemplate in the short time they had, so they decided to ignore it! Those involved did not have the tools to analyze the values, perceptions, and ethics involved. And the question was not mentioned at the press conference or afterwards.[3]

In the following, different views concerning values, their characteristics, and involvement in environmental risk decisions will be examined.

General Characteristics

The concept of values is a general as well as specific term, involving examples such as: aesthetic values, scientific values (accuracy, coherence), and ethical values (maximize honor, autonomy, self-determination, doing good for individuals, justice), as well as others. We are here interested in those values that are directly or indirectly involved in environmental risk decision making.

In general, values operate throughout the decision analysis process and often permeate that process. However, the values concerned citizens and leaders act upon are likely not representative of carefully worked out systems, and there may be differences between personal values and those of the community. Most find it difficult to say in detail what their own values are because in the U.S. and other Western countries there is no unified morality, and religious concepts have played a very small role in ethical theorizing. The Western democratic tradition puts great value on justice, fairness, equality, democracy (can technical values be reconciled with democratic ones — see Reference 4), autonomy, and responsibility. We believe that these are good values, and that societies (including our own) should be evaluated according to the extent they promote such values.

Other values we consider important include: health, quality of life, responsibility, truth, equity, stewardship, honesty, sanctity of the individual life, exceeding the "limits", dependence of all living things on each other, and spiritual and emotional balance.

In many administrative processes there is a requirement that facts and values be separated, although this may not be always possible. Important ethical and values questions can be distorted through the use of language of technical "experts" such as shoptalk that further complicates this separation. It is possible that by the translation of environmental problems into technical and scientific language the value questions are distorted or even lost.

Views of Values

We are all biased and this has important implications for environmental risk decision making. We may be biased because of our educational backgrounds and bring different values to the activity of environmental risk decision making. The scientist focuses on truth, the psychologists on feelings, the theologian or philosopher on the meaning of life, the journalist on what is news, the economist on allocation of scarce resources, an individual on NIMBY (not in my back yard), the attorney on winning, and so forth. This is not to be judgmental, but to acknowledge reality.

Scientists are taught to value scientific truths above other truths because ideally, scientific truths are usually never accepted until they have been publicly tested. In contrast, since the "truth" of ethical positions cannot be empirically verified in the same way and is therefore less "objective" than scientific truth, many scientifically trained people express open hostility to ethical discourse and value judgments. Ethical questions are often called "soft" or "fuzzy", in contrast with scientific questions and solutions that are supposed to be "hard". Some suggest that expert formulations of scientists are more rational and valid than the more intuitive, subjective, and thus irrational judgments of the lay public.[4]

New values in health and the environment have emerged since World War II, due to ideas such as freedom from illness, physical and mental fitness (exercising), control of infectious diseases with antibiotics, and a new focus on reproductive, developmental and immune diseases as well as degenerative changes.[5] There has also been an increased interest in consumer values due to changes like a shorter work week, more leisure time, and a greater role for the family. Prior to WW II the land between cities and the wetlands were something that no one wanted — now they are to be valued and protected.

Values in Environmental Risk Decisions

The development of the low-dose effects paradigm is based on a value judgment. The concepts involved of a linear dose–response curve with no-threshold for the shape and character of the curve are not based on any available scientific information or data. These concepts were developed by noting what would likely yield the highest estimates of risk for low dose exposures. In the absence of known data, this choice would cause regulation to err on the side of safety in setting standards. Sagan contends that too often this assumption is accepted as scientific fact. He states that scientists have a responsibility to separate fact from value judgments.[6]

That we are unwilling to experiment on humans but do so on animals is a value judgment that places other living things on a lower level. In setting standards for environmental contaminants we have a choice of using the average person or the most susceptible — this choice is value laden.

A possible organizing idea is the value of integrity. This proposal is systematically examined in a volume by Westra[7] that derives from a quote from Aldo Leopold from *The Land Ethic* — "A thing is right when it tends to preserve the integrity, stability, and beauty of the biotic community. It is wrong when it tends otherwise." This practical philosophical proposal is nonanthropogenic in its eventual direction, and involves cultural, ethical, philosophical, scientific, and legal aspects. The values involved in the idea of integrity include: freedom, health, the whole, harmony, biodiversity, sustainability, life, morality, and scientific reality.

The importance of human life is a value-laden concept. Is one human life sacred, or do we balance numbers and in the interest of efficiency save the largest numbers? Values associated with life and death are important in environmental risk decisions. With our societal denial of death we credit standards with saving life when it only lengthens life. We seem much more concerned with contaminants or health effects that shorten life as opposed to those that cause sickness. It is a value judgment that we think that contaminants that cause cancer are more important than those that cause neurotoxic, immunotoxic, or developmental effects. Is it ethically sound to allow exposures to rise to the level given by a standard?

It is a value judgment and perception of the public that estimates based on risk assessments are not believable because they do not trust the scientists that generate them.

There are other components in quantitative risk assessment where value judgments enter which include: uncertainty, no causal link or only a correlation, synergism or antagonism, latent period, morbidity vs. mortality, hormesis, threshold, comparing different health endpoints (or which are the more important?). Scientists often disagree on these issues.[8] In each of these cases, the risk assessor must make assumptions to complete the analysis — the choices are value laden.

One clearly value-laden decision is what is an acceptable risk? Or what is a safe level? Each of us has different levels of risk that we would find acceptable. There is no universal acceptable level. Some of the values that affect our individual decision on this question are: is it voluntary or involuntary, old or new, catastrophic or ordinary, known or unknown?[9]

In the "subjective" areas described by the channel model, values enter more obviously and in some cases are defaults similar to those in the "objective" areas, including:

Freedom: I do not care what the risk is; I am free to not use my seat belt, to smoke cigarettes, etc.

Equity: factory siting, waste sites, incinerators, etc., may be put near the poor and politically weak; or more generally, there is a conflict between private interest and public good

Trust: do not trust some scientists because they cannot even agree, e.g., emf fields; do not trust the government since some politicians are crooked

Quality of life: things that make my life better are good (hopefully by not damaging other living things)

Safety: err on the safe side by using the value-laden linear no-threshold dose–response curve assumption (or is it due to reaction to Rachael Carson's *Silent Spring*, Ernest Sternglass' *The Death of All Babies*, and the fallout debate?)

Stewardship: conservation of wetlands, trees, living creatures, a two-edged sword, gene-pool reduction, deforestation, species extinction

Natural is good: radon apathy, natural carcinogens, responsibility for environmental protection from toxic material and hazardous substances, sustainable development

Indoor air pollution: my home is my castle

Upstream or downstream: values differ with respect to position

Anthropocentric or biocentric: values differ by point of view

Too often default assumptions, such as nonthreshold, whatever happens to animals will happen to humans, most exposed or susceptible individual, are accepted as "science policy" or "expert judgment". Without careful scrutiny, these can lead to politically controversial results which are challenged as arbitrary rules that have no basis in either science or public policy.[10] By examining the value dimensions of this process we can get a better and more useful perspective concerning the environmental risk decision process.

Scientists make value judgments when they choose to research those problems with the largest funding levels or those most politically important. Choosing topics that would save lives would be an alternative value-laden decision. This lack of principle may be due to risk assessment being a new field and without a philosophical base.[11]

The value judgments of all involved in risk assessments and risk decisions have a strong effect on their nature, character, and outcomes. The value-laden approach is used widely in making risk decisions without much acknowledgment.

PERCEPTIONS (INCLUDING THE IDEA OF VALUES IN QUANTITATIVE RISK ASSESSMENT)

General

Perceptions are flavored by emotional feelings (such as fear, guilt, and embarrassment), limited by lack of educational background (e.g., they are quantitative in probability, uncertainty, reading graphs), steeped in biases (cultural, social, gender), confused by language (we hear what we want to, different connotations of words), and thus provide a block to the communication of facts in general and environmental risks specifically. "Actual, measurable risks are assumed to belong to the real world of hard, material things, whereas perceived risks are thought to lie in the domain of fallible human beliefs and intuitions" is a quote that sums up how too many view this situation.[10] Many people believe that what is really happening is not nearly as important as what we think or believe is happening.

Perceptions are deeply rooted in our feelings and emotional being as well as the cultural backgrounds in which each of us developed. "How people interpret a given set of facts about risk may depend on a host of variables, such as their institutional affiliations, their trust in the information provider, their prior experience with similar risk situations, and their power to influence the source of the risk."[10]

Perceptions are closely tied to values and for too many people the moral and ethical test is whether it feels right, and thus judgment is based too often only on feelings. "Our values, and therefore our actions, are closely tied in with our perceptions."[12] The perception is that the criterion is how we think we ought to be treated.

To be better able to understand the decision-making process, it would seem helpful to separate feelings, perceptions, scientific facts, and professional judgments. This is not necessarily to make a judgment about the various components and their relative weight in a decision, but merely to recognize the components and their role. Another spin on this question is to ask, why does the U.S. insist on making public policy on an "objective" basis instead of a value or cultural basis? It is not immediately clear what the basis should be; however it does seem desirable to recognize what the basis is.

The public suffers from a limitation in understanding in that some perceptions are inaccurate, risk information may frighten people, and strong beliefs are hard to modify. In this area of risk communication there has been research and thought. Some feel that the use of two-way communication is an important missing ingredient. Others observe that we seldom talk to each other; usually we talk past each other. One observation is that it might be better to reduce the use of words with negative connotation such as: death's uncertainty, regulation, rule, law, fear, embarrassment. It would be better to use positive thoughts such as: stewardship, quality of life, justice, freedom, and Mortimer Adler's six great ideas, viz., truth, goodness, beauty, liberty, equality, and justice. This leads to one final question: how important are opinion polls that show majority feelings? What role should these play in environmental risk decision making?

Many have observed that everything is connected to everything else. In that sense and in even a deeper sense, values, perceptions, and ethics are connected. On the other hand, no two people share the same perception of anything.

"Science has never been more successful nor its impact on our lives greater, yet the ideas of science are alien to most people's thoughts." This and other similar observations in Wolpert's volume, *The Unnatural Nature of Science,* show that there is a deep-seated fear of science.[13] "Science is perceived as materialist and as destructive of any sense of spiritual purpose or awareness; it is held responsible for the threat of nuclear warfare and for the general disenchantment with modern industrial society that pollutes and dehumanizes....The practitioners of science are seen as cold, anonymous and uncaring technicians." The central theme presented in this book is that many of the misunderstandings about the nature of science might be corrected once it is realized just how "unnatural" science is. He argues that science is not constructed on a common sense basis, and that it requires a conscious awareness of the pitfalls of "natural" thinking. This is consistent with the theme of many that "natural equals good". Scientists are seen as "meddling" with nature and callous to the ethical and social implications of major issues like nuclear weapons, genetic engineering, and similar issues.

Examples

The concept of quantifying perceptions and value judgments is a useful one in overall risk assessment so that the various contributions can be weighted according to their importance. Perceptions and value judgments have been analyzed quantitatively and those listed below were found to differ by one to two orders of magnitude: natural/manmade; ordinary/catastrophic; voluntary/involuntary; delayed/immediate; controlled/uncontrolled; old/new; necessary/luxury; and regular/occasional.[9]

There are problem areas in quantitative risk assessment and comparative risk assessment where no one seems to listen to the risk numbers or other scientific and technical information and depends almost primarily on perceptions, e.g.,

- Radon: people think, if it is natural and I cannot sense it, it cannot be bad
- Superfund: people think if I can smell it, it must be bad; tend to mistrust industry
- Nuclear power (fear of the bomb): Chernobyl, TMI
- Taking lead out of gasoline: people wonder, did we do the right thing for the wrong reason?
- Dioxin: this is touted as the most toxic chemical known, but not necessarily the most toxic to man
- Fluoride: people react with fear, ignorance, and lack of data
- Alar: people heard "children", and paid no attention to actual risk numbers
- Pesticides: the perception is that they are useful and needed
- For cancer, AIDS, *Legionella:* the overlay is fear of the unknown and helplessness
- For emf: overlay is hearing only about childhood leukemia
- Plutonium: half life in billions of years (metals last forever!)
- Global climate: we seem to see the effects only

A common theme is that we do not trust government, industry, scientists and other "experts" and feel helpless to argue against them.

The controversy concerning the use of animals in medical research and toxicological testing provides an example of a situation where neither side seems to be listening to the other. Animal rights activists appear to be concerned about the ethics and moral aspects and question the value judgments involved. They do not seem to have any trust in those who use animals in their studies. On the other side, the medical profession argues that animal studies have led to many major advances in our understanding of the human body and how to care for it. The American Medical Association has developed a large resource kit entitled "Medical Progress: A Miracle At Risk" for physicians to use in explaining to their patients the importance of the use of animals in providing a better and more healthy world for us. The question one must ask is whether either of these groups is listening to the other. The answer appears to be no. Each side has strong opinions and feelings that are understandable. Why are they not listening to each other?

A common theme is the problem of what to compare things to. For example, how should we view our responsibility to future generations? What are the responsibilities of our current generation to future generations in considering how to dispose of nuclear waste? One way to get perspective on this comparison is to accept that most of our nuclear waste was generated by the weapons program and this is part of our defense. The risk to future generations due to buried nuclear waste might be compared to the risk to current populations from unexploded munitions in past battlefields like those from WW I, WW II, Vietnam, etc. Another perspective on the dilemma of responsibility for future generations is that the investment of $1 today would be worth considerably more in 1000 or 10,000 years.

Most people tend to assign responsibility by focusing on the origin of a problem or on who or what has the power to alleviate it — for example, the focusing of public attention on the state of Boston Harbor led many to think that Governor Dukakis was the cause of pollution rather than an agent of treatment and control.[14] Responsibility comes in many forms, depending on our perception. We feel responsible to society at large and view some actions as our duty; we feel responsible to a group, friends, family, and ultimately, for ourselves.

We can observe that citizen protests can result from the perceived failure of government and industry to protect the health and safety of the people — e.g., putting nails and tacks on the highways to prevent burying of cattle with PBBs, digging trenches to prevent a landfill operation, etc.[15]

The problem of perceptions has entered the classroom in the conflict between environmentalists, who are encouraging teachers to impart more complex and controversial messages, and parents, who complain that children genuinely fear some of the more apocalyptic predictions about the fate of the Earth.

At a recent symposium organized by the Episcopal church, the audience responded to an invitation by a speaker to share some of their thoughts concerning their perceptions of environmental risk. Some of the ideas that were expressed included:

- People, the church, and politicians do not take the environmental problems seriously
- There is a lack of information and understanding about environmental problems
- There is apathy stemming from fear and despair
- Environmentalism is too confrontational, rather than cooperative
- There is a tension between preservation, conservation, and sustainable development
- There is confusion regarding individual and collective values.

All of these show a wide gap between the lay person and the reality of our environmental situation.

Another dimension of perceptions is denial. Why do some hazards pass unnoticed or unattended, such as: nuclear winter; the Sahel famine of 1983 to 1984; chemical pesticides; and problems of burning coal (as opposed to burning uranium). Other examples of this phenomenon include: asbestos, desertification, deforestation, gene-pool reduction, hazardous wastes, indoor air pollution, soil erosion, and toxic material.[16]

ETHICS

Introduction

Involvement of ethics in decision making is not a new idea. Boulding contended that ethics enters at two points:[17] first, in choosing the alternatives; and second, in ordering the alternatives. However, many risk decision makers have not included this dimension, which can be a serious mistake.

Many observers have noted the importance of ethics in risk decision making. Edward O. Wilson asks "Is Humanity Suicidal?"[18] He relates that: "My short answer — opinion if you wish — is that humanity is not suicidal, ... But the technical problems are sufficiently formidable to require a redirection of much of science and technology, and the *ethical issues* are so basic as to force a reconsideration of our self-image as a species." Another thoughtful observer, William Lowrance, in his volume entitled *Modern Science and Human Values,* commented that "rights are a fundamental, but in many ways insubstantial, basis for moral and *ethical* principles."[19] Another observation, from the area of bioethics, is that "Man's survival may depend on *ethics* based on biological knowledge."[20] This theme is further developed by Hassel in noting that personal salvation and human convenience have been pursued apart from planetary well being.[21] Many who think that science is ethically neutral confuse the *findings* of science, with the *activities* of science; data are neutral, but actions may not be.[37] All of these points suggest the importance of ethics and the lack of them in scientific and technical areas such as environmental risk decision making.

It seems that within the past decades ethics has moved out of the halls of academe and into the world of the marketplace because of controversies such as abortion, capital punishment, mercy killings, cloning, the treatment of animals, behavior of public officials, the morality of medical practice, and decisions of researchers — this new enterprise is often called "practical ethics". One observer claims that environmental ethics is different from these other types of applied ethics in that it involves the community, while other ethics involve the individual.[22] However, ethical thinking is becoming more involved in risk decision making, as evidenced by the Society For Risk Analysis including a section on ethics in their annual meeting and by the extensive literature listed in the bibliography of this chapter.

There is a need to involve ethical discourse in science and environmental risk decision making for the following reasons: it will assist in resolving potential conflicts, it will prevent default values based on supposedly value-neutral analysis, those making decisions need to know the value judgments imbedded in the information available, most scientific information contains uncertainty and may be easily thwarted by different value judgments, and these are often the normative principles that are actually used in the decisions.[25]

What Is Ethics?

Perhaps the earliest writings concerning ethics are those of Aristotle, recorded about 50 B.C. He was seeking to discover the good life for mankind: a life of happiness. To achieve happiness, he counseled finding the midground between excess and deficiency in what is often called the "Golden Mean" doctrine. Aristotle's ends or "finals" included: happiness, good, right, virtue, and blessed, which he considered as fixed. The ways to achieve these ends, the means, are changeable and cultural and include: beauty, bravery, courage, education, equality, health, honesty, honor, justice, lawfulness, learning, nobility, pleasure, political power, rationality, reason, victory, wealth, and wisdom. Whether a person finds himself or herself in a situation where an act is voluntary or involuntary is also an important determination of what is the proper course of action.

A hierarchy of ethics is shown schematically below[23]

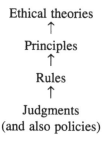

Ethical theories
↑
Principles
↑
Rules
↑
Judgments
(and also policies)

The test of the value of ethical theories includes, among other things, that they be clear, internally consistent, complete and comprehensive, simple, and able to account for the whole range of moral experiences (including our ordinary judgments). Also involved in the development of ethical theories are nonmoral values such as pleasure, friendship, happiness, knowledge, health, freedom from pain, and moral values such as good and right.

Webster's Dictionary defines *ethics* as questioning what is good and bad or right and wrong, and also terms it as a system of moral principles. Ethics should be distinguished from the social sciences, such as sociology and psychology, which attempt to determine why individuals or groups make statements about what is good, right, or obligatory. Some ethical systems that define good include: Aristotelian ethics, utilitarianism, Kantian ethics, natural rights, and Rawlisian contract theory.

An ethical theory attempts to decide what moral principles are correct. At a more abstract level, metaethics analyzes concepts like rights and duty. An ethical person is one who has any set of values and lives by them, has any set of values which are also shared by a group, and lives by a set of values which are universally valid. The term "ethicist" implies that the solution to value-laden problems is reducible to a series of questions which, using logical and codified methods of analysis, certain highly trained people are equipped to answer. An example of an ethical problem would be whether to tell those involved that the design of a toxic waste site that fulfills all the existing regulations could be negated by a local geological problem.[12]

Other ethical theories used in Western thinking include: situation ethics, which is a system that rejects rigidness and considers that right action is that motivated by love (agape); and existentialist thinking, where the value of choice is the courageous assertion of humanity and autonomy in a meaningless universe — one must choose for oneself.

Risk and Ethics

"In any analysis of a decision problem involving risks, ethical implications are introduced."[24] Because we are humans, our value systems are part of our decision-making process. Ethical concepts involved in such decision making include: value of life, getting what one wants, justice, equity, means and ends (if the ends do not justify the means, what does?).

In a risk decision, we ask if it is safe. In an ethical context, this is asking if it is right or good. Too often we ask the opposite in considering if something bad could happen and if so, how bad. To answer such an ethical question we need a way to determine what is bad, deleterious, undesired, wrong, unfair, unjust, and also what is good. The test is determined by the ethical system and definition of values therein and all this can get clouded by our perceptions. Another view would be to develop a base of information from the elements of risk/cost/feasibility, and overlay that with ethics, values, and perceptions. All too often the so-called quantitative and objective information is obscured by the overlay of values and perceptions.

Some observe that environmental decisions must be viewed primarily as ethical choices rather than as technically dictated conclusions. It is important in an age of increasing scientific complexity that interested parties attempt to understand the value positions and ethical issues that underlie scientifically derived policy choices. Experts and concerned citizens must realize that critical policy choices concerning environmental pollution and toxic chemicals are value judgments, matters of morality, and social and political judgments.[36]

The ethical interest impinges on decision making at two points. It impinges even at the first stage when we ask ourselves, "how do we come to know the range of possible alternatives", for there may be alternative ways of coming to know alternatives, and we have to make some kind of value judgment among them. At the second stage of the decision making process, in which the alternatives are subjected to value ordering, the ethical interest is clearly implied, for one of the major concerns of ethics is the evaluation of value orderings themselves. "Ethics, that is, is concerned with what might be called decision problems of the second degree, that is, decision about how decisions are going to be made, and according to what principles are they going to be made."[17]

Do risk decisions depend on ethics or do ethics depend on risk decisions? Or does decision making depend on risk, cost, feasibility, ethics, psychology, religion, politics, etc.? This point of view would reject the notion that risk assessment and risk management can be separated and thus recommends that the National Academy of Sciences "red book" on risk assessment methodology be rewritten to include these "subjective" aspects.

Since scientific, technical, and other specialists who may not be trained in ethics or value studies make judgments involved in risk decisions, it is important that those performing the risk assessment identify with clarity and precision uncertainties, assumptions, and ethical issues, as well as costs and other transscientific considerations involved.[36]

We need to involve ethical discourse in science and law because:

- It will assist in resolving potential conflicts
- It will prevent default values based on supposedly value-neutral analysis
- Those making decisions need to know the value judgments imbedded in the information available
- Most scientific information contains uncertainty and may be easily thwarted by different value judgments
- Ethics are often the normative principles that are actually used in the decisions[28]

Theology

In the related area of theology, one could wonder why environmental issues have not captured religious interest. Some suggest that this is due to their being too difficult and that they have multiple dimensions.[26] Another dimension is that technology has always found a solution to the problem — this is another kind of faith.

In a recent volume, *Models of God: Theology for an Ecological, Nuclear Age,* Sallie McFague claims that we need a new theology, or a reexamination of the one we do have.[27] I wonder *do* we really need a new theology, and is the one that she presents *really* new? My answer to both is NO. Perhaps what we need is a reaffirmation of our ethics. One of her contentions is that we need to adopt a holistic view that is more like the ecological view and that because of our ability to destroy all life with nuclear weapons we have a responsibility for all living things. Perhaps we need the holistic view simply because it is "right" or ethical. What is the reason that we should adopt the ecological view rather than the anthropocentric view? Her "models" are three — God as mother, lover, and friend. The corresponding values or ethics are justice, healing, and companionship. McFague asks if our nuclear capability shifts the human responsibility for the fate of the earth to God. She quotes from Jonathan Schell, "We have always been able to send people to their death, but only now has it become possible to prevent all birth and so doom all future human beings to un-creation." She further observes that to feel in the depths of our being that we are part and parcel of the evolutionary ecosystem of our cosmos is a prerequisite for contemporary Christian theology. Such a lack of attention leads at the very least to an attitude of unconcern for the earth that is not only our home but, if we accept the evolutionary, ecological paradigm, also the giver and sustainer of our lives in basic and concrete ways. If one were to practice Christian theology from the holistic perspective, it is evident that some significant changes from traditional models and concepts would be necessary for expressing the relationships between God and the world and between ourselves and the world.

We have too narrow a view of ethical issues in environmental matters because. humans are ecologically segregated and can exploit nature; we have forgotten natural history in light of human history; and we fail to recognize the relationships between humans and nature.[28] Nash advocates that we endeavor to make ethical values compatible with religious traditions.

There are references in the Bible that are pertinent and are discussed in more detail in *The Green Bible*.[29]

> God planted a garden in Eden in the east and there placed Man whom God created. Yahweh God caused to grow from the ground every kind of tree that is pleasing to see and good to eat, also the tree of Life in the middle of the garden and the tree of the Knowledge of Good and Evil. *Genesis* 2:8–9.

> For the destiny of humanity and animal is identical: death for one as for the other. Both have the same spirit: humans have no superiority over animals for all passes away like wind. Both go to the same place, both come from dust and return to dust. *Ecclesiastes* 3:18–20.

Environmental Ethic

"There is as yet no ethic dealing with man's relation to land and to the animals and plants which grow on it."[20] Also see Aldo Leopold's land ethic, where the central value is that "a thing is right when it tends to preserve the integrity, stability, and beauty of the biotic community. It is wrong when it tends otherwise."[30] As yet, there has been no similar environmental ethic put forward to guide our thinking in the area of environmental risk decision making. William O. Douglas, in his volume *The Three Hundred Year War*, observed that, "As a people we have no ecological ethic."[31]

If one would develop an environmental ethic, what would be involved? The following areas could be included:

- Oceans
- Future generations
- Space
- Animal rights
- Political boundaries
- Value of human rights

- Value of all life
- Resources
- Population
- Irreversibility
- Loss of species (flora and fauna)
- Preservation
- Other nonhumans
- Whose happiness is to be considered?
- What is safe?
- How can a balance be achieved?

Also in an environmental ethic we would like to include a list of moral virtues: thou shalt not erode, pollute, poison, make ugly, or irradiate the world. These are quite distinct from "thou shalt not kill, covet, steal, or deceive." Finally, an environmental ethic should include a place for values, perceptions, and ethics and a way to incorporate value judgments, keep them visible, and allow for change.

From the Presidents Council on Sustainable Development, the following should be considered: social equity, racial justice, population stabilization, improved quality of life, elimination of waste, reduced consumption, reduced poverty, and fairness in terms of rich vs. poor, equity, and sustainable development.

James Nash discusses the rights of nature in a chapter, "The Case for Biotic Rights."[32] A proposed bill of biotic rights includes the right to:

...participate in the natural competition for existence, health and whole habitats, reproduce their own kind without chemical, radioactive, or bioengineered distortions, fulfill their evolutionary potential with freedom from human-induced extinctions, freedom from human cruelty, flagrant abuse, or frivolous use, restoration, through managerial interventions, of a semblance of natural conditions, and a fair share of the goods necessary for the sustainability of one's species.[32]

DECISION MAKING

Introduction

The way we make decisions concerning the environment reflects the biases in our society, and that is acceptable if these are acknowledged.

Much more frequently than they may expect to, technical leaders and advisory committees find themselves drifting into — and then becoming immersed in — heated discussions of personal rights, or 'the natural', or obligations to future generations, or John Rawls's theory of justice, or the ethical dimensions of cost-benefit analysis. Often these considerations turn out not to be peripheral, but central. At issue may be the substance of decisions, or institutional or professional roles, or social procedure.[19]

The decision-making process is affected with the ethical dimension. It impinges at two points: first, when considering the values of the several possible alternatives; and second, in the ordering of those alternatives. Ethics involves the process of how decisions are made and according to what principles they are decided.[17]

Environmental Risk Decision Models: Values, Perceptions and Ethics

The existing models for environmental risk assessment do not contain any explicit mention of values, value judgments, ethics, or perceptions. However, these are often the main bases used in making such decisions.

For example:

- Alar was banned to protect *children*
- The linear, no-threshold dose response curve and the use of combined upper 95% confidence limits are based on *safety*, not science
- The Superfund program started with the idea that if I can sense it, it must be bad, while indoor radon has met with widespread apathy because it cannot be sensed, so why worry?
- The idea of zero discharge is based on the *sanctity of the individual*
- Forests and wetlands are preserved because of *stewardship*
- Nuclear power is avoided because of fear of *catastrophe*

In the specific area of risk assessment as it is involved in the environmental risk decision process there are numerous opportunities for value judgments. "Perhaps fifty opportunities exist in the normal risk assessment procedures for scientists to make discretionary judgments. Although scientists are presumed to bring to this task an expertise untainted by social values to bias their judgment, they are not immune to social prejudice, especially when their expertise is embroiled in a public controversy."[33]

The general theme here is to examine the place of values, value judgments, ethics, and perceptions in decision models. The hypothesis is that these characteristics are directly involved in current risk decisions, but that existing models do not include them. In some decisions, attempts are made to disguise these characteristics of values and ethics with other labels such as scientific or technical. Values and ethics seem like perfectly good ways to analyze, balance, and choose in the environmental risk decision-making process, and since they are widely used, why not acknowledge this and formally include them in the models?

Is the current and are the future environmental problems and decisions more complex and of a different character than those of the past? If so, then a new decision paradigm will be needed. Some have observed that the current environmental problems are characterized by levels of complexity and uncertainty never before experienced by any society.

Several models exist to describe the process of environmental risk decision making. Some of these will be presented and discussed here, including: an ideal model, the National Academy of Sciences "red book" model, a channel model, an overlay model, the cost-benefit analysis, and a proposed continuous model. These are not necessarily all the models that exist and are presented here to give an idea of the kinds that exist to focus thinking in the area of values, value judgments, ethics, and perceptions.

Ideal Model

The ideal situation is when all possible information is known about a situation, including the scientific and technical aspects; the health consequences of possible alternative actions and their alternatives; the exposure routes of all possible causes; the costs now and in the future; the social, political, and psychological consequences of all decisions and all other possible relevant information. Since this is not the case, and in general only fragments of the necessary information and data are available, it is folly to think that the ideal situation can ever be achieved. Decisions will have to be made with imperfect information and incomplete data. To keep perspective, it is well to use the perfect and ideal as a goal and develop methodologies that can help move closer to the ideal.

The National Academy of Sciences "Red Book" Model

The National Academy of Sciences regulatory decision model starts by combining hazard identification with dose–response assessment and then combining this with the exposure assessment to yield a risk characterization as shown below.

Hazard identification → Dose-response assessment
↓
Exposure assessment → Risk characterization

The regulatory decisions that emerge from this analysis use inputs from this risk characterization along with possible control options and nonrisk analyses such as economics, politics, and statutory and legal considerations as well as social factors.

Cost–Benefit Analysis

The idea of this analysis is to compare the benefits of a decision (such as preventing death or disease, reducing property damage, or preserving a resource) to the costs. This approach can be used to determine the best solution from among several options at the lowest costs. Any situation involves limited resources, and knowing how the costs and benefits compare is thought to be helpful. However, many of the benefits are difficult if not impossible to quantify, for example: the benefit of preserving a species, the aesthetic value of a forest, or how valuable it is to be able to boat and swim in a river or lake. Also many comparisons are difficult: the relative benefit of averting sickness or death, averting a cancer case, or a case of a birth or developmental defect. Almost all of the problem areas in cost–benefit analysis involve value judgments and thus this is an area that could be improved with the inclusion of value and ethical analysis.

In the past, many of the questions being raised here were addressed in what was called a cost–benefit study. Such an approach has its value; however some would disagree. "The world will end neither with a bang nor a whimper but with strident cries of cost–benefit ratio by little men with no poetry in their souls. Their measuring sticks will have been meaningless because they are not big enough to be applied to the things that really count."[34] Others have observed that examining risks and not also the costs and the benefits is like the Zen question of what is the sound of one hand clapping.

A Framework Model

It has been observed that risk assessment is an example of what some call a regulatory or mandated science.[35] This is one which tries to fill the gap between theoretical or laboratory science and making reliable and defensible regulatory or management decisions. Pure science is value free, and regulatory or mandated science is not. An alternative observation is that we all possess the keys to the gates of heaven and the same keys open the gates of hell.

Classical risk assessment involves two stages: factual judgment which is free of values and evaluation which is value laden. The classical model separates risk estimating from managing risk. That factual can be separated from normative (value laden), descriptive from prescriptive, risk assessment can be value free even though it is dominated by human judgment in the face of uncertainty. The classical model does not acknowledge the role of value-based judgment. Values can feed back between risk assessment and risk management without anyone realizing this.

The alachlor controversy is an example of the breakdown of the classical risk assessment model for decision making.[35] It was not a conflict between those who accept the verdict of the risk assessment and those who do not. It is also not a conflict between those who understand the objective risks and those who are guided by subjective perceptions. It is a political debate among different value frameworks, different ways of thinking about moral values, different concepts of society, different attitudes towards technology, and different ideas about risk taking.

Authors who analyzed the alachlor controversy concluded that: "A more realistic model of risk assessment, one that is sensitive to the role of values in the estimation of risk, is urgently needed."[35] They recommend a frame that includes the acknowledgment of the interconnections between the scientific and social policy elements.

The components of their framework model include:

- Attitude towards technology (positive or negative)
- Uncertainty (statistical, lack of knowledge, incomplete knowledge, methods to use)
- Risk taker or risk-adverse
- Causality (including confidence)
- Burden of proof, who has it and what are the criteria
- Rationality
- Voluntariness (John Stuart Mill's liberalism) or social order

The principal lessons learned from this analysis and proposal are **not** that we need to start a global debate on the meaning of rationality, the merits of technology, or the importance of voluntary risks — these issues are too broad. However, these are among the value issues that need to be addressed by risk assessors. "Sensitivity to the biases that are introduced by broad attitudes concerning rationality, technology and the liberal state should bring recognition by risk analysts that their activity is not, as they imagine, neutral and value-free."[36]

A Channel Model

There are several elements or "channels" that can be used to move from an environmental risk decision problem to possible solutions. Several of these are shown in Figure 1 as horizontal elements moving from the problem to the solution.

The model is arbitrarily separated into two areas: the so-called "objective value" elements such as risk, cost, and feasibility, and the "subjective value" elements like social, political, psychological, and safety elements. It is suggested that, although seldom explicitly mentioned, all of the elements involved in environmental risk decisions involve values, perceptions, and ethics.

All too often a decision concerning an individual environmental problem is made using only a few or even only one of the many elements shown above. In these cases, other horizontal channels depicted in the model above are known, but are ignored or overlaid with what the decision maker knowingly or unknowingly thinks are more important values. There appears to be no element that does not involve a value or ethical dimension. For example, the relative value of cancer, neurological, developmental, or immune endpoints as well as the relative importance of mortality and morbidity are value decisions made in the risk assessment process. Also the choice of model to describe the flow of a contaminant through the environment (and default assumptions) and the

Problem	Values or Value Laden Components	Solution or Decision

Environmental Problem Requiring Solution ⤷	**Objective, Hard or Quantitative**
	  Quantitative Risk
	  Comparative Risk
	  Cost
	  Feasibility
	
	Subjective, Soft or Qualitative
	  Social
	    Prejudice
	    Justice
	    Equity
	    Freedom
	    Trust (scientist, government, media)
	    Responsibility
	    Blame
	  Quality of Life
	    Job security
	    Self image
	  Safety (error on safe side)
	  Political (power)
	  Religious (e.g., stewardship)
	  Ethics (standards of moral values
	  Psychological (feelings)
	    Fear
	    Embarrassment (ignorance)
	    Guilt
	    Helplessness
	    Security
	  Life (prolong)
	  Judical (let someone else decide)
		⌐↓
		
		Decisions and Policies

Figure 1   Channel environmental risk decision-making model. (Figure courtesy of B. R. Cothern.)

extrapolation of a dose–response curve into the unknown involve values and perceptions. Discounting is a value-laden decision concerning estimating cost.

An Overlay Model

As a variation of the channel model where it was observed that each horizontal element was value laden, one can think of the horizontal elements as being value free. In that case, values, ethics, and value judgments are added as an overlay to the analysis. By adding the values at the end, one can easily lose sight of the critical features of a problem and focus almost completely on the value or ethic. An example of this approach is the use of the value of zero risk.

To the uninitiated or uninformed (or those who do not appreciate or understand the complexities), the most desired decision concerning environmental risk is the one that would result in no risk, zero risk, or zero discharge. Witness the recent laws in Massachusetts and Oregon using the idea of toxics use reduction (TUR) — they rest on a simple argument that the use of every toxic chemical should be reduced or eliminated. Other attempts to effect zero risk include the Delaney Amendment for pesticides in food, effluent guidelines for discharges into water, resistance to de minimus regulations (or any other minimization approach), and resistance to fluoridation of water, to name only a few.

To overlay information concerning an environmental problem with a value such as zero risk prevents perspective, and this simple-minded approach prevents any understanding of the risks actually averted or the cost of doing so.

Continuous Model

In the model proposed here, values, perceptions, and ethics enter the process in several places and do so continuously. These elements are inserted by many different individuals in the form of assumptions or defaults at different places in the overall process. These individuals include: scientists (physical, biological, and social), economists, attorneys, politicians, regulators, engineers, managers, and many other professions. Few of these individuals are trained in the use of values and ethics.

What is needed is a model that includes inputs at each stage from those trained in the use of values, value judgments, ethics, and perceptions such as that shown in Figure 2.

This model or paradigm is presented as one view or quick snapshot of a continuously changing process.

The first step is to gather the known information and put it onto a "common table" so that it can be intercompared, weighed, and balanced. All too often, only some of the existing information is assembled. However, in the sense of an ideal model, let us assume that much of the existing data and information is assembled, including inputs from all sectors, viz., government (federal, state, and local), regulators, industry, acadame, labor, public, journalists, environmentalists, and any other stakeholders in potential decisions. The risk assessment input includes: scientific and technical data and information concerning exposure and health effects to humans and other living things, qualitative and quantitative risk assessments as well as comparative risk assessments.

Options are generated in the next phase by considering the information and data gathered in the common table phase. These options are affected by the values of those generating them. The options are scrutinized to determine what information is missing, and research is instituted to develop it. At the same time, the values impinging on this decision are considered in a process that could be described as "YES, BUT...," and new and different inputs are selected and added and existing ones subtracted, due to value judgments created in the process of developing options. The first tension in the process is generated by the conflict between the values and value judgments of the decision makers involved.

The third step is to balance and chose among the options. In this process, values and value judgments affect the choices, and some options are discarded as unacceptable. An important step here is to test and check the options.

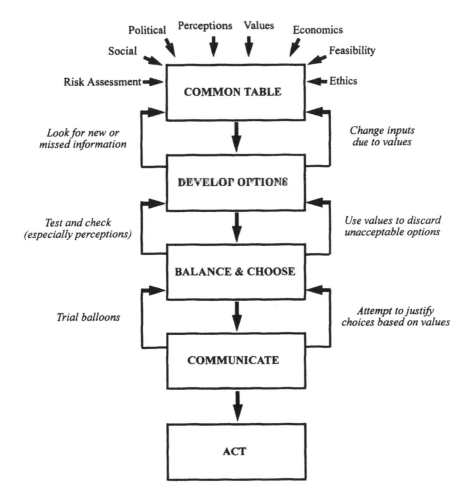

Figure 2 Snapshot in time of the continuously moving process of environmental risk decision making. (Figure courtesy of B. R. Cothern.)

The option chosen is then floated as a trial balloon to test its acceptability, and usually an attempt is made to justify the choice based on values and value judgments. Some of the possible values used include (this is a partial list): catastrophic vs. ordinary effect, dependence of all living things on each other, cancer vs. a noncancer endpoint, equity, emotional and spiritual balance, exceeding limits such as sustainability and physical ability of the earth, fairness, health, honesty, justice, protection, quality of life, relative value of humans and other living things, natural vs. man-made, nonsecrecy, privacy, responsibility, safety, sanctity of an individual life, severity of the effect, stewardship, and voluntary vs. involuntary.

The final acts are to communicate the decision and test it again, and lastly to act.

Conclusions

Others before have seen the wisdom of acknowledging the place of values in environmental risk decision making. In the conclusion of one these proponents' recent papers, the following statement sums the situation better than I can:

> In conclusion, environmental decisions must be viewed primarily as ethical choices rather than as technically dictated conclusions. It is important in an age of increasing scientific complexity that interested parties attempt to understand the value positions and ethical issues that underlie scientifically derived policy choices. Experts and concerned citizens must realize that crucial policy choices concerning environmental pollution and toxic chemicals are value judgments, matters of morality, and social and political judgments.[36]

One overall objective is to use the value of honesty and ask that the values, value judgments, and ethical considerations used in environmental risk decisions be expressed and discussed. To a scientist, Brownowski's comment that "Truth in science is like Everest, an ordering of the facts" is a most important value.[37] It is a conclusion of this line of thinking that we should unmask the use of values in environmental decisions and challenge decision makers to clearly state how they are using values.[38]

The Big Picture

Who in this world, country, locality, etc., thinks in the larger sense, about the big picture, or where it all is going? Who thinks beyond today, and perhaps tomorrow? Gore[39] and Meadows[44] suggest such an approach in their recent books.

We all have emotional blocks as well as skill deficiencies that will need to be overcome if comparative quantitative risk assessments and analyses of the overall or big picture are to be fully achieved. The fears and anxieties can be faced and the lack of quantitative thinking can be overcome with education.

Another barrier to developing a new view or a new paradigm is discussed by Howard Margolis in a volume entitled *Paradigms and Barriers,* in the context of scientific beliefs.[40] The same idea applies to a barrier of the mind in seeing and understanding the big picture of environmental risk decision making. He analyzes patterns of thinking and cognition in the area of science with the thesis that physical and mental habits are the same. He argues that the critical problem for a revolutionary shift in thinking lies in the robustness of the tacit habits of the mind that conflict with the new ideas. The current thinking in the area of risk decision making could use some change of habit in attempting to see the big picture.

Few people perceive a need to rank environmental problems, either on ordinal or cardinal scales. Some alternative reactions are that the "scientific" community understands that kind of thing, that it is too technical and, anyway, the government will protect us from such complex and unpleasant things. Many think that anything that can cause a deleterious problem should be eliminated, so why worry if one is worse than the other? Another component of this way of thinking is there are no bodies on the floor (at least no obvious ones) as a consequence of environmental problems, so why all the fuss?

It is important to try to develop a decision with knowledge of the biases and values of the decision maker. To help in this endeavor, we can use the concept of the veil of ignorance:

> ...parties situated behind this wall do not know how the various alternatives will affect their own particular case and they are obligated to evaluate principles solely on the basis of general considerations ... no one knows his place in society, his class position or social status, his fortune, strength, psychology, particular circumstances of their own society, and no information as to what generations they belong.[41]

This is a way to judge if a procedure, principle, ethic, etc., is fair. In the effort to develop an environmental ethic, "a modified Rawlisian theory of justice generates a relatively strong environmental and animal rights ethic."[42] The idea here is to understand the biases that the decision maker has.

Perhaps a more serious impediment (for all of us amateurs in environmental analyses) to being able to rank environmental problems quantitatively is that few think they have the training in quantitative and analytical thinking to be able to achieve this broad perspective. Few have complete knowledge about probability, statistics, uncertainty, and logarithms in or out of school. Further, most do not realize that they use this knowledge in everyday life. Many would argue that it takes too much time to learn about these concepts and that they have more important things to do. It seems paradoxical that the very concepts that would help in getting perspective about life are rejected because we have too little perspective about their value.

What are the potential consequences of continuing the status quo and not trying to get the big picture?

- We may lose our bearings without knowing the bigger picture
- The population could exceed the food supply
- Public health could deteriorate
- We could through diminished species upset the natural balance
- We could exceed the limits of spaceship earth
- We could change the global climate and reduce the amount of food available

One paradigm for the understanding of the big picture is an examination of seven world views or fragments of world views, all of which are sufficient to trigger a constraint on exploitation and development:[43]

- Judaeo-Christian stewardship
- Deep ecology and related value systems; rights to nonhuman nature
- Transformationalist/transcendentalism; wild nature has spiritual value because experience of nature can transform human perception and value
- Constrained economics; based on human economics
- Scientific naturalism; Darwin, emphasis on the natural
- Ecofeminism; gender domination
- Pluralism/pragmatism; practical problems solving

The history of environmental problem solving can be understood as a range of paradigms involving a constellation of beliefs, values, and techniques all requiring a changing array of data and information to achieve solutions. One analyst observes five paradigms of environmental management in development.[45] His paradigms and their dominant imperative are listed below:

- Frontier economics; progress as infinite economic growth and prosperity
- Environmental protection; trade-offs as in ecology vs. economic growth
- Resource management; sustainability as necessary constraint for green growth
- Eco-development; codeveloping humans and nature
- Deep ecology; antigrowth, constrained harmony with nature

What good is a world view, anyway? Without a view of the big picture we cannot put our individual problems and lives into proper context. If we cannot see the larger picture, we do not know the forces and will likely never know the causes of individual situations. It may not be best to make all decisions, or even many decisions from the larger view. However, without that view we can easily make the wrong decisions on the regional, local, or individual level.

A useful perspective on the importance of the big picture can be derived from analyzing five detailed discussions of examples of failed decision making in the federal government, viz., The Bay of Pigs, Pearl Harbor, Korean War, the Watergate cover-up, and the Vietnam War.[46] Also included in the analysis were two counterexamples: the Cuban missile crisis (same cast as for the Bay of Pigs) and the making of the Marshall Plan. Janis[45] provided some suggested recommendations, which are paraphrased here, as they apply to environmental risk decision making: all assumptions involved in a decision must be critically evaluated, there is a need for an impartial overview of all channels between a problem and potential solutions, there should be an independent review by different analysts, it is helpful to test-market the decision, and finally one should realize that a consensus may not be achieved and that in such cases one needs to accept that and also carefully examine all the contributing reasons from the perspective of a broad overview. One of the pitfalls that can cause these "groupthink" problems is the failure to challenge all the assumptions and rationalizations. Following are some of these in the area of environmental risk decision making that need challenging:

- Quantitative risk assessment is all we have
- Have to use the existing information
- Something is better than nothing
- *Yes* we know the default assumptions may not be valid, *but* that is all we have
- Denial that value judgments are involved and the claim that it is a scientific basis
- If we do not project a smooth face of unanimity, the public will think we do not know what we are doing

It would be a useful exercise to try to list all the assumptions and rationalizations and then critically examine each for validity, importance, and value.

Risk

The concept of uncertainty in risk assessment is too often used to cloud important issues. One observer contends that "uncertainty is always present in science-based decisions and what causes controversy and divisions among the interested parties is determining the amount or nature of the uncertainty that can be tolerated in order to resolve the issue...."[47] Further, she contends that the divorce between risk assessment and risk management is "uneasy" and is a result of "bad management and unscientific assessment." Whatever the cause, it is a process that is constantly being reexamined. Until the technique of risk assessment is more mature, it should be subject to question and change and constant reexamination.

In an analysis of risk assessors, Shrader-Frechette sees two different groups: cultural relativists who overemphasize the role of value judgments in risk assessment, and the naive positivists who claim risks can be objectively measured.[48] She proposes a middle position of scientific proceduralism where risks are not purely objective, but the role of values is put into more perspective.

One of the thorniest problems in quantitative risk assessment is how to compare different health endpoints. When comparing a neurological effect to one of the immune system, to a birth defect, to a cancer death, invariably we hear the comment "apples and oranges." "Some skeptic is always protesting that the health, environmental, energy, and other decisions I will describe impossibly amount to 'trying to compare apples and oranges' — to which one must respond that life always requires comparing apples to oranges, and to kumquats, ambulances, and aircraft carriers as well. Difficult and frustrating though it may be, it is unavoidable."[25] Also, consider the *New Yorker* cartoon showing that the common elements between apples and oranges are: edible, warm color, round shape, similar size, contain seeds, grow on trees, good for juice, names begin with a vowel, similar pesticide treatment, and unsuitable for most sports.

There are two systems used in establishing causation for chronic health risks: accept the animal data or use indirect evidence such as structure activity and negative epidemiology. The choice is value laden.[49]

CONCLUSIONS

Lowrance[49] gave the following guidelines for scientists concerning the impact of the intertwining of facts and values in developing public policy:

> Recognizing that they are making value judgments for the public, scientists can take several measures toward converting an arrogation of wisdom into a stewardship of wisdom. First, they can leaven their discussions by including critical, articulate laymen in their group. Second, they can place on the record their sources of bias and potential conflicts of interest, perhaps even stating their previous public positions on the issue. Third, they can identify the components of their decisions being either scientific facts or matters of value judgment. Fourth, they can disclose in detail the specific bias on which their assessments and appraisals are made. Fifth, they can reveal the degree of certainty with which the various parts of the decision are known. Sixth, they can express their findings in clear jargon free terms, in supplementary non-technical presentations if not in the main report itself.

Values, perceptions, and ethics permeate and impact directly on environmental risk decisions. To ignore their impact would be folly. What is needed is recognition of the role they play, a different view of the current reality this view provides, and a movement toward getting the big picture. All aspects of environmental risk decision making involve values and value judgments. These appear as assumptions, default positions, and policies.

Are environmental risk decisions too complex? Are there too many details, pressures, and components for any one person or group of people to understand, weigh, and analyze? Is it possible for even a single issue to be understood enough to be explained to those involved, the public and interested parties? All of this demands that a simplifying model be developed that will facilitate understanding and thus, a safer and more healthy environment.

Ethical issues pervade environmental risk decision making. The new area of practical ethics is an important component of decision making that helps to guide toward right decisions in protecting justice, equity, and the value of life, to mention only a few aspects.

Is it time to develop an environmental ethic? Or is the situation so complex that a set of ethics needs to be developed in several areas? The answer to these questions could go a long way toward simplifying our understanding of environmental problems.

It seems to be essential to work toward grasping the big picture in a more complete and better way. The use of modeling could help here. The continuous model discussed here is a step in that direction.

If we are going to deal with reality we must be more honest. The value of truth should be a central one in environmental decision making. These values should be central to the development of a new paradigm in the field of environmental risk decision making.

BIBLIOGRAPHY

1. D. Coleman, The Group and the Self: New Focus on a Cultural Rift. *New York Times,* December 30, 1990, p. 37, 41.
2. F. Capra, *The Turning Point: Science, Society, and the Rising Culture,* Bantam Books, New York, 1988.
3. G. Taubes, *Bad Science: The Short Life and Weird Times of Cold Fusion,* Random House, New York, 1993.
4. D.J. Fiorino, Technical and Democratic Values in Risk Assessment, *Risk Anal.,* 9:3, 293–299, 1989.
5. S.P. Hays, From Conservation to Environment: Environmental Politics in the U.S. Since World War II, in *Environmental History: Critical Issues in Comparative Perspective,* K.E. Bailes, Ed., All University Press of America, Lanham, MD, 1985.
6. L. Sagan, A Brief History and Critique of the Low Dose Effects Paradigm, *BELLE Newsletter,* University of Massachusetts, Amherst, MA, Vol. 2, No. 2, 1993.
7. L. Westra, *An Environmental Proposal for Ethics: The Principle of Integrity,* Rowman and Littlefield, Lanham, MD, 1994.
8. C.R. Cothern and D.W. Schnare, The Limitations of Summary Risk Management Data, *Drug Metab. Rev.,* 17:145–159, 1984.
9. D. Latai, A Risk Comparison Methodology for the Assessment of Acceptable Risk, Ph.D. thesis, Massachusetts Institute of Technology, Boston, January 1980.
10. S. Jasanoff, Bridging the Two Cultures of Risk Analysis, *Risk Anal.,* 13:123–129, 1993.
11. M.R. Greenberg, H. Spiro and R. McIntyre, Ethical Oxymora For Risk Assessment Practitioners, *Accountability in Research,* 1:245–257, 1991.
12. A.S. Gunn and P.A. Vesilind, *Environmental Ethics for Engineers,* Lewis Publishers, Boca Raton, FL, 1987.
13. L. Wolpert, *The Unnatural Nature of Science,* Harvard University Press, Cambridge, MA 1993.
14. S. Iyengar, *Is Anyone Responsible?* University of Chicago Press, Chicago, 1991.
15. K.S. Shrader-Frechette, *Risk and Rationality,* University of California Press, Berkeley, 1991.
16. R. Kasperson and J. Kasperson, in D.G. Mayo and R.P. Hollander, Eds., *Acceptable Evidence,* Oxford University Press, New York, 1991.
17. K.E. Boulding, The Ethics of Rational Decision, *Manage. Sci.,* 12:161–169, 1966.
18. E.O. Wilson, Is Humanity Suicidal?, *New York Times Magazine,* pp 24–28, May 30, 1993.
19. W.W. Lowrance, *Modern Science and Human Values,* Oxford University Press, New York, 1986.
20. V.R. Potter, *Bioethics: Bridge to the Future,* Prentice Hall, Englewood Cliffs, NJ, 1971.
21. D.T. Hassel, Ed., *After Nature's Revolt,* Augsburg Fortress Press, Minneapolis, 1992.

22. E. Shirk, New Dimensions in Ethics: Ethics and the Environment, *J. Value Inquiry,* 22:77–85, 1988.
23. J. Beauchamp and A. Childress, *Principles of Biomedical Ethics,* Oxford University Press, New York, 1983.
24. R. Keeney, Ethics, Decision Analysis and Public Risk, *Risk Anal.,* 4:117–129, 1984.
25. D.A. Brown, Environmental Ethics; After the Earth Summit: The Need to Integrate Environmental Ethics Into Environmental Science and Law, *Dickinson J. Environ. Law and Policy,* 2:1, 1–21, 1992.
26. James Nash, private communication.
27. S. McFague, *Models of God: Theology for an Ecological, Nuclear Age,* Fortress Press, Philadelphia, PA, 1987.
28. J.A. Nash, *Loving Nature: Ecological Integrity and Christian Responsibility,* Abingdon Press, Nashville, 1991.
29. S.B. Scharper and H. Cunningham, Eds., *The Green Bible,* Orbis Books, Maryknoll, NY, 1993.
30. A. Leopold, *A Sand County Almanac and Sketches Here and There,* Oxford University Press, New York, 1968.
31. W.O. Douglas, *The Three Hundred Year War: A Chronicle of Ecological Disaster,* Random House, New York, 1972.
32. J. Nash, in Earth Rights and Responsibilities: Human Rights and Environmental Protection, a Symposium, *Yale J. Int. Law,* Vol. 18, No. 1, 1993.
33. W.A. Rosenbaum, *Environmental Politics and Policy,* CQ Press, Washington, D.C., 1991.
34. N. Cousins, The fallacy of cost-benefit ratio, *Saturday Review* 6, no. 8:8, 1986.
35. C.G. Brunk, L. Haworth, and B. Lee, *Value Assumptions in Risk Assessment: A Case Study of the Alachlor Controversy,* Wilfrid Laurier University Press, Waterloo, Canada, 1991.
36. D.A. Brown, Superfund Cleanups, Ethics and Environmental Risk Assessment, *Boston College Environ. Affairs Law Rev.,* 16(2), 181–198, 1988.
37. J. Brownowski, *Science and Human Values,* Perennial Library, Harper & Row, New York, 1972.
38. M. Sagoff, *The Economy of the Earth,* Cambridge University Press, Cambridge, U.K., 1988.
39. A. Gore, *Earth In The Balance: Ecology and the Human Spirit,* Houghton Mifflin, New York, 1992.
40. H. Margolis, *Paradigms and Barriers: How Habits of Mind Govern Scientific Beliefs,* University of Chicago Press, Chicago, 1993.
41. J. Rawls, *A Theory of Justice,* Harvard University Press, Cambridge, MA, 1971.
42. B. Singer, An Extension of Rawls' Theory of Justice to Environmental Ethics, *Environ. Ethics,* pp. 217–232, Fall 1988.
43. B.G. Norton, *Toward Unity Among Environmentalists,* Oxford University Press, New York, 1991.
44. M.E. Colby, World Bank Discussion Paper #80, The World Bank, Washington, D.C., 1990.
45. I.L. Janis, *Groupthink,* 2nd ed., Houghton Mifflin, Boston, MA, 1982.
46. E. Silbergeld in *Acceptable Evidence,* D.G. Mayo and R.P. Hollander, Eds., Oxford University Press, New York, 1991.

47. K. Shrader-Frechette in *Acceptable Evidence,* D.G. Mayo and R.P. Hollander, Eds., Oxford University Press, New York, 1991.

48. D. Byrd and W. Gawlak, The Rules of the Game: What Recent Rulings Say About Courts' and Regulators' Differing Approaches to Establishing Causation for Chronic Health Risks, in *Analysis, Communication and Perception of Risk,* B.J. Garrick and W.C. Gelker, Eds., Plenum Press, New York, 1991.

49. W. Lowrance, *Of Acceptable Risk: Science and the Determination of Safety,* Kaufman, Los Altos, CA, 1976.

SECTION II

Issues in Environmental Risk Decision Making

4

INTRODUCTION TO ISSUES IN ENVIRONMENTAL RISK DECISION MAKING

Scott R. Baker

A number of issues emerge in the discussion and analysis of values and value judgments. In the area of environmental risk decision making these include, among others, health and safety, pollution prevention, uncertainties, fallibility, biases, models such as the one that separates risk assessment and risk management, public participation and concerns, and intergenerational equity. This partial list represents the topics discussed in this section by authors from a variety of disciplines and professions.

If scientists cannot suppress their subjective values and beliefs as influences, then how can their honest scientific opinions be protected from bias? To try and integrate these ideas one can use: collective opinion, peer review, collegial interaction, skepticism, criticism, sharing of research data, serving as mentors, and postpublication review. Concerning the risk assessment paradigm in particular, one can observe that in classical risk assessment there is no opportunity for admitting the subjectivity of the risk assessor. The separation between risk assessment and risk management was intended to prevent the corruption of the risk assessment process by the values of the risk manager. However, the paradigm does not guard against subjectivity entering the risk assessment process. For example, the goal of safety leads to default positions that consistently overestimate risk.

In a similar way, several authors point out that the "red book" from the National Academy of Sciences separated risk assessment and risk management and included a two-way communication between them, but the common misinterpretation is to put all value considerations into risk management, with no communication to risk assessment; thus there is no opportunity to discuss public values in the risk assessment process. Our environmental regulatory system has done little to enable regulators to respond to public concerns and

1-56670-131-7/96/$0.00+$.50
© 1996 by CRC Press, Inc.

virtually nothing to recognize the essential role of community values in the shaping of environmental solutions.

The perceptions and feelings of the public about environmental risks include those of being unfair, uncontrollable, untrustworthy, and involving the reality of fear, anger, and suffering. Some authors observe that the public are interested in a number of risk questions such as: what is an acceptable risk?; what are the "facts"? (but often we have too few); what is the balanced view?; how can we build trust?; and can we not sensationalize insignificant problems? The importance of these and related concepts, major contributions to risk decision making, need to be consciously acknowledged.

Uncertainty raises ethical questions that no amount of science can answer, such as what to do until science provides certainty, who should have the burden of proof in demonstrating that an individual contaminant poses a risk, can a piece of scientific information be ethically neutral or unchallengeable? We cannot all afford technical experts, and the lack of scientific information should not be used as an excuse for failing to take cost-effective action. These questions need to be addressed as a regular element in environmental risk decision making.

Ethical issues are often masked or hidden by current risk assessment practices — for example: the burden of proof, distributive justice; race, color, creed, social status, communicative rationality of the general public; and the implied conclusion is that it is "safe" to pollute as long as it is within a given standard. Without considering these issues directly, environmental risk decision making is incomplete.

5   INDUSTRY'S USE OF RISK, VALUES, PERCEPTIONS, AND ETHICS IN DECISION MAKING

P.J. (Bert) Hakkinen and Carolyn J. Leep

CONTENTS

The First Principle: Scientifically Sound Risk Assessment 74
The Second Principle: Public Participation 75
The Third Principle: Risk-Based Priority Setting 78
The Fourth Principle: Flexible, Cost-Effective Approaches to Risk
    Management 80
Summary 81
References 81

This chapter provides an overview of a risk-based approach to environmental management and examples of how industry integrates information about risk with public perceptions and values in making management decisions. Well-designed approaches to risk assessment, complemented by flexible, cost-effective risk management practices, have a crucial role in reconciling what are sometimes viewed as competing demands of job-creating growth and responsible environmental stewardship. The risk-based approach to environmental management is built around the four major themes shown in Table 1. First, a scientifically sound risk assessment is essential for environmental decision making. Second, the public must be involved in the risk assessment and management processes. Third, resources should be focused on problems where the greatest risk reduction can be achieved. Fourth, environmental regulations should be performance oriented and allow companies to make risk reductions in the most cost-effective manner.

These themes illustrate a number of values common to industry. Industry values scientific strength; excellence and continuous improvement in health,

1-56670-131-7/96/$0.00+$.50
© 1996 by CRC Press, Inc.

Table 1 The Four Major Themes Associated with a Risk-Based Approach to Environmental Management

1. A scientifically sound risk assessment is essential for environmental decision making.
2. The public must be involved in the risk assessment and management processes.
3. Resources should be focused on problems where the greatest risk reduction can be achieved.
4. Environmental regulations should be performance-oriented and allow companies to make risk reductions in the most cost-effective manner.

safety, and environmental performance; public trust and understanding; and getting the biggest "bang for the buck" — focusing efforts on opportunities for getting the greatest risk reduction in the most cost-effective way. The rest of this chapter will provide examples of how industry puts each of the above principles into action, incorporating these values.

THE FIRST PRINCIPLE: SCIENTIFICALLY SOUND RISK ASSESSMENT

As noted above, a scientifically sound risk assessment is essential for good and responsible decision making by industry and others. While the information used in an initial risk assessment might be incomplete (i.e., not yet containing complete information on all aspects of the hazard identification, dose response, and/or exposure assessment), it should contain information considered to be scientifically sound, based on expert judgment.

When should industry be doing risk assessments? Risk assessment work at the beginning of a project is ideal, and allows a focus on identifying opportunities for changes and improvements, along with a possible comparison of alternate approaches, before a project is far along in development. This early emphasis helps develop an optimal finished product, with any changes based on the outcomes of risk assessments made as early and as easily as possible. A letter to the editor of a journal several years ago discusses this in greater detail.[1]

As an example of the above, the philosophy of the Procter & Gamble Company (P&G) is that its consumer products will be safe for all recommended uses and reasonably foreseeable uses, and will be perceived as being safe by consumers and others. This includes very early work by P&G toxicologists and others to assess which potential components (e.g., starting materials and solvents) and possible competing technologies have the lowest possible number of actual and perceived human and environmental safety issues. This approach results in proactive risk assessment and management by involving toxicologists and other experts early in the product and technology development cycle to help achieve safe and perceived-to-be-safe products.

The approach used by P&G and many other companies is what could be called a tiered approach to risk assessment. An initial risk assessment is based on available public and/or company data to support initial development activity and to identify key data needs and areas of uncertainty to be addressed later. The tiered approach leads to an increase in sophistication in risk assessment activity as judged by toxicologists and other experts to be needed to support a particular level of development and could ultimately lead to development of extensive hazard identification, exposure assessment, or other information if those types of information are unavailable at the beginning. Key in this activity is the judgment of the risk assessment experts about how much information is needed at any given time in the development cycle, along with expert judgment about when enough risk assessment-related work has been done to support commercialization and the resulting potential widespread human and environmental exposures that might occur.

This early involvement by risk assessment experts and the use of a tiered approach extends to the workplace (worker safety) and possible manufacturing plant emissions. As above, early involvement can lead to changes pertaining to safety and possible worker and environmental exposures when they are easiest to make and when potential competing chemicals, manufacturing processes, and environmental control processes can be compared. Also, as above, a tiered approach can be utilized.

THE SECOND PRINCIPLE: PUBLIC PARTICIPATION

The public must be involved and/or considered in the risk assessment and management processes. For example, as noted above, the philosophy of P&G includes that its consumer products will be perceived as being safe by consumers and others. This makes understanding potential public perceptions a key part of the development work for a new product and a part of the proactive risk assessment and management approaches noted above. This ideally results in development of technologies and products with the lowest possible number of actual and perceived human and environmental safety issues.

How can public perceptions of chemicals and consumer products be assessed? In 1993, the Chemical Manufacturers Association published a document based on research sponsored by P&G.[2] This document demonstrates how consumer perceptions of risk can be quantified and examined, presents examples of consumer risk perception data for a wide range of consumer products, and discusses the importance of assessing and considering perceptions of risk as part of the development of a product. The 47 household chemical products covered in this document are shown in Table 2. Each of these products was evaluated for the 15 topic areas shown in Table 3.

The above work provides examples of what toxicologists and others in industry are or could be doing in the initial phases of development of a new product or technology. They can consider public perceptions early in the development process, based on how their past experience indicates certain toxicology and perhaps other information might be perceived or by gathering

Table 2   Risk Perception Data for the Following Types of Household Chemical Products Are Available

1. Aspirin	25. Mothballs
2. Non-aspirin pain relievers	26. Dandruff and medicated shampoos
3. Children's aspirin	27. Adult, non-medicated shampoos
4. Powdered laundry detergents	28. Baby shampoos or "no tears"
5. Liquid laundry detergents	formulas
6. Hypoallergenic laundry detergents	29. Fluoride toothpastes
7. Soaps for baby clothes	30. Mouthwash
8. Hard surface cleaners	31. Home permanent preparations
9. Ammonia	32. Hair styling preparations
10. Oven cleaners	33. Hair spray
11. Automatic dishwasher detergent	34. Deodorants and antiperspirants
powders	35. Bulk fiber laxatives
12. Automatic dishwasher detergent	36. Hair removers
liquids	37. Glass cleaners
13. Hand dishwashing liquids	38. Fabric softener liquids
14. Chlorine bleach	39. Fabric softener sheets
15. Non-chlorine bleach	40. Detergents for fine washables
16. Deodorant soaps	41. Bathroom cleaners
17. Hypoallergenic or special bar soaps	42. Hair dyes
18. "Beauty soaps"	43. Lemon-scented products in general
19. "Pure soaps"	44. Perfumed (non-lemon scent)
20. Antacids	products in general
21. Lotions and creams	45. Products containing enzymes
22. Antidiarrheal products	46. Products containing phosphates
23. Disinfectants	47. Products containing ingredients
24. Spot and stain removers	made from biotechnology

From Neil, N., P. Slovic, and P.J. Hakkinen, "Mapping Consumer Perceptions of Risk," Chemical Manufacturers Association (1993).

data from consumers. This can be done initially without going to the public for input and permits early considerations of possible changes that could eliminate or at least minimize any perceptions of risk.

Actual input from the public can come from the type of research noted above, or from the use of focus groups, examination of information gathered from consumer test questionnaires or market research interviews, and eventually from letters or phone calls (e.g., many companies now have toll-free telephone numbers consumers can use to call companies to ask questions, praise a product, or to express concerns).

Public participation is a key component of the Chemical Manufacturers Association's Responsible Care®* initiative, which requires each member company to continuously improve its performance in health, safety, and environmental quality and to speak openly with the public about any concerns. Table 4 outlines key elements of Responsible Care®; Table 5 describes the six Responsible Care® Codes of Management Practices.

This two-way communication can involve several audiences, including employees, customers, and the general public. The Product Stewardship Code is one of the six Responsible Care® Codes of Management Practices. Product

* Registered trademark of Chemical Manufacturers Association, Washington, D.C.

Table 3   The 15 Perceived Risk Topic Areas Evaluated for the Household Chemical Products Shown in Table 2

1. RISK TO ADULTS: To what extent are most adults at risk of personal harm from this type of product?
2. RISK TO YOUNG CHILDREN: To what extent are young children (under age 5) at risk of personal harm from this type of product?
3. RISK TO THE ELDERLY: To what extent are the elderly at risk from this type of product?
4. ENVIRONMENTAL EFFECTS: To what extent does this type of product or its ingredients pose a risk of harmful effects to the environment?
5. PREVENTIVE CONTROL: To what extent can accidents or harmful effects involving injury to a person associated with this type of product be prevented?
6. SERIOUSNESS OF INJURY: If an accident or unfortunate event involving this type of product occurred, to what extent are the harmful effects (to a person) likely to be mild, or serious?
7. SEVERITY OF CONSEQUENCES: If an accident or unfortunate event involving this type of product occurred, to what extent are the consequences likely to include loss of life?
8. CONTROL OF SEVERITY: Once an accident or unfortunate event involving this type of product has occurred, to what extent can proper action reduce the severity of the consequences of that accident or unfortunate event?
9. KNOWN TO THOSE EXPOSED: To what extent are the risks associated with this type of product known to members of the general population who are using the products?
10. MEDIA INFORMATION: To what extent do people need to be informed by the news media of potential hazards of this type of product or its ingredients?
11. INFORMATIVENESS: If a serious accident or injury involving this type of product occurred, to what extent would the mishap serve as a "warning signal" for society providing new information about the probability that similar or even more destructive mishaps might occur in the future?
12. OLD OR NEW: To what extent is this type of product old or new?
13. NECESSITY OR LUXURY: Do you consider this type of product to be a necessity or a luxury (for society in general, not just for you personally)?
14. BENEFITS: How beneficial is this type of product to you and your family?
15. REGULATION: How important is it that the labeling of this type of product be controlled by federal regulation?

From Neil, N., P. Slovic, and P.J. Hakkinen, "Mapping Consumer Perceptions of Risk," Chemical Manufacturers Association (1993).

Stewardship in the development of new products and technologies includes having toxicologists and others examine appropriate ways to communicate and manage any unavoidable possible risks (e.g., skin or eye irritation from accidental exposures to a cleaning product). This could lead to changes in package design to reduce the potential for accidental exposures and the addition of cautionary labeling and first-aid statements. Public participation via consumer feedback will help confirm the judgments of company experts about the effectiveness of package design and label wording, with changes made to packaging and labeling if judged to be needed.

Another example of Responsible Care®-related public participation involves member company interaction and involvement in Local Emergency Planning Committees (LEPCs) as part of a community's emergency response planning process. Also, Community Advisory Panels (CAPs), independent voluntary groups of individuals who have made a commitment to meet with

Table 4   Key Elements of Responsible Care®

1. GUIDING PRINCIPLES: Statements regarding health, safety, and environmental quality upon which management practice codes are based. The principles recognize both public concerns and the industry's desire for self-improvement.
2. CODES OF MANAGEMENT PRACTICES: Responsible Care® is defined and implemented through a series of management practice codes. Each code states the intended results and defines, in a qualitative way, what is expected of Member and Partner Companies. Codes aim to encourage companies to stretch themselves to achieve higher levels of performance.
3. PUBLIC ADVISORY PANEL: A fundamental component of Responsible Care®, this panel is composed of informed citizens and environmental and community leaders from across the country. It helps ensure the public's concerns are understood and that actions are taken that respond to those concerns.
4. MEMBER SELF-EVALUATION: Each Management Practice Code requires a Member/Partner to conduct an annual self-evaluation. This assists company management to determine whether a change in implementation is necessary. The evaluations also assist CMA in gauging overall industry progress and in identifying areas where additional resource materials are needed.
5. MEASURES OF PERFORMANCE: Code Performance Measures are essential to enhancing the overall credibility of the industry's health, safety, and environmental performance improvement efforts. CMA will use aggregate Member and Partner results to demonstrate the industry's progress.
6. MANAGEMENT SYSTEMS VERIFICATION: Verification of the management systems CMA Members and Partners are employing to implement Responsible Care® are viewed as critical. Such process will involve external participation and lead to certification.
7. EXECUTIVE LEADERSHIP GROUPS: Regional meetings of Executive Contacts that provide an opportunity for companies to meet and discuss their progress and share experiences with Responsible Care® implementation.
8. MUTUAL ASSISTANCE: CMA Members and Partners coming together regionally, at the coordinator and practitioner level, to share successful practices.
9. PARTNERSHIP PROGRAM: Spreading Responsible Care® beyond the CMA Membership to other companies that take ownership or possession of chemicals is a priority. Through the Partnership Program, non-CMA members and their associations can be involved in this important performance improvement initiative.
10. OBLIGATION OF MEMBERSHIP: Participation in Responsible Care® is an obligation of CMA membership. Each Member Company Executive Contact must sign the Guiding Principles; communicate that commitment to all employees; instruct management to make good-faith efforts to implement the Codes of Management Practices; participate in the self-evaluation process and meet the expectations of the initiative; and use the Responsible Care® name and "hands" logo in accordance to CMA's guidelines.

plant managers on a regular basis to discuss issues of mutual interest, can complement LEPC activities. The CAPs serve as ways for open and honest communications to occur with companies, with the interactions ideally leading to constructive solutions to any issues.

THE THIRD PRINCIPLE: RISK-BASED PRIORITY SETTING

There is a growing awareness that, while we are spending more and more money on environmental controls, it is not being spent in the most cost-effective

Table 5   Responsible Care® Codes of Management Practice

COMMUNITY AWARENESS AND EMERGENCY RESPONSE CODE (CAER)
• Designed to assure emergency preparedness and to foster community right-to-know.
PROCESS SAFETY CODE
• Designed to prevent fires, explosions, and accidental chemical releases.
DISTRIBUTION CODE
• Designed to reduce the risk that the transportation and storage of chemicals poses
  to the public carriers, customers, contractor and company employees, and to the
  environment.
POLLUTION PREVENTION CODE
• Designed to promote industry efforts to protect the environment by generating less
  waste and reducing pollutant emissions.
EMPLOYEE HEALTH AND SAFETY CODE
• Designed to protect and promote the health and safety of people working at or
  visiting CMA Member Company sites.
PRODUCT STEWARDSHIP CODE
• Designed to promote the safe handling of chemicals at all stages — from initial
  manufacture to distribution, sale, and ultimate disposal.

manner to address the most serious problems. Recently, a bipartisan group of mayors from 114 cities and towns across the U.S. called upon the U.S. Congress to "assure that environmental protection investments are made where they accomplish the greatest good."[3]

Current U.S. environmental policy was built largely in reaction to public concerns, not on a strong, scientific analysis of which hazards present the greatest actual risks. The result, as U.S. EPA's Science Advisory Board noted in a widely quoted study, is that regulatory attention often has been focused on less significant risks while, overall, our environmental protection efforts, "have been ... less effective than they could have been."[4] A risk-based alternative for setting priorities would make environmental management efforts more effective.

A first step in achieving risk-based priority setting is to develop a scientifically sound means of ranking relative risks. This would incorporate the best available scientific information, while also including public perceptions and concerns to lead to broad-based public acceptance. A second step would be to include considerations of feasibility, cost-effectiveness, and other impacts of actions that might be undertaken to address the different risks. The purpose of this step is to identify the risks that present the best opportunity for the greatest risk reduction, thereby achieving the "biggest bang for the buck".

An industry example of risk-based priority setting comes from the Chemical Manufacturers Association's Responsible Care® Pollution Prevention Code. Member companies establish priorities, goals, and plans for waste and release reduction based in part on an assessment of relative risks and also on overall community and employee concerns. For instance, the Dow Chemical Company conducts risk assessments on Toxics Release Inventory emissions to help prioritize opportunities for emissions reductions.

THE FOURTH PRINCIPLE: FLEXIBLE, COST-EFFECTIVE APPROACHES TO RISK MANAGEMENT

Part of the process of identifying the best opportunities for risk reduction involves finding the most cost-effective approaches to risk management. Use of the phrase "cost-effective approaches" here is not intended to serve as a call to reduce environmental protection brought about by regulations. Rather, it involves an effort to apply to environmental management the same principles of total quality management that have been applied to many other aspects of business.

Why is this needed, and how can it be accomplished? Current environmental laws and regulations have a "command and control" focus that was effective in achieving major improvements in air and water quality over the past 20 or so years. A new approach is needed to make the smaller, incremental improvements being achieved today in the face of resource limitations. However, current environmental laws make this difficult, since they emphasize highly prescriptive, media-specific regulations, which place obstacles in the way of flexible, cost-effective regulations.

Is there a good example of a cost-effective approach to environmental management? A hopefully convincing example is a rather unique pollution prevention study conducted cooperatively by U.S. EPA and Amoco Corporation at Amoco's Yorktown, VA, refinery. Among the goals of the Amoco/EPA project were to:

- Determine the types, amounts, and sources of emissions that the refinery releases to air, land, and water;
- Develop options to reduce these releases;
- Determine the benefits, impacts, and costs of different options to select the most effective ones; and
- Identify factors that encourage or discourage pollution prevention initiatives.

A key finding was that, if the company had been free to pursue a flexible, performance-oriented approach to pollution prevention, 90% of the emissions reductions required under applicable regulations could have been achieved for 20 to 25% of the cost of meeting specific requirements of the regulations. Had a performance-oriented approach to emissions reductions been possible, releases at this refinery could have been reduced at an average cost of $500 a ton compared to $2100 a ton average cost under the current prescriptive command and control regulations.[5] Other findings were that better data can improve environmental management decisions and that current practices discourage innovative solutions to complex environmental problems. This project demonstrated the value and challenge of government/industry partnerships and their use in identifying ways to set priorities for making environmental improvements.

SUMMARY

This chapter has provided an overview and examples of a risk-based approach to environmental management involving four major themes. First, a scientifically sound risk assessment is essential for environmental decision making. Second, the public must be involved in the risk assessment and management processes. Third, resources should be focused on problems where the greatest risk reduction can be achieved. Fourth, environmental regulations should be performance oriented and allow companies to make risk reductions in the most cost-effective manner.

REFERENCES

1. Seiler, F. A. "On the Use of Risk Assessment in Project Management," *Risk Anal.,* 10(3): 365–366 (1990).
2. Neil, N., P. Slovic, and P. J. Hakkinen, "Mapping Consumer Perceptions of Risk", Chemical Manufacturers Association (1993). For a copy, contact Chemical Manufacturers Association, Publications Fulfillment, 2501 M Street, NW, Washington, D.C. 20037.
3. Fink, T. January 15, 1993 letter to members of Congress from Tom Fink, Mayor of Anchorage, Alaska, cosigned by 113 other mayors.
4. "Reducing Risk: Setting Priorities and Strategies for Environmental Protection," U.S. Environmental Protection Agency Report SAB-EC-90-021 (1990).
5. "EPA, Amoco Project Reduces Emissions, Highlights Benefits of Risk-Based Actions," *Chem. Reg. Rep.,* 17(25): 1088 (1993).

SUMMARY

6
REGULATING AND MANAGING RISK: IMPACT OF SUBJECTIVITY ON OBJECTIVITY

Scott R. Baker

CONTENTS

Introduction 83
Risk Assessment, Objectivity, and Subjectivity 85
Beliefs and Values as Influences on the Process of Balancing Costs
    and Benefits 88
Personal Values and Beliefs of Scientists 90
Thwarting Subjective-Driven Bias in Science 91
Conclusion 91
References 92

INTRODUCTION

The nature of subjective influences on science and their use in environmental policy is the focus of this chapter. The hypothetical case study of a fictitious company, presented below, describes common circumstances under which a risk assessor must, in the absence of sufficient data, commingle objective scientific measurement with subjective opinion.

Case study: Ajax Widgets, Inc. has been producing consumer products for 20 years. During that time, the Company has relied on extensive use of bulk solvents in its manufacturing processes. At times in the past, solvents were spilled or discarded, so that there was widespread soil and ground water contamination. Ajax agreed to clean up the contamination under a negotiated consent decree with the U.S. Environmental Protection Agency (U.S. EPA). As a first step, an extensive investigation was undertaken to delineate the extent and amount of contamination, and its threat to human health and the ecology. The study was used to select the appropriate cleanup strategy. Remediation of the site cost $10 million.

The study was directed by a hydrogeologist. Based on his knowledge of the physiography of the site, he designed a scheme for soil and ground water data collection that provided chemical concentrations used in the health risk assessment. The risk assessor, who was not consulted prior to environmental data collection, had to rely on the results of chemical analyses on only two ground water samples, taken at one monitoring well in the vicinity of what could be, sometime in the future, the site of private drinking-water wells used by potential future residents near Ajax's plant. The risk assessment predicted that cancer risk would be significant and unacceptable from drinking well water contaminated with one solvent. The solvent's presumed toxicity to humans is based on limited studies in laboratory animals. The result of unacceptable human health risk led to the U.S. EPA opinion that Ajax should clean up the contamination.

In conducting the health risk assessment, where necessary information was lacking, the risk assessor made several prudent simplifying assumptions, relying as much as possible on conventional procedures prescribed in U.S. EPA guidance documents. These assumptions presumed that:

- The ground water monitoring data represented chemical concentrations in the drinking water wells
- Ground water chemical concentrations were log-normally distributed, for purposes of averaging the results of all of the sample analyses
- There would be residents present in the area sometime in the future who would drink the contaminated ground water
- The residents would drink 2 liters of the contaminated water each day, every day, for a presumed residence time of 30 years
- During that 30-year period, the chemical concentration in the ground water would remain unchanged from the concentration determined in the monitoring well
- The body weight of all individuals drinking the contaminated water for 30 years would be 70 kg
- All input values used to calculate health risk (daily water consumption, body weight, residency time, ground water chemical concentration, toxic potency of the chemical) were conservatively assumed to be 95% upper confidence limits of the measured or assumed values. Accordingly, the resulting risk estimate was itself conservative — a 99.9999+% upper confidence limit, based on compounded conservative input values, some measured, some assumed

In this case study, the risk assessor was willing to subjectively accept that, for purposes of estimating human health risk from drinking the ground water:

- The uncertainty associated with the number of monitoring samples was tolerable
- The indicated ground water concentration of chemical (95% upper confidence limit on the mean) could be an underestimate of the real concentration*
- Someone in the future will drink the contaminated ground water
- An individual will drink 2 liters of that water every day for 30 years
- The chemical's concentration in ground water will not change in 30 years, based on blind faith and disregard for past experience indicating the opposite at other sites
- Cancer in experimental animals is a sufficiently reliable predictive indicator of cancer in humans to use in costly cleanup decisions

It is clear from this case study that the risk assessor was biased toward being highly protective, based on the absence of sufficient information. The degree to which the risk assessor was protective was strongly influenced by his or her beliefs and values. Proponents of the risk assessor's approach might claim that his tendency to be protective was justified in the absence of information and in the presence of uncertainty. Critics of his analysis might claim that, for the same reasons, he was being unreasonable and *overly* protective.

RISK ASSESSMENT, OBJECTIVITY, AND SUBJECTIVITY

The risk assessment–risk management paradigm, presented in Figure 1, was proposed by the U.S. National Academy of Sciences in 1983[1] and has been widely adopted by the U.S. government as a decision-making framework. Assessment and management of human health risk is a methodical process. While the process ideally strives for objectivity, in reality it contains a lot of subjectivity. A great deal of effort has been expended on developing risk guidelines, criteria, policies, and procedures in an attempt to promote objectivity and consistency in the risk assessment and management process. Because of this emphasis, very little attention has been paid to whether risk analysts influence the process with their subjectivity, and to what degree such subjective influence occurs.

The architects of the risk assessment–risk management paradigm were concerned about the prospect of the value-laden risk *management* process corrupting what they hoped would be a value-limited, risk *assessment* process, by influencing its outcome. Therefore, it is no accident that risk characterization, the nexus of the venn diagram in the risk assessment–risk management paradigm, can be influenced only in one direction, through the science of risk assessment, as illustrated in Figure 1.

* Log normally distributed environmental monitoring data generally produce statistical measures (i.e., means and associated 95% upper confidence limits) that are lower than statistical measures based on normally distributed environmental data.

RISK ASSESSMENT

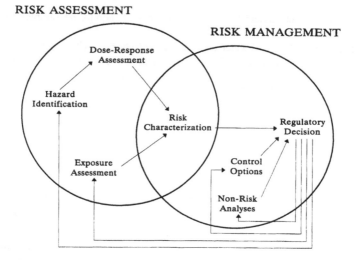

Figure 1   The risk assessment–risk management paradigm.

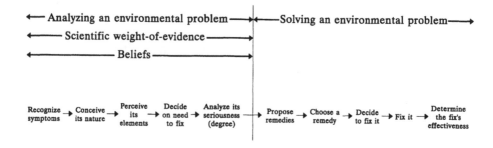

Figure 2   Characterizing an environmental problem and developing a solution.

However, the paradigm does not guard against subjectivity entering a supposedly objective risk assessment process. Characterizing an environmental problem and developing an appropriate solution to it can be broken down into a progression of several steps (see Figure 2). The main features of the progression include identification of the problem's elements, analyzing their seriousness, choosing the best remedy, and implementing it. Each step in the process leads to a conclusion, with uncertainty, based on "best professional scientific judgment."* The objective scientific elements of "best professional judgment" (see Figure 3A) have been codified in a number of risk assessment procedural guidance documents that establish the scientific standards for risk

* "Best professional judgment," in turn, is based on knowledge and acquired wisdom applied to objective scientific observation and subjective belief.

A. Objective Elements

Certainty is high	Scientific weight of evidence	Certainty is low
Statistically positive	Results prove the hypothesis	No correlation
In more than one species	Study is reproducible	Only the most sensitive species
Strong epidemiologic association	Human evidence	None
According to accepted scientific standard	Quality of study	Elements missing
Proper measurement techniques used	Data precision and accuracy	Do not know data variability
At more than one site in the body	Body sites showing health effect	One
Physiologically significant	Adversity of health effect	Reversible change in chemical level
Critical organ is affected	Severity of health effect	Modest biochemical change
Published in peer-reviewed journal	Study is authoritative	Unpublished report
Multiple routes of interest produce effect	Routes of administration	One route, not the one of interest

B. Subjective Beliefs

Certainty is high	Scientific weight of evidence	Certainty is low
Estimated risk is correct	One in a million risk	A useful probability estimate
Risk is based on upper-bound inputs	Conservatism in risk assessment	Risk is based on average inputs
Protect human life at all costs	Value of human life	Based on actuarial statistics
Protect earth at all costs	Social and environmental conscience	Nonharmful decrement acceptable
Acceptable for all	Quality of life (human condition)	Unacceptable for some
Chemicals will not hurt	Invincibility	Any chemical will impair health
Can control destiny	Vulnerability	Fate is in the hands of others
Little regard for the plight of others	Sensitivity to the human condition	Unfortunate should have better
Things are fine as is	Activism	Progress requires a strong voice
There is no environmental injustice	Empathy	Environmental inequities should be fixed
Environment is sustainable as is	Concern for future generations	Left alone, quality of life will decrease
Nature is in harmony with all religion	Spirituality	Basis for nature is strictly biological

Figure 3 Competing objective information and subjective beliefs. A. Objective elements. B. Subjective beliefs.

assessment, most notably the U.S. EPA's Risk Assessment Guidelines.[2,3] Taken together, these objective scientific elements form the scientific weight of evidence that a chemical is or is not a health hazard. The elements can be tested, challenged, debated, disproved through experimentation, or otherwise examined, based on fact in a scientific consensus-building process.

Superimposed over the scientific weight of evidence, yet still a part of "best professional judgment," are the subjective elements listed in Figure 3B. The subjective elements are often covert, hard to notice, influencing objectivity by introducing scientific bias, examples of which were illustrated in the case study.

The decision that an analyst makes on each *objective* element is a statement of certainty about confidence in the information, based on the scientific weight of evidence. The following example, drawn from Figure 3A illustrates the point:

Objective decision:*	The results prove the hypothesis
Certainty (confidence in the decision):	The results are statistically positive

Converse

Objective decision:*	The results do not support the hypothesis
Uncertainty (no confidence in the decision):	There is no statistical correlation in the results

* Based on scientific weight of evidence.

Similarly, the decision that an analyst makes on each *subjective* element is also a statement of certainty about confidence in the information, but based on morals, values, and belief. From Figure 3B:

Subjective criterion:	Risk of one-in-one-million
Certainty (confidence in the belief):	The risk is acceptable to me

Converse

Uncertainty (no confidence in the belief):	The risk is useful only as a useful probability estimate

All of the objective and subjective elements, taken together and integrated, form the complex basis of decision making, with personal bias superimposed over fact.

BELIEFS AND VALUES AS INFLUENCES ON THE PROCESS OF BALANCING COSTS AND BENEFITS

The staunchest of environmental activists might advocate that any environmental health hazard is unacceptable and should be prevented at all cost. Conversely, a strong proponent of economic development might caution against

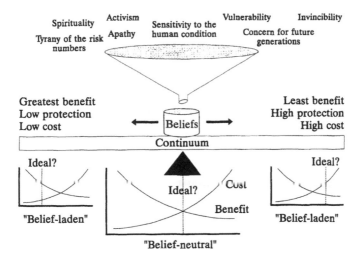

Figure 4 How personal beliefs influence the health risk decision-making process.

overreacting to environmental contamination and might advocate minimizing pollution control. For purposes of illustration, these two individuals can be considered to represent opposite extremes of a continuum, illustrated in Figure 4, that runs between:

1. A baseline level of relatively low health protection for relatively low cost. The greatest amount of pollution and risk reduction (greatest benefit) per dollar is achieved at this level of protection.
2. High health protection for relatively high cost. Dollar for dollar, pollution reduction to achieve high health protection occurs only with a high economic premium and is not cost-effective. The least amount of pollution and risk reduction (least benefit) is achieved in this case.

Most points of view fall in between. Where an individual's point of view falls on the continuum is influenced by the beliefs, presented in Figure 3B, that he or she holds. For practical reasons, a consensus-seeking conciliator might argue that the fairest and most objective approach is to conduct a cost–benefit analysis, resulting in the general graphical illustration in Figure 4. Where the cost and benefit curves intersect is a reasonable approximation of optimal balance between the two competing forces of cost and benefit, devoid of subjective beliefs. This "belief-neutral" position is ideal to the consensus seeker. In the cases of the two activists cited above, their "belief-laden" positions, illustrated in Figure 4, shift appropriately to the right or left of the "belief-neutral" position.

Practically speaking, risk assessment issues are resolved and risk management decisions are made via one of two processes,[6] as illustrated in Figure 5:

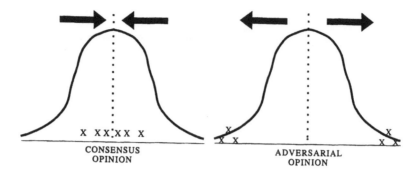

CONSENSUS OPINION   ADVERSARIAL OPINION

Figure 5 Range of expert judgment. (From Goldstein, B.D. "Risk Assessment/Risk Management Is a Three-Step Process: In Defense of EPA's Risk Assessment Guidelines." *J. Am. College Toxicol.* 7(4):543–549, 1988. With permission.)

1. Through consensus building, with a tendency toward belief neutrality. In this case, agreements among interested parties are identified and fortified. Differences of opinion are circumscribed and attempted to be resolved. Scientists operate according to this process.
2. Through adversarial conflict, with a tendency toward belief imposition. In this case, there are few (if any) agreements. Polarized views and opposing differences of opinion are called out and emphasized. Lawyers, environmental activists, and other individuals with strongly held beliefs about the environment in one direction or the other operate according to this process.

While the consensus-building and adversarial approaches are very different, their end results are often the same: a collective opinion or decision that is toward the central tendency of where most people would like to see that opinion or decision rest.

PERSONAL VALUES AND BELIEFS OF SCIENTISTS

Scientists are trained to be objective empiricists in their approach to science. They formulate present-day testable hypotheses based on the conclusions of hypotheses tested in the past, and their own conclusions are used in the future to postulate and test still more hypotheses. Yet even for the most disciplined of scientists, it is difficult to separate subjective personal beliefs, as extensions of self, from objective experimental observation and interpretation. The subjective criteria presented in Figure 3B can lead to two interesting scientific traits:[4,5]

- **The scientist as fallible human being.** Scientists are trained to be opinionated about epistemological science. As thinking individuals, they have philosophical, religious, cultural, political, and economic values, beliefs, and opinions that will invariably shape their scientific judgment. In and of themselves, personal values and beliefs are not harmful to science unless they introduce bias. Then, values and beliefs are critically harmful and, in the extreme, can result in prejudiced, unfair, and unrealistic risk-management decisions. A blatant example of where bias becomes serious involves attempts to link certain biological or behavioral traits with overtly racist or sexist views.

- **The scientist as self-deceiver.** It is easy for scientists to be duped by their own beliefs, to fool themselves into "seeing" what they expect to see (i.e., seeing what is not there), and to fail to notice what they believe should not be there. Problems and the methods to investigate them are often not well defined. Experimental techniques are pushed beyond the limits of quantification to the limit of detection. The sought-after signal (positive test result) is often hard to distinguish from noise (background).[*] The opportunity for error is high, such as terminating a toxicity experiment too early because the data conform to the hypothesis (e.g., the chemical is a carcinogen), when continuing it might reveal that the scientific weight of evidence for carcinogenicity is equivocal. Scientists call this effect a false positive or false negative result. One is led to wonder how much of the "false" should read "subjective."

THWARTING SUBJECTIVE-DRIVEN BIAS IN SCIENCE

If scientists cannot suppress their subjective values and beliefs as influences, then how can their honest scientific opinions be protected from bias? One effective way is by forming collective opinion through consensus building (common to scientists; anathema to lawyers), peer review, collegial interaction, skepticism, criticism, sharing of research data, serving as mentors, and post-publication review. Risk assessments used in the regulatory setting, such as in the U.S. EPA's rule-making process, are subject to administrative review, formal public comment, and scientific review by the U.S. EPA's Science Advisory Board. While each of these activities produces its own brand of subjectivity and bias, the result of collectivization of opinions is (1) at best a purging of biases, or (2) a normative balance of biases, or (3) at worst (hopefully) a recognition and acceptance that the outcome is biased to some measured degree.

Making decisions without knowing that biases are influencing them can lead to serious errors (e.g., racial or sexual biases). We should strive to avoid unrecognized biases by purging them through the processes described above.

CONCLUSION

The degree to which the risk assessor tends to be protective is strongly influenced by his or her beliefs and values. In the absence of information, the

* Which is one major reason why toxicologists often conduct chemical dose-response studies in rodents at doses much higher than background or what would be measured in real life.

risk assessor tends to be overly protective, with subjective decisions superimposed over objective criteria. The decision that a risk assessor makes on each *objective* element is a statement of certainty about confidence in the information, based on the scientific weight of evidence. While a decision that a risk assessor makes on each *subjective* element is also a statement of certainty about confidence in the information, it is based on morals, values, and belief. Belief in the degree of protection that should be afforded to the public runs along a continuum between (1) a baseline level of relatively low health protection for relatively low cost and (2) high health protection for relatively high cost. Most points of view fall in between. Where an individual's point of view falls on the continuum is influenced by the beliefs. A "belief-neutral" position is ideal to the consensus-seeker. More extreme "belief-laden" positions at either end of the continuum are suited to individuals with strongly held views in one direction or the other. Risk decisions that have subjective components can best be protected from damaging bias through consensus-building activities.

REFERENCES

1. U.S. National Academy of Sciences. *Risk Assessment in the Federal Government: Managing the Process* (Washington, D.C.: National Academy Press, 1983).
2. "The Risk Assessment Guidelines of 1986." Office of Health and Environmental Assessment, U.S. EPA Report 600/8-87/045 (August 1987).
3. "Part VI. U.S. Environmental Protection Agency, Guidelines for Exposure Assessment," *Federal Register,* Friday, May 29, 1992, pp. 22888–22938.
4. National Academy of Sciences. *On Being a Scientist.* Committee on the Conduct of Science (Washington, D.C.: National Academy Press, 1989).
5. Institute of Medicine. *The Responsible Conduct of Research in the Health Sciences.* Committee on the Responsible Conduct of Research (Washington, D.C.: National Academy Press, 1989).
6. Goldstein, B.D. "Risk Assessment/Risk Management Is a Three-Step Process: In Defense of EPA's Risk Assessment Guidelines." *J. Am. College Toxicol.* 7(4):543–549, 1988.

7 BACK TO THE FUTURE: REDISCOVERING THE ROLE OF PUBLIC HEALTH IN ENVIRONMENTAL DECISION MAKING

Thomas A. Burke

CONTENTS

Introduction 93
Why a Public Health Approach? 96
    Comparing the Approaches 97
    The Tools and Health Endpoints 98
    The Agencies — Outputs, Actions, and Solutions 98
Public Values and Decision Making 99
Measures of Success 99
Recommendation: A Combined Approach 99
References 101

INTRODUCTION

The incorporation of community values into environmental decision making has become somewhat of a battle cry in the environmental regulatory and risk management community. Spurred by the environmental justice movement, the revolt of states and municipalities against unfunded mandates, and the movement toward an inclusionary comparative risk approach, regulators at every level of government are recognizing that successful environmental policies are dependent upon the recognition and integration of public values.

While this may seem like "motherhood and apple pie", in fact we are currently saddled with a prescriptive regulatory infrastructure which has systematically ignored or discouraged the consideration of public values in the decision process. Although most of our major environmental laws were originally the product of heartfelt environmental values, the regulatory process has evolved with blinders to address media-specific or pollutant-specific issues with little regard for public values and few successful mechanisms for stake-holders to participate in the decision process.

Ironically, the codification of the quantitative risk assessment process in the 1983 National Academy of Sciences "red book"[1] may have contributed to a departure from the consideration of public values in environmental decision making. Agency interpretations, perhaps more aptly called misinterpretations, of the "red book" paradigm have led to concerted attempts to separate the scientific risk assessment process from risk management and its inherent influences of politics, economics, and social values. In fact, the "red book" called for a conceptual distinction between risk assessment and risk manage-ment, but stressed that this should:

> ...not imply that they should be isolated from one another; in practice they interact, and communication in both directions is desirable...[1]

Unfortunately, the common misinterpretation has led to an isolated approach to risk characterization and contributed to a departure from traditional public health methods.

To illustrate the exclusionary nature of the regulatory process I would like to describe a personal experience in "public outreach". Eleven years ago I was involved in quite a memorable public meeting. The setting for the meeting was a hot summer night at a community center in a public housing project in the Iron Bound section of Newark, New Jersey. The purpose of the meeting was to report to the community on a major Superfund emergency remedial action and investigation which was being conducted following the discovery of 2,3,7,8-TCDD, dioxin, at an abandoned herbicide manufacturing facility in their neighborhood. Levels of dioxin as high as 50,000 ppb were detected on-site, and neighborhood sampling indicated offsite migration well in excess of the Times Beach trigger level of 1 ppb.[2]

Several federal and state agencies were present at the meeting. We arrived in the glare of television cameras and under the guard of state and city police because of concerns from the Mayor's office about the potential for violence. The Mayor himself insisted upon moderating the meeting, sensing the frustra-tion and anger of the community. His fears were well founded. Shortly after the meeting began, a violent scuffle with the police erupted when a community activist, unhappy with the process, refused to relinquish the microphone.

The contaminated facility sat on the dividing line between a Portuguese and African American neighborhood. The segregation of the community, not unlike many urban communities nationwide, only served to underscore the challenge of understanding the fears and expectations. The wishes of the two groups could not have been more different. The residents of the project wanted relocation — a ticket out. The homeowners would not consider leaving; they just wanted the site cleaned and an explanation of the potential risks to their health.

In fact, no one got what they wanted from the public meeting. Rather than an exchange of ideas to address a community problem, it became a forum to describe what the government could do, given the constraints of the regulatory and budget process. Most of the community requests could not be honored. No relocation, no long-term community health study, no way within existing laws to remove the dioxin contamination from the site. The frustration of the residents was matched only by the disappointment of the government officials. There were no satisfactory solutions to this difficult, frightening problem.

Today, the site remains contaminated, a monument to the failed expectations of Superfund. A costly, risk assessment-driven, off-site clean-up was conducted. However, the contamination was merely containerized and placed back on the site. Most of the questions asked that night remain unanswered. The debates about the risks of dioxin continue, and the public remains on the outside of the decision process.

Although this example may seem extreme, the fact is that our environmental regulatory system has done little to enable regulators to respond to public concerns and virtually nothing to recognize the essential role of community values in the shaping of environmental solutions. The national environmental laws have divided our environment, and our environmental agencies, into compartments — air, water, soil, waste. For each of these compartments, specific pollutants of concern have been identified, and a risk-based approach has been developed as the basis for a complex permitting and enforcement system.

While many positive environmental results have been realized, the risk-specific technocratic approach has Balkanized our national efforts and limited the ability of our agencies to respond to many complex environmental issues. This is not news. The trend was recognized by the EPA Science Advisory Board in their landmark report, *Reducing Risk: Setting Priorities and Strategies for Environmental Protection*:

> Because most of EPA's program offices have been responsible for implementing specific laws, they have tended to view environmental problems separately ... and questions of relative seriousness or urgency have remained unasked. Consequently, at EPA there has been little correlation between the relative resources dedicated to different environmental problems and the relative risks posed by these problems.[3]

Another vexing result of the risk-specific approach has been the demise of the role of the public health community and the use of traditional public health approaches in environmental decision making. This is not to say that environmental protection is not a vital public health function; indeed it remains a cornerstone of our national public health efforts — environmental agencies are public health agencies. However, the growing trend toward risk-specific regulation and enforcement has resulted in fundamental changes in the organization and practice of environmental health.

The decline of the public health approach began with the first Earth Day and the rise of the federal EPA as the primary regulator of the environment. From their historic roots in public health departments, environmental responsibilities evolved to become the purview of environmental regulatory agencies at both the state and federal levels. Driven by increasing statutory mandates, the new agencies organized along environmental media to enforce the laws.

The diminishing role of public health was highlighted in the 1988 Institute of Medicine Report, *The Future of Public Health,* which concluded that:

> The removal of environmental health authority from public health agencies has led to fragmented responsibility, lack of coordination, and inadequate attention to the health dimensions of environmental problems.[4]

On the state level, "mini-EPAs" have become the dominant model for environmental protection. Currently, public health departments retain the lead responsibility for environmental programs in only eight states.[5]

The trend away from the public health approach is seen not only in the organizational charts, but also in the budget. While overall government expenditure on environmental regulatory and remediation programs has never been higher, support for evaluating and addressing public health concerns is almost nonexistent. For example, support for the Agency for Toxic Substances and Disease Registry's public health assessment effort under Superfund is well below 1% of the annual budget for remediation and enforcement activities.[6] On the state level, 80% of environmental expenditures go to regulatory activities. When natural resource protection efforts are included only, $0.08 of every state level environmental program dollar goes to support public health activities.[5]

WHY A PUBLIC HEALTH APPROACH?

Is it coincidence that the declining role of *public* health has mirrored the decline in *public* involvement in addressing environmental issues? Perhaps it is time to go "back to the future" and rediscover the public health approach to environmental problem solving.

The public health approach offers a valuable model for the integration of values, perceptions, and ethics into environmental decision making. Historically, public health strategies have been shaped by two factors: the level of scientific and technical knowledge, and the content of public values and popular opinion.[4] The strong influence of public values and acceptability has shaped public health initiatives from the cholera epidemics to the anti-tobacco movement.

Unlike the risk assessment approach, public health has not seen the need to separate the assessment of risks from risk management considerations. By its very nature, public health is population based, and shaped by the integration of public values with science. Public health strategies to reduce risk are shaped

Table 1   Comparison Of The Risk Assessment And Public Health Approaches

	Goal: The Protection Of Public Health	
	Risk Assessment	**Public Health**
Driver	Single hazard	Population health impact
Approach	Pollutant specific	Population based
Process	Hazard ID	Assessment
	Dose response	Policy development
	Exposure assessment	Assurance
	Risk characterization	
	Risk management	
Health endpoint	Limited (cancer)	Diverse/multiple
Tools	Toxicology, modeling	Epidemiology, surveillance
Agencies	Federal/state	Local/community
Outputs	Standards, regulations	Intervention strategies
Actions	Permitting, enforcement monitoring	Outreach, screening, treatment
Solution	High tech/pollutant specific	Low tech/broad based
Public values	No mechanism for inclusion	Essential component of intervention strategy
Decision making	Top down	Bottom up
Success	Regulatory compliance	Community health improvement

by the biological basis, technical feasibility, and public support and acceptability of interventions. Table 1 presents a categorical comparison of the risk assessment-based and public health approaches to environmental decision making. This table is not intended to be all-inclusive, but rather to point out several of the basic differences in the two approaches.

Comparing the Approaches

The fundamental goal of both approaches is the protection of public health. However, there are a number of clear contrasts between the two methods. As shown in Table 1, the risk assessment approach is generally driven by a single hazard and is pollutant specific in its approach. This contrasts with the population based approach of public health which is driven by population health impacts.

The risk assessment approach is based upon the four-step "red book"[1] paradigm to provide the metric for characterizing risk and guiding risk management. This contrasts with the more robust core functions of public health: assessment, policy development, and assurance, as defined by the Institute of Medicine.[4] Ellen Silbergeld has described the contrast between the two approaches to assessment. Quantitative risk assessment seeks to identify the toxicological properties of a specific substance, while public health assessment is based upon an evaluation of the prevalence of disease and the health status of a population.[7]

The policy development and assurance components of the public health process are analogous to risk management as described in the "red book", although assurance implies a continual evaluation of interventions and a continuation of public support.

The Tools and Health Endpoints

Several other aspects of the two approaches also provide a sharp contrast. While there may be some degree of overlap in the analytical tools (i.e., epidemiology), risk assessment is largely based upon the principles of toxicology and is dependent upon the results of animal testing. This limits the health endpoints which have been considered. As a result, risk assessment-based regulatory approaches have largely been dominated by efforts to control carcinogens. In contrast, the public health approach is based upon surveillance of community health indicators such as morbidity and mortality. Although the approach may lack the sensitivity to identify specific etiologic agents in the environment, it provides an ongoing indicator of community health for a broad range of health endpoints.

The Agencies — Outputs, Actions, and Solutions

The agencies involved in the two approaches also offer a sharp contrast. The traditional public health approach is based upon a strong local/community health infrastructure to provide health status data and implement programs. The risk assessment approach is structured by national environmental laws which established a strong federal/state partnership of regulatory agencies with little participation or financial support for health agencies at the local or community level. The exclusion of local agencies from the decision-making process contributes to the exclusion of public values and has fueled the current backlash against federal "unfunded mandates" to address environmental risks.

The end results or abatement strategies of the risk assessment approach are also quite different from the public health approach. The output of the risk-based process are often high technology, source-specific standards or regulations. These regulations are implemented through an enforcement strategy shaped by permits and monitoring. In contrast, public health strategies often involve a low technology community-based approach to identify and reduce risks through education, screening, and medical intervention.

The management of risks from lead in the environment provides an illustration of the two approaches. The EPA regulatory approach included a national ban on leaded gasoline, the establishment of a national standard for lead in drinking water, and limiting the lead content in plumbing fixtures and pipes. In contrast, the Centers for Disease Control public health approach has established a strategy implemented by health agencies at the local level to: educate communities and medical care providers about the hazards of lead, identify those at high risk through screening, and define an approach for medical management of the highly exposed.

PUBLIC VALUES AND DECISION MAKING

As I have stated previously, and illustrated through the Newark dioxin example, the risk assessment-based regulatory approach offers little or no opportunity for the incorporation of public values into the decision-making process. The current approaches to risk characterization do not provide an adequate mechanism for weighing social values, perceptions, or ethics. In addition, the rigid structure of the major statutes provides little flexibility for the consideration of community concerns. Decision making is predominantly "top down", the product of federal laws implemented through state and regional agencies.

On the other hand, the public health model *does* offer a mechanism for the consideration of public values. The traditional community-based approach offers a more "bottom up" decision-making process. Many programs are implemented through a "block grant" approach which enables communities to determine the prevention strategies most appropriate to meet their needs. Because many of these interventions are nonregulatory in nature, their success is dependent upon community acceptability and local political support. Therefore, inclusion of public values has been essential to the development of effective public health programs.

MEASURES OF SUCCESS

As a former environmental regulator and a former state health commissioner, I have a somewhat unique perspective on the contrasting measures of success. Success in the risk-assessment driven regulatory arena is measured by environmental compliance. Numbers of permits written, percentage of monitoring results below standards, violations reported, and fines collected are the standard tools to track success. These are generally effective measures of activity, but say little about the benefits to public health.

The standard measures of success for public health programs are generally direct measurements of community health status. For example, a food protection program would be evaluated by its effectiveness in reducing reported cases of food-borne illness. Similarly, a tuberculosis control program is judged by its effectiveness in reducing the number of cases in a community. While direct measurements of health benefits of environmental programs can be elusive, it should be a goal for defining success.

RECOMMENDATION: A COMBINED APPROACH

Each approach clearly has its strengths and weaknesses. The purpose of this paper is not to advocate one over the other. Nor is a return to the past health agency-based organization of environmental services advocated. Rather, a combined approach to environmental health which draws upon the strengths of

Table 2   Combination of the Risk Assessment and Public Health Approaches

Goal: The Protection of Public Health	
Combined Risk Assessment/Public Health Approach	
Drive	Individual hazards and population health impacts
Approach	Population based to identify pollutants of concern
Process	Quantitative and public health risk evaluation
	Characterization of health and social aspects of risk
	Policy development/ risk management
	Assurance/evaluation
Health endpoint	Broad range to prevent adverse effects and address community concerns
Tools	Toxicology, modeling, epidemiology, surveillance
Agencies	Federal, state, local, community
Outputs	Intervention strategies including standards, regulations, education, and nonregulatory guidance
Actions	Permitting, enforcement, monitoring, outreach, screening, and treatment
Solution	Combined high tech/pollutant specific and low tech/broad based
Public values	Included as an essential component of intervention strategies and prevention
Decision making	Shared and inclusionary — agencies and community
Success	Community health and environmental improvement

the two models is recommended. The characteristics of a combined approach are presented in Table 2.

The combined approach offers an opportunity to enhance the role of public values in the decision-making process and draw upon a well-established community-based network of public health agencies. Such an approach would not require a "reinvention" of government, but rather a rediscovery of the historical roots of environmental protection. By adopting a combined approach, environmental decision making might move beyond the pollutant and media-specific approach toward a population-based approach to community health. Such an approach would also enhance the assessment of risks through the consideration of actual measurements of morbidity and mortality in the population. Organizationally, the approach could improve the responsiveness of decisions to community needs through improved involvement of state and local health agencies. This would also shed new light on "outside the Beltway" concerns in environmental regulation.

The solutions to environmental issues can also be strengthened by a partnership of public health and environmental agencies. The previously mentioned national strategy for lead could serve as a model which provides a continuum of prevention from the source of pollution to the exposed individual.

Perhaps the most important benefit of the rediscovery of public health would be the opportunity for greater public involvement in the decision-making process. The community-based perspective of public health can provide a vehicle for the inclusion of public values and enable stakeholders to participate in the shaping of environmental solutions.

In conclusion, environmental decision making is currently at a critical crossroads. High costs and low credibility have undermined support for traditional regulatory strategies. The interface between science and public values is being revolutionized through the environmental justice movement and the comparative risk approach to priority setting. Future success may well hinge on our ability to incorporate public values and restore lost trust. Hopefully, the rediscovery of public health will play an important role in the restoration of that trust.

REFERENCES

1. National Research Council, 1983. *Risk Assessment in the Federal Government: Managing the Process.* National Academy Press. Washington, D.C.
2. Kimbrough, R.D., H. Falk, P. Stehr, and G. Fries. "Health Implications of 2,3,7,8-tetrachlorodibenzodioxin (TCDD) Contamination of Residential Soil". *J. Toxicol. Environ. Health* 14:47–93, 1984.
3. United States Environmental Protection Agency. 1990. *Reducing Risk: Setting Priorities and Strategies for Environmental Protection.* Washington, D.C.
4. Institute of Medicine. *The Future of Public Health.* Washington, D.C., National Academy Press, 1988.
5. Burke, T.A., N. Tran, and N. Shalauta. Identification of State Environmental Services: a Profile of the State Infrastructure for Environmental Health and Protection. A report prepared for Bureau of Health Professions, Public Health Branch, Health Resources and Services Administration, U.S. Public Health Service. Rockville, MD. 1994.
6. Agency for Toxic Substances and Disease Registry. 1989. ATSDR Biennial Report to Congress: Vol. 2, U.S. Department of Health and Human Services, U.S. Public Health Service. Atlanta, GA.
7. Silbergeld, E.K. "A Proposal for Overcoming Paralysis in Improving Risk Regulation". In *Regulating Risk: The Science and Politics of Risk.* T.A. Burke, N.L. Tran, J.S. Roemer, and C.J. Henry, Eds. Washington, D.C.: ILSI Press, 1993. pp. 45–47.

8 TELLING THE PUBLIC THE FACTS — OR THE PROBABLE FACTS — ABOUT RISKS

Victor Cohn*

We live in a time when the planet we live on and all its inhabitants are at risk. Almost no reporters or editors, whether the chroniclers of politics, business, economics, science, medicine, or the environment, can ignore this. Politics, economics, science, health, the state of the environment — every piece of news about risk involves all these elements. Given the difficulty of the subject, or subjects, and their complex interactions, it is little wonder that the chairman of a National Research Council committee on the communication of risk — John Ahearne, former chairman of the Nuclear Regulatory Commission, past vice president of Resources For the Future — has said, "Right now everyone seems frustrated. Government officials and industry managers think that the public just does not understand. The public is tired of failed promises and of being treated in a condescending manner. Scientists and engineers are distressed because the media and public misinterpret their complex research."

We of the media can indeed do better — far better. There has indeed been much misinterpretation or poor reporting. There has also been much clear and correct interpretation and reporting on risks of all kinds. There needs to be more. All systems will continue to grow, and even reducing risks beg risks. Incinerating garbage creates air pollution. Closing a nuclear plant increases the demand for fossil fuels. A smaller car burns less fuel, but is less safe in a crash. Spending huge sums attacking the wrong risks runs the risk of neglecting far greater ones. Over-worry about slight risks can create destructive paranoia.

* Victor Cohn is former science editor of the *Washington Post.* He was a fellow during 1993–1994 at Georgetown University and is currently a fellow of the American Statistical Association. He is the author of *NEWS & NUMBERS: A Guide to Reporting Statistical Claims and Controversies in Health and Other Fields* (Iowa State University Press) and — the volume from which most of this chapter is abridged — *Reporting on Risk* (published by the Media Institute, 1000 Potomac Street NW, Suite 301, Washington, D.C., 20007).

1-56670-131-7/96/$0.00+$.50
© 1996 by CRC Press, Inc.

It is clear that we must choose between risks.

It is clear that the good life we like has environmental costs.

It is clear that we may no longer safely count on a standard of living based on destroying the environment.

I believe that pointing out all these facts is part of the environmental story. Pointing out that there is no free lunch, that affordable electricity, food, and daily living all put someone at risk, is an essential part of reporting risk.

All reporting of risk and the environment should be fair and objective, yet I believe there can be no true reporting without these facts and values. This does not mean reporting should be propagandizing. That is what so-called journalists do under tyrannies. However, it takes values — fairness, decency, thoroughness, responsibility — to create responsible reporting.

Whether we will it or not, journalists have in practical fact become part of the regulatory apparatus. What we choose to report and how we report it helps set the agenda for regulating *all* health risks, for no kind of reporting touches human emotions and fears more closely.

This is a heady responsibility, and it is properly a responsibility not only of journalists, but of all who communicate to the public, including the executives, scientists, and public relations officers and others who speak for the government, industry, and environmental organizations.

Morris A. (Bud) Ward, director of the Environmental Health Center of the National Safety Council, has written that "'risk communications' to many has come to represent environmental public relations at its worst: accept 'acceptable risk' or, worse yet, delay, delay, delay." Those who so offend are not the only offenders. Drs. Jean Mayer and Jeanne Goldberg of Tufts University wrote of food problems: "Time and time again, industry takes the defensive, trying to protect its turf when it should strive for the public interest. Consumer groups tend to leap to conclusions with the hysteria of someone shouting 'fire' in a crowded theater."

It is the responsibility of reporters of risk to seek out and report the truth, or as close as we can come to the truth. If the public is to learn to live with unavoidable risk — and we should not ask more — then, said Chris Whipple, technical manager for risk and health science, Electric Power Research Institute, and a member of the National Research Council committee on risk communication: "Openness is the surest policy ... public perception is based as much on trust of the risk managers as it is on any numbers ... that trust is essential if risk managers would manage successfully ... based on pragmatism." There is a much better chance of long-term success, if all the facts, assumptions, and uncertainties are made public.

Reporters must try to report all the real dangers, if they are indeed dangers, in proper perspective, avoiding either dangerous hysteria or dangerous apathy.

Good reporting on the risks of life is not easy. Regarding virtually every risk, we hear one thing from one "side", one thing from the other. To quote Ron Kotulak of the *Chicago Tribune*: "Information is power, and because of that industrialists, politicians, bureaucrats, administrators, doctors, scientists, and activists often try to control it or shape it or otherwise use it to promote their private goals ... or to serve a particular cause or ideology."

So pesticides are good, pesticides are bad. Radon in our homes causes lung cancer. Low levels are harmless. Nuclear power is dangerous. Nuclear power is the least harmful source of energy.

Why is this?

- Even the most honest and most eminent scientists disagree when the results of their studies are in conflict, or evidence is incomplete.
- Organizations — corporate or activist — typically arrive at conclusions that serve their own causes or interests. In 1982 the newspaper ad of a major oil company (not Exxon) said "OIL SPILLS: LESS THREAT NOW" and asserted that "oceanic pollution from oil transportation has abated, thus insuring the continued health of the seas. That's progress of which we are understandably proud."

Well, oceanic pollution from oil transportation had abated. Then came the June 1989 *Exxon Valdez* spill off the shore of Alaska.

The main offenders on all sides are scientists for government, industry, and environmental groups who "make sweeping judgments on the basis of incomplete and hence inadequate data", stress their own opinions, and omit or minimize conflicting evidence, a Twentieth Century Fund task force said. The same organizations' leaders and public relations officers may sincerely do the same.

- Or they may do so insincerely.
- Politics often leads to lies or distortions or at least to decision ruled by a rigid political philosophy rather than fact.
- The greatest risks are not always the risks that the public worries about — and that the media emphasize.

Critics therefore accuse the media of fostering "chemophobia" and "cancerphobia." Which happens first, the public perception or the media emphasis? It is hard to say. Yes, the public gets its information from the media, but the media mainly react to someone's concerns.

We of the media are also fathers, mothers, residents of a town and neighborhood, consumers of food, water, and air. The residents around the Love Canal toxic waste site were worried sick over neighborhood illnesses well before the media discovered them.

Still, one must admit, the great risks to society are not always reflected either in public fears or TV pictures and newspaper headlines. The public worries greatly about pesticides and nuclear radiation, but worries little about heavy drinking or wearing seat belts or driving 70 miles an hour. A sign on a Wisconsin tobacco shop said: "Cigarettes, 2 Packs for $3.09. We Sell Radon Test Kits."

In one survey, League of Women Voters members ranked nuclear power at the top of "events contributing to death". So did college students. Yet the country's greatest killers are smoking, drinking alcohol, motor vehicle accidents, and handguns, in that order, according to two authors in the *Cancer Bulletin*. A study for the Carter Center of Emory University said two out of three U.S. deaths and one hospitalization in three are linked to six personal health risks: tobacco, alcohol, injury, high blood pressure, over-nutrition (as measured by obesity and high blood cholesterol), and gaps in primary medical care.

"In the battle with environmental hazards, our resources are limited and we cannot afford to squander them", Dr. Stephen Thacker of the U.S. Centers for Disease Control (CDC) told an environmental conference. "Insignificant problems are sensationalized while significant problems are ignored..." CDC's Dr. Vernon Houk said injuries are leading cause of death "yet we have only $20 million to prevent them, and zero to prevent lead poisoning in our children".

What about the pesticide issue?

The Food and Drug Administration said in late 1989 that 96% of fruits, grains, vegetables, and dairy products analyzed in its annual food safety survey either had no pesticide residues or had residues within permissible limits. A few more were in technical noncompliance, and only 1% had levels above legal limits.

Dr. Elizabeth Whelan, president of the American Council on Science and Health (which says most environmental health risks are exaggerated) says "...the margins of [food] safety are enormous" and "...there is no scientific evidence that residues in food from regulated and approved use of pesticides have ever been the cause of illness or death in either adults or children." (She has also admitted, however, that this may not be the case among agricultural workers.)

She has written — must I say that I disagree? — that the central problem may be that "...the eastern press, which dominates network news and wire-service coverage, tends to be considerably more politically liberal, anti-corporate, anti-free enterprise, anti-establishment, and anti-traditional, as compared with the rest of the nation's population", and that these media extremists are "rejecting the traditional goals of American society and are out to dismantle it and replace it with some poorly defined alternative structure."

The eminent Dr. Bruce Ames, inventor of the widely used Ames test for carcinogens and mutagens (mutation-causing chemicals), is less tendentious, but strongly agrees that pesticides have received a bum rap. He has labeled as myths the ideas that "cancer rates are soaring" or soaring due to pesticides. Ames (in collaboration with Lois Swirsky Gold) has said:

- "Americans ingest in their diet at least 10,000 times more weight of natural pesticides than they do of man-made pesticide residue. These natural 'toxic chemicals' ... serve as protection against fungi, insects and animal predators."
- "The important issue is not merely to identify those chemicals that are carcinogenic or teratogenic (producing birth defects) [but] to put our efforts into discovering and eliminating the important causes of human cancer [and] the major preventable risk factors ... such as tobacco, dietary imbalances, hormones, and viruses."

Others tell us that our times are "the safest of times", that life expectancy is at an all-time high, and that our food and water supplies are the safest in history, all this despite the enormous increase in use of chemicals.

Without question, there are greater perils to humankind than pesticides or other chemicals. Part of the undue fear of cancer epidemics may have been caused by estimates made by some scientists that "70 to 90 percent" of all cancers might be "related to", though not solely caused by, "environmental factors". It would have been more accurate to have said "behavioral" factors, since by various generally accepted estimates, tobacco use is associated with 25 to 30% of all cancers; alcohol use, 3 to 5%; diet 30 to 60%; sexual behavior, 7%; radiation, 3%; food additives, perhaps 1%; man-made environmental (mainly chemical) pollution, 1 to 3%; and job exposures, 4 to 6%.

Regarding cancer, the latest U.S. figures at this writing (as adjusted for changing age distributions in the population) are these:

- Cancer incidence, meaning the overall probability of getting cancer, has indeed been increasing — by an average 1.2% yearly or by 22.4% between 1973 and 1991. Even eliminating the tragic toll of lung cancer, largely the effect of smoking, there was a 20.6% increase.
- The greatest increase took place in those over 55 — not unexpected since many cancer-causing agents take many years to take effect — but there was also a slow but steady increase of 0.5% yearly in younger people.
- The cancer death rate — much debated when people try to decide whether or not there has been a cancer epidemic — has increased by an average 0.4% yearly; by 1991 it was up 6.9%. Remove lung cancer's effect, and the overall rate decreased by 2.9%. For under-55s, the death rate dropped by 16.1% between 1973 and 1991. For those 55 and older, death rates increased (by 3.7% for those aged 55 to 64 and by 16.1% for those 65 and older).

Is the increased incidence due to chemicals or pollution? Or is there just better detection and diagnosis? Have there been lower death rates in some cancers because of earlier and better treatment? Or perhaps because many tiny cancers that would never develop into harmful tumors are being detected early and counted in the statistics, making death rates appear lower? The answers are unclear, but given the overall increases "something," or many "things", are going on that demand greater scrutiny and better long-range study.

Contrary to the views of Ames and others, many scientists believe that:

- Even if the effect on human physiologies is small, it is foolish to add man-made toxins to any natural burden, especially dyes or food colorings that serve no major purpose.
- Studies almost universally deal with the effects of some single toxin, but in real life we are exposed to the unknown cumulative and possible synergistic effects of many toxins. We know that some combinations — exposure to tobacco smoke and asbestos or radon, for example — are deadly.
- Even small added burdens may cause or help promote unacceptable numbers of cancers. Perhaps many more years must pass before we realize it, if we ever do. No one predicted a lung cancer epidemic when men, and then women, started smoking. There have been some recent, little understood, possibly environmentally caused increases in asthma and some neurodegenerative diseases.
- Toxins may particularly affect the most vulnerable: children, the aged, the chronically ill.
- FDA and EPA monitoring of foods, air, and water is far from complete. And EPA *does* rank pesticides high among possible carcinogens. EPA scientists and managers in May 1989 ranked "problem areas" on the basis of cancer risk. Tied for first: worker exposure to chemicals and indoor radon. Next in order came pesticide residues on food, other indoor air pollution, consumer exposure to chemicals, hazardous air pollutants, depletion of stratospheric ozone, and inactive hazardous waste sites.

The public is not an unimportant participant in judging all this. There can be logic, or at least legitimate personal choice, in public acceptance of some risky activities or technologies, while the public may reject others that are less risky.

A Rutgers University environmental risk and risk communication group (Professors Peter M. Sandman, Michael Greenberg, and David B. Suchsman) has said:

For risk assessment professionals, 'risk' is some kind of multiplication of probability (how often does it happen) times magnitude (how bad is it when it happens)...

For the rest of society ... mortality matters. But so does whether the risk is involuntary or coerced, whether the effect is immediate or delayed, whether the hazard is natural or industrial, whether the people imposing it are trustworthy or untrustworthy, etc.... If you are injured while skiing, your response to the injury will be very different from your response to, say, the discovery of buried industrial wastes in your backyard....

People know what feels involuntary, unfair, uncontrollable, untrustworthy....
[They know] the reality of human fear, anger, and suffering.

Among other factors that affect risk perception, according to recent studies: the "…why take a chance? factor, the feeling that the direst of the doomsayers have often been right, the fear of the unknown vs. the known, the feeling that the victims may be our children, the belief that a normally 'safe' technology (e.g., nuclear power) can kill thousands if the worst happened and — very much — the feeling that 'we are no longer in control'."

We have again and again been told that some technology was "safe", only to learn otherwise later. No one, at least no one in charge, adequately prepared us for a Three Mile Island (TMI) or Chernobyl, an *Exxon Valdez* or toxic waste dumps by the hundreds. We constantly hear of new hazards on a global scale. We know that birds and fish have died or been spoiled as foods.

We have seen disasters — TMI and Chernobyl — occur when managers or operators were careless, ill trained, or complacent. Underlying the space shuttle *Challenger* disaster, says John Ahearne, was "complacency … a disbelief that this technology was really hazardous." Now, he says, it is "not only the general public" which is "skeptical of technologists who present an argument based on trust me, I know best."

"The public…" concludes Peter Sandman, "pays too little attention to hazard, the experts pay absolutely no attention to outrage." Risk communication, he says, must include an "effort to alert people to risks they are not taking seriously enough", an effort to "reassure people over risks they are over-reacting to", and an understanding that the public's feelings must become part of the equation.

Alas, our job as reporters would be easy if we just find out the "facts". Our greatest problem may be that there often are none, or very few, or few the experts agree on.

"Having to piece together a story from multiple sources, even recalcitrant ones, is hardly new to journalists", wrote Professor Fischoff of Carnegie-Mellon University. "What is new about many environmental stories is that no one knows what all of the pieces are or realizes the limits of his understanding…. There may be no scientific expertise anywhere for measuring the long-term neurological risks of a new chemical." Or, for that matter, the neurological or other effects of an old one.

What about risk numbers? "In almost 99 percent of the cases, we do not know," Thomas Burke of the New Jersey Department of Health has reported. "We have no idea, really, how to interpret a biological assay [to tell] a community what their risks are."

Even where they exist, studies of environmental health effects are often weak, in part because there are seldom enough cases to stand out against the normal case load or background. Or effects may take so many years to develop that no one can accurately ascribe any cause.

This is not to say that scientists — with proper support — could not do better. Paul Portney of Resources for the Future has said that "we currently do a disgraceful job of collecting, analyzing and disseminating information about environmental conditions and trends."

None of this, of course, is any excuse for bad reporting. The Rutgers University group (including two journalism professors) in their comprehensive manual, *Risk Communication for Environmental News Sources,* describe us not very flatteringly, though they are no easier on sources who clam up, lie, or obfuscate, and they emphasize that not all journalists are the same and some posses "extraordinary" knowledge. Focusing mainly on unspecialized reporters, and pointing out that reporters do their jobs with "extraordinary constraints of time and space", they conclude that:

- "Who, what, when and where come well ahead of 'how dangerous' in the journalist's list of important questions.... For the most part [journalists] will not sit still for a lecture on the fundamentals of risk assessment."
- "Journalists are often woefully ignorant of the basics of environmental science and engineering."
- Reporters' goals are too often only "accuracy and balance rather than 'truth'," and objectivity equals accurate reporting of what sources said and did ... accompanied, ideally, by fairness or balance" in emphasis. Scientists seek methodological rigor, the trend of evidence, and the consensus of experts "in the unending search for truth", while journalists tend to look for the "other side."
- "[The] balanced treatment conveys ... the notion that one side is as credible as the other.... But often one side represents mainstream scientific thought, while the other is a minority of dissident voices.... The scientific establishment, like any establishment, sometimes turns out systematically wrong, [but] the scientific majority usually turns out closer to the mark than the mavericks. It is deceptive to present such opposing positions as if they had equal support."
- "Reporters pay more attention to alarming than to reassuring information.... One [reason] is the reporter's desire to come back with a story that will appeal to the editor and thus earn 'good play'."
- "Typically, reporters give short shrift ... to views too extreme to be credible [or that are] 'oddball' ... [but] also pay relatively little attention to [moderate] positions usually seen as too wish-washy to make good copy.... William Lowrance of Rockefeller University complains that 'articles often present the absolutely most polar views [but] the issues are usually not black and white'."
- "Most risk controversies are newsworthy more because of the controversy than because of the risk. Serious risks that are not highly controversial — driving without a seatbelt, for example — rarely receive much media attention.... A risk that is new (as indoor radon was in 1985) or extraordinarily dire (as AIDS continues to be) merits coverage even without conflict. But [many risks] soon disappear from public view."

They quote a Dow Chemical executive who tells how 75 reporters from all over the country attended a 1983 news conference at which Dow responded to a national furor over dioxin by announcing an extensive study program — but "in the four years since.... I doubt if we have had half a dozen calls from reporters ... to find out if we did what we said we would do."

Professor David Sachsman examined news sources' influence on San Francisco Bay environmental coverage by 28 daily newspapers, 6 TV stations, and 8 radio stations with independent news operations. He found just 15 environmental reporters among them, and said many journalists relied on "official" or public relations sources for lack of time, rarely questioning "the reliability of 'official' spokesmen."

Dr. Patricia Murphy was assigned by the Environmental Protection Agency to investigate a midwest cluster of ALS cases (amiyotrophic lateral sclerosis or "Lou Gehrig's disease") that were attributed for a time to use of a sewage-derived fertilizer. The charges did not stand up. A state epidemiologist told reporters there was "no statistically meaningful difference" between the local and national death rates, but one newspaper chose to say, "City ALS Death Rate Higher Than U.S.," and another said, "[No] Epidemic at the Moment."

John Ullmann, the first executive director of Investigative Reporters and Editors, has said: "Often we are talking about stories that can scare people to death, if we do not do them well. I mean sometimes we are a walking time-bomb ourselves, because we are so ignorant."

In a study commissioned by the Environmental Protection Agency, Harold Sharlin examined the 1983–1984 controversy over the pesticide ethylene dibromide (EDB), whose residues were found in breads, cake mixes, and other foods. He gave major national newspapers the highest marks for accuracy, but said local newspapers tended to miss or misuse technical concepts. He said the weakest medium was television, especially local TV, with its brief, dramatic visual material and stress on "clear, definitive answers" creating a bias not obvious to viewers.

EDB may or may not have been a serious threat. There were no known deaths, however. This did not stop a tabloid in a major eastern city from screaming in a headline: "Killer Muffins Found on Grocery Shelves".

Professor Fishchoff, in many ways a defender of press performance, wrote: "[Its] performance in the reporting of risk issues is not always up to the standards found in other topic areas. Most news organizations would not tolerate sports or business reporting by reporters who do not understand the subject.... The same is not always true of ... the technical and social dimensions of risk messages."

Journalist Cristine Russell tells of the chemical companies in Port Arthur, Texas, which in June 1989 wanted to be among the first to present their toxic-emissions data to the public. "After briefing public officials, they found themselves faced with a press luncheon almost devoid of the eagerly anticipated reporters.... Where had they gone?... The hottest story in town was a more immediate environmental threat: the U.S. Forest Service had gone to court over plans by the Rainbow Family of Living Light to camp over the Fourth of July holiday weekend. Some members ... reportedly cavorted in the nude.... Industry gave a party and nobody came. Said [one industry speaker], 'Do you think if we took our clothes off the press would come and hear about hazardous chemicals?'"

Sharon and Kenneth Friedman of Lehigh University have written that despite much good reporting at the 1979 Three Mile Island nuclear accident, "Research has shown that most reporters ... lacked sufficient background.... They did not know how a nuclear plant worked, had no understanding of radiation terminology, and did not know which experts to seek out for help."

When it comes to emotionally charged issues, or trying to make "page one" or the "evening news", the best among us sometimes overstate. Some years ago I decided, tongue only partly in cheek, that there are only two kinds of medical stories: "New Hope" and "No Hope". The same is true of stories about environmental and other risks. The in-betweens got buried or lost. New Hope and No Hope get the play.

When it comes to risks, it is mainly No Hope that gets the play. We are thus often accused of perpetrating the "Scare of the Week."

We are very far from the whole story, of course. At a news conference presenting the National Research Council's 1989 report on risk communication, John Ahearne said: "[One] problem area we identified was the tendency to blame journalists as a central cause for risk communication difficulties. Contrary to a view often held in government, industry and technical circles, we do not believe that journalists are a significant independent cause of risk communication problems."

This report found:

- There are instances in which media coverage has favored extreme positions, but it is wrong to say the media always exaggerate health risks.
- Extensive media coverage of the TMI nuclear accident found the balance between supportive and negative statements to be more reassuring than alarming.
- "Journalists need to understand how to frame [both] the technical and social dimensions of risk issues," but when an expert "is not communicating effectively, that usually is due to inability or unwillingness." (The Rutgers professors expand on this point as bluntly as in the language they use to describe inadequate journalists.)
  - "Sometimes [sources] do not really want the information reported, even accurately, fearing perhaps that the public would be inappropriately frightened. If there is no ongoing public controversy, many sources see little reason to upset the applecart with publicity.... In the midst of controversy, many are reluctant to fuel the fire ... These may be unwise preferences — we think they usually are — but they are widespread."
  - "[Many sources] enter their first interview already angry or contemptuous about what they regard as media hostility, sensationalisms, or stupidity. Their chip-on-the-shoulder attitude may lead to excessive caution, or

to off-putting condescension, or even to a refusal to grant the interview at all ... [all] self-defeating reactions ... [Such sources] will resist even learning from experience. Each bad interview is further evidence of the ill will or incompetence of reporters, never of their own need to acquire more skill as a source."

In a searching analysis of our ways and our failures — and one that by no means advocates namby-pamby reporting — Dorothy Nelkin of New York University has written that:

- "While reporting about risks is often problematic, the most serious problem may be less one of bias, inaccuracy, and sensationalism than the reluctance of journalists to challenge their sources of information."
- "Journalists must try to convey understanding as well as information ... the social, political, and economic implications ... the nature of the evidence that forms the basis for decisions, and the limits as well as the power of science...."
- "'The scientific method' confer[s] a standard of objectivity.... Understanding the methodology and the evidence behind a scientific statement is critical to evaluating its value."

How can we all do better? How can you who are scientists, statisticians, epidemiologists, environmental officials help reporters report the facts — or the best "facts" you can muster — on environmental problems and controversies? How can you help the public understand these and make decisions?

Because the public is confused by constantly conflicting "they say's," your most important contribution may be to speak candidly about the strength and weakness of your, or anyone's, statistics and studies. Tell all you can about:

- The certainty of the uncertainty of knowledge and findings, how all you can say is the best estimate at the moment.
- The necessary use of probability to decide on action in the face of uncertainty.
- The importance of the power of large vs. small numbers in accessing studies.
- The ever-present danger of biases — unconsidered factors — distorting results.
- The ubiquity of variability, how two studies seldom produce just the same result.
- And the fact that there is a hierarchy of studies, with varying degrees of probable validity.

All these things tell us that one study or set of statistics seldom proves anything, that we need to look at a consensus of studies, and often lean on a consensus of the best informed, most neutral authorities.

A wise counselor once said, "...if you would have public confidence, confide in the public." There is no better advice.

THE URGENT NEED TO INTEGRATE ETHICAL CONSIDERATIONS INTO RISK ASSESSMENT PROCEDURES

9

Donald A. Brown

CONTENTS

Introduction 116
Risk Assessment Reforms 116
Risk Assessment Procedures and Scientific Certainty 117
Risk Assessment Decisions Made in the Face of Scientific
   Uncertainty Must Be Understood to Raise Important Ethical
   Questions 119
   The Use of Worst-Case Scenarios 121
   The Duty to Take into Account Most Recent Toxicity Studies 122
   Failures in the Application of Animal Data to Humans 122
   The Inappropriate Use of Fate and Transport Models 122
   Cascade of Conservative Assumptions that Overstate the "Real"
   Risk 123
Risk-Based Decisions Often Hide Other Controversial Ethical
   Questions 125
   The Problem of Burden of Proof 125
   The Problem of Distributive Justice 126
   The Problem of Communicative Rationality 126
   The Problem of Risk-Based Decisions Authorizing Pollution in
   Certain Situations 126
Risk Management Decisions Based on Cost-Benefit Analysis Also
   Hide Other Important Ethical Issues 127
Conclusion 128
References 129

INTRODUCTION

We have truly reached the age of risk assessment in environmental affairs. Presently, the U.S. Environmental Protection Agency is using risk assessment procedures in environmental programs to set environmental priorities, standards, and clean-up levels. In June 1992, in Rio de Janeiro at the Earth Summit, the nations of the world adopted Agenda 21, which urged the use of risk assessment procedures in hazardous and nuclear waste decisions throughout the world. In a short time, risk assessment procedures have become a very important tool of environmental managers worldwide.

Along with the increased popularity of risk assessment there is growing controversy about some approaches and methods in the use of certain risk assessment procedures by the government agencies. In response to this controversy, several legislatively mandated reforms of risk assessment procedures are currently under serious consideration in the U.S. Congress. This legislation has two goals: (1) to increase the use of risk assessment in environmental affairs as a priority setting mechanism; and (2) to reform perceived tendencies in current risk assessment practices to overstate the "real risk". Both of these goals stem from a desire to cut unnecessary costs of environmental regulation, in the first instance by making sure that environmental regulation focuses on the more important risks while not wasting money on trivial ones, and in the second case through assuring that the risk quantification does not overstate the "real" risk. A common theme of these reform efforts is that better risk assessment procedures can be assured by putting risk assessment procedures on a sound scientific footing and that regulatory efforts should be directed to reducing high risks while ignoring insignificant risks.

This paper argues that the most important reform needed in risk assessment procedures is to integrate ethical considerations into risk assessment procedures. It is premised on the notion that there are controversial ethical questions that remain mostly hidden and unexamined throughout risk assessment methodology, and that current reform efforts have the potential to exacerbate serious ethical problems with the use of risk assessment procedures.[1]

RISK ASSESSMENT REFORMS

Two legislative proposals dealing with risk assessment are currently under serious consideration. Amendments to the Clean Water Act and the Safe Drinking Water Act have been introduced that call for expanded use of risk assessment procedures, reform several perceived abuses of risk assessment methodology, and change the basis of the degree of environmental protection to one based on cost–benefit analysis (CBA).

Clearly, the objective of these reforms is to limit the scope of environmental regulation so that money is not wasted on environmentally trivial regulations and to assure that any risk assessment is to the maximum extent possible based on hard science rather than on untested speculation. Implicit in these goals are notions that: (1) ethically "neutral" scientific reason is capable of, and should ground, any quantification of the risk; (2) risk assessment not based on sound science is irrational; and (3) rational regulations are those for which the benefits of the regulation exceed costs of regulation.

These notions suffer from the following problems:

1. Risk assessment quantification must be made in the face of pervasive scientific uncertainty;
2. Decision making in the face of uncertainty raises ethical questions that no amount of science or comparative analysis can answer by ethically neutral analytical methods;
3. Risk assessment-based decisions typically raise controversial ethical questions that are often hidden in the risk assessment quantifications;
4. Cost–benefit analysis as a basis for risk-based decision making is an ethically controversial approach to public policy that has a tendency to hide important ethical questions and issues.

For these reasons, the most important reforms that should be made to risk-based decision making on environmental matters are provisions that require that ethical considerations are integrated into risk-based decisions.

RISK ASSESSMENT PROCEDURES AND SCIENTIFIC CERTAINTY

Calls for reform of risk assessment procedures assume that risk assessment procedures can be put on a sound scientific basis, yet risk assessment-based decisions must be made in the face of pervasive scientific uncertainty, a limitation of risk assessment that is not likely to change in the foreseeable future. Scientific uncertainty exists in risk assessment because:

1. Epidemiological data relating dose rates to human disease do not exist for most contaminants;
2. Extrapolating dose–response results from animals to humans requires the selection of untested assumptions;
3. Effects of exposure may take years or generations to materialize for chronic diseases;
4. Human experimentation is excluded on ethical grounds;
5. Experiments must assume some dose rates, thereby giving no information about other dose rates; and
6. Exposure assessments must rely upon complex models that attempt to describe how pollutants may be transported through air, water, and soil and thereby create exposure opportunities to humans and animals.

Scientific uncertainty is often found in many of these component steps of a risk assessment because in many of these areas, either a scientifically sound theoretical basis has not yet been developed, or empirical data is inconclusive. Because theory is weak or data are incomplete, risk analysis must rely upon making many assumptions for which there is no *a priori* scientific basis that compels the choice of that assumption.

Much of the literature on risk assessment and scientific uncertainty deals with uncertainty in determining the probability of causing cancer in humans through exposure to toxic substances at various dose levels, that is, problems 1 through 5 above. However, often the greatest source of scientific uncertainty in most environmental risk assessment quantifications is created by the need to predict how toxic substances will behave in the environment, and thereby create potential exposures to humans, animals, or the environment, problem 6 above. This part of risk assessment is generally referred to as the problem of fate and transport of toxic substances.

Fate and transport calculations must rely on theoretical models of the way toxic substances move in air, soil, water, and mixtures of other substances. Because extreme degrees of complexity characterize "real-world" conditions, attempts to quantitatively model consequences of many human activities on regional or local environmental systems or to model even the simplest of ecosystems have met with very limited success. Environmental models are fraught with uncertainty because mathematical relationships between "real-world" variables described in the model are not known, or because models must greatly simplify "real-world" conditions to make the model usable.

Moreover, data needed to run or test the model are difficult or impossible to obtain. Although important data are sometimes theoretically obtainable, they are often practically unavailable because of the extraordinary high costs of obtaining the data. For example, it often costs as much as $10,000 to drill a properly cased monitoring well and $1000 per sample to test for the presence of certain toxic substances in the well. These samples may be necessary to adequately determine the extent of groundwater contamination when attempting to calculate the risk posed by the contamination. Environmental models used in risk assessment therefore suffer from two kinds of uncertainty: (1) theoretical uncertainty, and (2) informational uncertainty.

If risk assessment techniques are to be used more widely as priority-setting mechanisms, calls to limit the use of risk assessments to those which are grounded on strong scientific basis are unrealistic. Risk assessment quantifications should be understood to be heuristic tools that are often based on unproven assumptions about potential harm rather than quantitatively accurate descriptions of harm. As a heuristic tool, risk assessments can still be used as priority setting mechanisms and to assist decision makers in other ways in the regulatory process.

RISK ASSESSMENT DECISIONS MADE IN THE FACE OF SCIENTIFIC UNCERTAINTY MUST BE UNDERSTOOD TO RAISE IMPORTANT ETHICAL QUESTIONS

In day-to-day decision making about risk, environmental protection controversies are often thought of as technical–instrumental problems. To solve such problems, scientists or other technically trained personnel use scientific procedures to develop the "facts" about a particular environmental danger and to describe measures that can be taken to prevent or remediate environmental degradation. For example, if risk associated with toxic substances is viewed primarily as a technical–instrumental problem, science needs to determine whether certain substances create toxic risks to humans or the environment (a question of scientific fact) and, if so, what steps can be taken to mitigate against any adverse environmental effects (an instrumental question of means). Since these questions are about "facts" and "means", according to dominant conventional wisdom, they are best answered by experts who use "value-neutral" scientific procedures as analytical tools to find answers.[2]

Conversely, environmental risk controversies can be understood as problems that most fundamentally raise ethical questions, questions about what is the "right" thing to do. For example, which environmental amenities "should" we protect or what "should" we do in respect to the environment when the technical "facts" about consequences are uncertain? Under an ethical lens, the most important questions about potential risk controversies might be: (1) what "should" we do about potential toxic substances before science can specify consequences with certainty?; or, (2) who should have the burden of proof in demonstrating that a particular chemical poses a risk?

Of course, environmental problems usually raise *both* complex technical–instrumental questions and thorny ethical questions (inextricably embedded in scientific reasoning due to the technical and practical need to make assumptions). However, if we allow technical–instrumental discourse to dominate our risk assessment decision-making discussions, several consequences follow.

First, positions taken by decision makers that are justified on scientific grounds, but have actually been based on the values of the decision makers, appear to be compelled by "neutral" technical reasoning, and therefore are not subject to public scrutiny.

Second, the values of those stake holders and institutions that command technical resources can determine the nature of action that needs to be taken and make it appear that this action is compelled by scientific reasoning. Because technical expertise is very expensive, only those who can afford to pay technical experts can participate in the public policy discourse. Therefore, if technical discourse is allowed to dominate public policy decisions, those who do not have large financial assets may be effectively disenfranchised from discussing matters that should be understood to be moral or political in nature.

Third, political action initiated to protect the environment before science can describe precise cause-and-effect relationships between a proposed action and its effect on the environment can be attacked as irrational because it is without a scientific basis that compels action. The above-referenced legislative reforms of risk assessment emphasize the preference to base risk assessment assumptions on validated scientific evidence. Yet the need to make assumptions in risk assessment follows from the absence of sound scientific understanding of cause and effect and practical limitations in obtaining data. That is, assumptions that guide risk assessment calculations need to be made for both scientific and practical reasons. Given the magnitude of scientific uncertainty embedded in risk assessment calculations, because proposed risk assessment reforms acknowledge that governments must be sensitive to the economic consequences of risk assessment decision making, there is likely to be increased pressure on decision makers to withhold government environmental action that has significant economic consequences in situations lacking a strong scientific factual basis.

In the face of scientific uncertainty, government officials must decide whether they will err on the side of a false positive, generally referred to as a class I statistical error, or a false negative, a class II statistical error. That is, government officials faced with uncertain evidence of potential harm cannot avoid deciding either on the side of human and environmental protection — a decision that in the light of better information may prove that the decision imposed unnecessary costs — or deciding to avoid imposing unnecessary costs — a decision that in the light of better information may be understood to have allowed human or environmental injury. If the decision maker waits until the scientific proof is in, then he or she is making an ethical judgment that favors the status quo.

Moreover, because scientists are taught to be silent in absence of sound scientific proof, if there is urgent need to take action to prevent environmental destruction where scientific proof is not conclusive, scientific norms may be inconsistent with ethical principles. Thus, the scientific norm that a scientist refrain from speculation in the absence of proof may conflict with the public policy need to protect humans or the environment for future generations. Decision making in the face of scientific uncertainty must be understood to raise ethical and normative questions, yet the proposed reforms would further entrust environmental decisions to scientists and thereby give the question of norms over to part of culture which is often called "non-normative." Therefore, although decision makers need to have the best science to inform them about consequences of decisions, they must understand that norms that insist on high levels of scientific certainty may be inconsistent with valid public policy objectives.

Therefore, if we let neutral scientific language dominate our discourses on risk assessment-based decisions under the proposed reforms, we are likely to see less environmentally protective government action. Clearly, the proposed reforms call for an emphasis on scientific information and decision making at the expense of values-related thinking. If scientific discourse is allowed to dominate the public policy discussions about environmental risk problems, there will be expanded opportunities for obstructive behavior for those who oppose government action.

Approaches to risk assessment that demand higher levels of scientific certainty before taking action conflict with the "precautionary principle" adopted by the nations of the world at the 1992 Earth Summit in Rio de Janeiro. The "precautionary principle" states that where there is reasonable basis for concern about serious or irreversible environmental damage, the lack of scientific certainty shall not be used as an excuse for failing to take cost-effective action.

Obviously, the desire to put risk assessment calculations on the best scientific footing found in the above-referenced legislative reforms follows from the conclusion of the proponents that government has often based risk assessment calculations on unnecessarily protective assumptions that waste money. The literature attacking government risk assessment usually points to several different types of unreasonable assumptions. These include assumptions which: (1) rely on highly improbable exposure scenarios, such as the assumption that humans will be exposed to contaminated groundwater at the rate of 2 liters per day for 70 years; (2) fail to take into account recent studies on toxic substances; (3) fail to understand limitations or proper application of animal studies to human exposures; (4) use unnecessarily conservative fate and transport models to predict movement of substances in the environment; and (5) cascade the use of conservative assumptions so that final calculations grossly overstate the "real" risk. Although it is admitted that governments should apply the best scientific understanding to risk-based decisions, these criticisms suffer from the failure to consider the following ethical and practical considerations.

The Use of Worst-Case Scenarios

Government is often criticized for basing assumptions on unrealistic worst-case exposure scenarios in risk calculations, such as the assumption that persons using contaminated groundwater *will* be exposed continuously for 70 years and consume 2 liters of water per day during that period. It is admittedly improbable that persons will be exposed to the same source of water for an entire lifetime. However, as long as it is possible that persons be exposed for a lifetime, an assumption that may not be valid for all risk calculations but appropriate for a small subset, the goal of protecting for the entire period of exposure is arguably sound public policy. Moreover, effective regulation may sometimes require government agencies to adopt crude but administrable decision strategies that do not incorporate a high degree of scientific sophistication. Along this line, it may be impossible for government to determine "actual" future exposure scenarios. A legislative mandate requiring government to base risk assessment calculations on "actual" future exposures, if possible to implement, so complicates the risk calculation process that it may make it administratively unmanageable. In addition, the idea that government should regulate only at the level at which actual harm is likely to result is based upon the dubious ethical notion that persons should have the right to pollute up to a level which causes actual proven harm.

The Duty to Take into Account Most Recent Toxicity Studies

Government is often accused of failing to take into account recent studies on properties of toxic substances when performing risk calculations. Yet studies of toxic substances often conflict with each other, requiring decision makers to carefully analyze and weigh evidence. It is therefore important for the government to carefully and systematically review the accumulating scientific evidence. The mere presence of scientific studies that challenge existing government guidance on toxicological properties of substances is not a necessary and sufficient condition for changing positions when the government is charged with protecting public health.

Failures in the Application of Animal Data to Humans

Government is often attacked for the inappropriate application of animal toxicological data to human exposures. One variant of this criticism is the notion that only evidence generated from human exposure, i.e., epidemiological evidence, should be accepted as scientifically valid evidence of human toxic effects. However, human experimentation is not appropriate for ethical reasons. Furthermore, reliable human epidemiological evidence from persons who have actually been exposed is not available for most substances because of statistical and other methodological limitations.

Another variation on the criticism of extrapolation of animal toxicological studies to humans is the failure to adjust for the type and site of tumor production in animals, given different physical and biochemical reactions of humans. This criticism argues that animal toxicological studies should not be applied to humans until the properties of a toxicant's movement through the body to target organs is understood. However, such studies may greatly delay or make prohibitively expensive the use of the animal studies, further shifting the burden of proof to the government and away from the producer or polluter of the toxic substance.

The Inappropriate Use of Fate and Transport Models

A common criticism of risk assessment calculations is that scientifically inappropriate assumptions are made in the fate and transport models used in risk assessment calculations. However, as explained in more detail above, environmental models used in risk assessment suffer from two kinds of uncertainty: (1) theoretical uncertainty, and (2) informational uncertainty. Because of this uncertainty, the models can always be attacked on scientific grounds if scientific certainty is the norm for regulatory rationality. However, for practical reasons of cost and administrative ease, along with a lack of understanding of certain environmental mechanisms, the government must use models that necessarily simplify "real-world" conditions. If absolute predictive accuracy of a model is the criterion of its acceptability, no fate and transport model will satisfy the criterion.

Cascade of Conservative Assumptions that Overstate the "Real" Risk

Probably the most common attack on government's use of risk assessment is that persons performing risk assessment cascade conservative assumption upon conservative assumption so that "real" risk is grossly overstated. These criticisms often focus on admittedly conservative assumptions sometimes made in risk assessment, such as the assumption that persons will be exposed to the substance continually for 70 years. However, implicit in the charge that risk assessments overstate magnitude of the risk is the notion that all or the vast majority of the assumptions made in risk assessments are conservative, that is they err on the side of protection, and that the risk calculation does not also include nonconservative assumptions which tend to understate the risk and thereby tend to balance the conservative assumptions. Yet almost all risk assessments fail to consider the cumulative or synergistic effects of substances that a person is exposed to even though there are no scientific grounds for such exclusions. Risk-based calculations usually assume that the only source of exposure will be the specific substance under consideration at its measured level of concentration, even when other substances creating threats from exposure are present. The failure of risk assessments to consider cumulative or synergistic impacts of multiple substances is one of many examples of non-conservative assumptions commonly used in risk assessment. As one commentator has concluded:

> Risk assessors often respond to scientific uncertainties by adopting conservative safety-oriented positions on some important issues while they use *best-current-scientific guess, middle of the range, methodological-convenience, or least-cost treatments on other material issues.* EPA and other agencies have never explained the scientific or policy rationales underlying these inconsistent treatments of uncertainty, and risk managers may not recognize that substantial inconsistency exists.[3]

Another commentator has concluded:

1. There is no such thing as "actual risk";
2. Conservatism is inherently no more or less biased than alternative approaches;
3. Only some conservative assumptions are gratuitous;
4. Not all the inferences that are made are in fact conservative;
5. *A cascade of truly conservative steps may still yield a reasonable estimate of risk.*[4]

In this writer's experience, although many of the assumptions made by risk assessors are conservative and therefore tend to be environmentally protective, others are not conservative and therefore may drive the quantification of the risk in such a way that the number may understate the actual risk.[5]

Additional non-conservative assumptions are often embedded in risk assessment calculations because there are many variables in risk assessment calculations that are derived from cost-driven policy considerations rather than toxicological sciences. These policy considerations are usually referred to as matters of "risk-management" as distinguished from risk assessment calculations. For example, risk assessment quantifications used at sites contaminated with hazardous substances ultimately make assumptions about whether the risk assessment should be calculated based upon the current proximity of nearby people or whether the calculation should assume that the site will be cleaned up so that all future land uses are possible. In the latter case, the risk assessment will be calculated so that water directly beneath the site will become the point of the calculation rather than water in wells at significant distances from the site. Because assessing risk in a groundwater system at considerable distances from the source of pollution allows the risk assessor to consider factors that allow for attenuation of pollutants between the source of pollution and the well, a decision to quantify the risk in an off-site well is a non-conservative assumption built into the risk assessment. In this way, the decision about where to measure the point of compliance in the groundwater is an important non-conservative assumption often hidden in the risk assessment calculations.

Additionally, a risk manager may have to choose between protecting people from contaminated soil by leaving some soil behind a fence or requiring that all the contaminated soil be completely removed or otherwise eliminated. These kinds of decisions are often made in the course of quantifying the risk, although the non-scientific nature of these policy judgments is rarely disclosed in the quantification of the risk.[6] These quantifications appear to be scientifically based, yet they depend on answering a question that cannot be answered by science alone. Because the nature of such a question is inherently prescriptive, rather than descriptive, it should be understood to be a question of ethics or norms. However, because it is embedded in the risk assessment calculation, it mistakenly appears to be a scientifically compelled consideration.

Most remarkably, risk assessment calculations, despite frequent charges that risk assessors compound conservative assumptions that overstate risk, rarely look at risk of threats to the environment, so-called ecological risk, even when being performed under laws that require protection of both human health and the environment (such as the Superfund law). Ecological risk calculations are almost without exception not performed in risk assessments dealing with environmental clean-up because of the lack of understanding of the effects of toxic substances on ecosystems. Like the joke about the drunk who looks for his lost keys under the light post because that is where the light is, risk assessment calculations usually only quantify the magnitude of the risk on human health because risk assessors know how to do these calculations. Disturbingly, ecological risk assessment is not performed even in situations where the law requires it, primarily because the science of ecological risk assessment is in its infancy. Calculation of a toxic substance's movement through air, water, and soil to compare resulting concentrations at points of exposure with human health-based numbers, despite the need to face numerous significant scientific uncertainties, is a relatively straightforward engineering problem. On the other hand, attempts to calculate ecosystem-wide impacts is much more technically ambitious. This is so because efforts to model the relationship between human activities and ecosystems have not met with success because ecosystem dynamics are so extraordinarily complicated that they make quantification of ecosystem-variable interrelationships virtually impossible.

RISK-BASED DECISIONS OFTEN HIDE OTHER CONTROVERSIAL ETHICAL QUESTIONS

From the above discussion it is clear that because of pandemic scientific uncertainties that must be faced in risk assessment, reforms that aim at putting risk assessment on a scientific footing that avoids normatively based assumptions are doomed to failure. In addition to the normative questions raised by scientific uncertainty, risk assessment also raises many additional ethical questions that are often hidden in risk assessment calculations or go unaddressed in discussions about reforms.

The Problem of Burden of Proof

Because of the pervasive nature of scientific uncertainty in risk assessment calculations, whether the government or those who have responsibility for the introduction of hazardous substances into the environment have the burden of proof for conclusions about risk is an extraordinarily important ethical and public policy issue. Yet burden of proof issues often get lost in discussions about risk assessment policy. Most proposed reforms would put the burden of proof on government to accurately quantify the risk. Insisting on high levels of scientific proof before government may take action based on risk assessment calculations creates a burden that government may not be able to bear. Such a rule may prevent protective government action where there is a reasonable basis for concern, but where science is uncertain about the consequences of risk-creating activities. The burden and standard of proof that should be required in risk assessment-based regulatory action is an ethical question, not a scientific question. If we let rigid standards of proof dominate risk assessment-based actions, we are making ethical choices that may be inconsistent with other ethical objectives such as protecting those who have not consented to be exposed to a risk.

The Problem of Distributive Justice

The questions of who must bear the risk and who receives the benefits from current industrial policies is central to any question of environmental justice. Yet current risk assessment practices and proposed risk assessment reforms do not require that risk assessments identify what members of society will be affected by risk assessment policies. Risk assessment policies often create uneven effects of decisions across subgroups. For instance, migrant farmers may be exposed to pesticides picking oranges in concentrations thousands of times higher than the concentrations that city dwellers are exposed to. In addition, poor people or racial minorities may assume a greater burden of harm than those who receive the benefits from actions affecting the environment. Therefore, risk assessment reforms should require that the distributive aspects of risk-based decisions are considered and disclosed.

The Problem of Communicative Rationality

Some proponents of risk assessment reform suggest that it is irrational for a member of the public to oppose a risk which is smaller than risks accepted by the general public from such common activities as driving or falling down steps. Yet such an approach fails to differentiate between risks that one accepts and risks that have been imposed on someone without their permission. Similarly, some proponents of risk assessment reform argue that to determine levels of risk acceptability in public policy, the policy maker should look to risks that appear naturally, such as the risk of ionizing radiation at high altitudes. However, such an approach fails to distinguish between risks that humans can do something about and those that we cannot control.

Because some proponents of reform perceive risk rationality to be largely a matter of risk comparison, policy makers are encouraged to see risk acceptability problems as problems of risk "communication". Under such an approach, risk "communication" training is often focused on explaining risk comparisons while ignoring important ethical issues, such as problems of distributive justice or whether the risk has been accepted consensually or not. Because the proposed reforms would make risk rationality a matter of comparing the magnitude of risks without considerations of other ethical questions, reform legislation should be amended to require that risk comparisons include information on whether the risks have been accepted, what subgroups would be exposed to potential risks, and whether risks that are being compared are risks that humans can do anything about.

The Problem of Risk-Based Decisions Authorizing Pollution in Certain Situations

Risk-based decisions can become a justification for allowing pollution where other rational approaches to public policy might prohibit or minimize the pollution below levels deemed acceptable by risk calculations. For instance, under some environmental laws, polluters are expected to install best available technology regardless of government's ability to show that a certain level of health risk will be exceeded if the technology is not installed. This approach greatly reduces issues of complexity in regulatory decision making and controversies entailed by scientific uncertainty.

In a similar vein, several states prohibit any degradation of groundwater and have expressly rejected the notion of giving potential polluters the right to pollute to risk-based levels. The public policy basis for this approach is premised on two considerations. First, groundwater once polluted is extraordinarily difficult and expensive to clean up. Second, toxicological understanding of harm to humans or the environment may change in the future so that levels that were once deemed to be safe are later believed to create harm. Because of these concerns, allowing degradation of groundwater to certain risk assessment-determined levels may encourage degradation of valuable groundwater resources. Therefore, using risk assessment as the basis for all regulatory decisions, an approach taken by the reform legislation, may not be appropriate public policy for all issues.

RISK MANAGEMENT DECISIONS BASED ON COST–BENEFIT ANALYSIS ALSO HIDE OTHER IMPORTANT ETHICAL ISSUES

The proposed reform legislation requires the use of risk assessment calculations in conjunction with CBA to make environmental decisions. CBA is an ethically controversial approach to public policy which appears to many to be an ethically neutral way of deciding questions of policy. Economists and others have attempted to develop a variety of analytical techniques, including CBA, that transform public policy questions into analytically answerable value-free, neutral, technical questions.[7] However, most philosophers believe such an approach is equivalent to squaring a circle because each analytical technique takes an unarticulated ethical position which excludes other viable ethical approaches. For example, CBA assumes something similar to utilitarianism as its ethical position.[8] Thus employing a CBA when making a risk management decision is tantamount to taking a position on an ethical question. Many philosophers view CBA approaches to public policy decision making to be inappropriate because utilitarianism rests on the dubious ethical position that moral rights can be balanced with expected utilities. Many philosophers believe that utilitarian approaches must be tempered by rights theories or other principles of justice. When CBA becomes the basis for a risk management decision, that an ethical approach has been implicitly taken is usually not understood by the analyst or is hidden in the jargon of the analysis.

The utilitarian approach raises ethical problems that cannot easily be answered from within a utilitarian system. A utilitarian, for instance, must decide which alternatives will be entertained in the utilitarian calculus, which consequences of a given action will be considered, whose assessments of harms and benefits will be allowed, and what time scale will be used in assessing those consequences. The utilitarian framework, therefore, often rests upon imprecise judgments independent of, and prior to, the utility calculus itself.

Moreover, the utilitarian approach contained in CBA is a strongly anthropocentric one which attempts to reduce value to a single measure to allow the economist to optimize the benefit to society through the CBA-based decision. The measure of value in CBA is the willingness of humans to pay for environmental entities in market situations. Plants, animals, and other environmental entities are in this way treated as commodities available for human consumption. By making willingness to pay the measure of value, CBA approaches are antithetical to ethical positions that hold that certain plants, animals, and environmental entities have intrinsic value.

Treating environmental entities as commodities is contrary to one longstanding tradition in natural resources and environmental law, the public trust doctrine. The public trust doctrine prevents government from treating public trust resources as commodities through a recognition of the government's duty to conserve these resources as a steward for the citizenry. This duty of stewardship includes a constraint on government's ability to sell these resources. In this way the public trust doctrine implicitly recognizes that public trust resources have a value that transcends the market value of these resources.

CBA also ignores distributional effects of the decision and therefore is often inconsistent with concepts of distributive justice. Similarly, CBA often does not take into account pollution which transcends natural boundaries. CBA rarely takes into account the rights of future generations or non-human species. When future benefits are considered by CBA, they are most often valued by a process called discounting, which assumes that future impacts are less valuable to present decision makers than immediate impacts. In this way CBA undervalues arguable duties that present generations have not to harm future generations.

Although more sophisticated utilitarian approaches are capable of dealing with some of the problems mentioned above, all too frequently the value analyses one actually finds in the environmental public policy debates are oversimplified utilitarian calculations that more sophisticated utilitarians would likely reject.

CONCLUSION

Although it is admittedly important to continue to enhance our analytical ability to make mathematical estimates of risk, it is also critically important to develop the ability and procedures to identify the many values questions that are often central to making a risk-based decision but that are often hidden in the risk assessment jargon. For that reason, it is important that any legislation on risk assessment include the following:

1. The law must expressly adopt the "precautionary principle". That is, legislation should state that where there is a reasonable basis for concern about threats to human health and the environment, lack of full scientific certainty should not be a basis for prudent protective action.
2. Although the government should be initially reponsible for performing risk assessments, the burden of showing lack of harm should shift to the proponent of a regulatory activity in the face of scientific uncertainty. The public should be protected against arbitrary government action by procedures that: (a) guarantee the public's right to comment on risk assessments; (b) create a duty of government to respond to significant comments.
3. Risk assessments must identify distributional, intergenerational, and, when appropriate, international effects of risk based decisions.
4. Not all environmental legislation should be based on risk assessment and cost-benefit analyis. Risk assessments should be used as heuristic tools rather than a prescription for environmental regulation. CBA similarly should be used as an analytical tool rather than as a prescriptive rule. When CBA is used in conjunction with risk-based decisions, how human health or evironmental entities were valued should be expressly identified in the decision. It is critical that public trust resources be protected from valuation techniques that would determine their value by market-based methods.

REFERENCES

1. For an excellent discussion of the ethical dimensions of risk assessment, see, Cranor, C.F. *Regulating Toxic Substances: A Philosophy of Science and the Law,* Oxford University Press, New York (1993); and Shrader-Frechette, K. *Risk and Rationality,* University of California Press, Berkeley (1991).
2. Although risk assessors and others often assume that scientific procedures can be "value-neutral", all scientific endeavors entail values choices about definition of problems or the metaphysical assumptions embedded in the scientific practice.
3. Latin, H., Good Science, Bad Regulation and Toxic Risk Assessment, *Yale J. Regulation,* 5, 89 (1988) at 94.
4. Finkel, A.M., Is Risk Assessment Really Too Conservative? Revising the Revisionists, *Colum. J. Env. L.* 14 427, 431 (1989).
5. The following represent experiences of the writer. Risk assessors in calculating a safe dose for a carcinogen will assume that the individual human receptor is the maximum exposed individual (MEI). The MEI, it is assumed, will be exposed for a 70-year period to the carcinogen and may drink 2 liters per day of the contaminated water. This, of course, is a very conservative assumption because few people are threatened with exposure from a carcinogen for 70 years. However, in the same series of calculations on which the safe dose is estimated, the risk assessment quantification will often will be based on the following non-conservative assumptions: (1) The risk assessor will often assume that the highest soil or groundwater measurement represents the highest concentration of that carcinogen found at the site even though the number and placement of the monitoring points are limited because of the cost of monitoring and it is very

possible that higher levels of contamination can be found at the site at locations not monitored; (2) the risk assessor will often not calculate the toxicological synergistic effects that may occur when the carcinogen of concern is mixed with other toxic substances found at the site; (3) the risk assessor will often assume that contaminated groundwater will move through geologic rocks very slowly while the contamination is taken up by the rocks by assuming a certain permeability for the aquifer rocks and a certain propensity of the contaminants to be bound up by the rocks through which the contaminated groundwater travels. In this way it is assumed that the contaminated groundwater will become dilute when mixed with other non-contaminated groundwater; however, groundwater may move very quickly and remain undiluted if the groundwater flows through open fractures rather than the semiporous rocks or if the contaminants are mixed with fluids which will not allow the contaminants to become bound up with the rock material; (4) the risk assessor will often assume that the calculation of the risk to human receptors should be calculated at places where people are located at the time the calculation is made. In other words, to determine the risk to humans at a particular site, the risk will be assessed based upon the location of existing drinking wells. This assessment may understate the risk to persons who move closer to the site in the future; (5) the risk assessment may assume that the dose defined as safe will represent the total dose that the human receptor will be exposed to in cases where it is possible that the human receptor may be receiving additional doses from other sources including ambient or background levels; (6) the risk assessment may be based upon animal data where humans are more sensitive to the carcinogen than the animal species that produced the carcinogenic response; (7) the risk assessment may be based on an extrapolation from animals to humans that is based on mass or surface area of the animals and a human which may underestimate the safe dose to an actual human; (8) the risk assessment may assume an average human response of a typical human receptor when certain individuals that are exposed may be unusually susceptible to the carcinogen.

6. Many environmental professionals assert that risk management procedures should be separated from the more scientific risk assessment procedures so that interested parties can identify the transscientific issues that have been considered in a risk management calculation. However, in actual practice risk management decisions are often not kept separate from risk calculation, and it is therefore sometimes difficult to determine the policy considerations that have affected the "neutral" calculation of the risk.

7. For a discussion of why cost–benefit analysis and other economic analytical techniques cannot transform environmental problems into technical problems which avoid ethical questions, see M. Sagoff, "The Economy of the Earth." (1988).

8. M. Sagoff, "The Economy of the Earth," at 104–111, Cambridge University Press, New York (1988).

THE PROBLEM OF INTERGENERATIONAL EQUITY: BALANCING RISKS, COSTS, AND BENEFITS FAIRLY ACROSS GENERATIONS*

10

Bayard L. Catron, Lawrence G. Boyer, Jennifer Grund, and John Hartung

CONTENTS

Introduction 131
The Problem of Intergenerational Equity 132
    The Nature of Our Obligations 133
    The Motivation Problem 134
    Uncertainty 134
Comparing Present and Future: Discounting 136
Suggested Principles for Intergenerational Equity 139
A Suggested Decision Model for Incorporating Intergenerational
    Equity Concerns into Risk Decision Making 142
Conclusion 145
References 146

INTRODUCTION

As humankind's ability to affect the far future increases, the issue of the *fairness* of imposing risks and burdens on the lives of future, unborn generations is accentuated. Associated with risks are benefits and costs which are

* This chapter is based on work done for a project conducted by the National Academy of Public Administration (NAPA) and sponsored by the U.S. Department of Energy (DOE) through Battelle Pacific Northwest Labs. The interpretations, opinions and recommendations contained herein are those of the authors and do not necessarily represent the views of any of these organizations. Note that, as of this writing (September 1994), NAPA has taken no position on any of these issues, and has not made any recommendations to DOE. The authors wish to thank Bob Zahradnik for assistance and an anonymous reviewer for helpful comments on an earlier draft of the paper.

1-56670-131-7/96/$0.00+$.50
© 1996 by CRC Press, Inc.

increasingly uncertain over long time periods. When benefits and costs are uncertain and asymmetrically distributed in time, particularly across generations, traditional efficiency-based management methodologies often fail to produce equitable and/or acceptable solutions. Nowhere is the equity concern in risk management and decision making more evident or more important than with issues of intergenerational equity, as exemplified by such issues as climate change, ozone depletion, and nuclear waste disposal. The ethical challenge is this: how to balance risks, costs and benefits fairly across generations.

The chapter is presented in five sections: (1) The Problem of Intergenerational Equity; (2) Comparing Present and Future: Discounting; (3) Suggested Principles for Intergenerational Equity; (4) A Suggested Decision Model for Incorporating Intergenerational Equity Concerns into Risk Decision Making; and (5) Conclusion. The first section highlights the broad agreement in the literature that present generations have obligations to future generations, while noting that there is less agreement about the basis of such obligations (rights or justice, for example), or about the specifics of the obligations. Then the "motivation problem" (how to motivate the present generation to fulfill its obligation) and the vast uncertainty associated with the far future are discussed. The second section provides a discussion of "discounting," the usual economic technique for making temporal comparisons, focusing especially on intergenerational considerations. Because intergenerational discounting appears inadequate, and is believed to be ethically inappropriate for several reasons, the third section proposes a set of intergenerational equity principles that can help make trade-offs between generations in decision making and priority setting. The various elements of a suggested decision model are summarized in the fourth section, and the concluding section offers some general comments about the limitations of making systematic trade-offs in our political culture.

THE PROBLEM OF INTERGENERATIONAL EQUITY

Three key elements of the intergenerational equity problem are discussed in this section:

- **The Nature of Our Obligations.** What obligations do current generations have to future ones? For example, is our obligation not to impose great hardship and suffering on them? Or, is it, more strongly, to provide the opportunity for a standard of living at least as high as our own? What do these obligations require of us in the present?
- **The Motivation Problem.** How can we motivate the present generation to fulfill its obligations to the future?
- **Uncertainty.** How can we take future generations adequately into account when we do not and cannot know their interests, values, priorities, etc.?

The Nature of Our Obligations

There is a broad consensus in the literature that current generations have some obligations to the future. At a minimum, according to Annette Baier (1984), "...we are obligated not knowingly to injure the common human interests they like all of us have — interests in a good earth and good tradition guiding us in living well on it without destroying its hospitability to human life." Moreover, as Daniel Callahan (1980) says, we should not jeopardize the possibility of future generations exercising those fundamental rights necessary for a life of human dignity. Edith Brown-Weiss (1990) goes further, arguing that "each generation is entitled to inherit a robust planet that on balance is at least as good as that of previous generations."*

What is the basis for our obligations to the future? Is there some rational moral grounding, for example, in the concept of justice or the rights of future generations? There is less agreement on this point. Brown-Weiss (1990) maintains that "intergenerational rights have greater moral force than do obligations." If future generations have rights, a strong basis for our obligations to invest and sacrifice for them would be provided — but is it morally and constitutionally possible for unborn generations to have rights?

Several arguments support the idea of the rights of future generations. Baier (1980) argues that future generations have rights, due to great advances in knowledge and power, including our powerful capacity to affect their welfare for better or worse. Kristin Shrader-Frechette (1988) favors a contractarian basis for the rights of future generations; she suggests that precedents set in intragenerational disputes can provide insight into intergenerational conflicts. Going further, Richards (1983) would extend constitutional rights to future generations (and would also establish new forms of legal rights to protect natural objects and resources).

There are also significant arguments that future generations do *not* have rights. Ruth Macklin (1980) argues that future generations do not have rights because rights are properly ascribed to *actual,* not *possible* persons. However, she does say we have moral duties to future persons as long as we believe they will exist in the future. Similarly, Richard De George (1981) argues that "only existing entities can have rights", although we have an obligation to continue the human race. Both agree that, although future generations do not have rights, we in the present have obligations to them.

In contrast, Hubin (1976) claims that we do not have obligations to the future so much as we have responsibilities that flow from generally accepted principles of justice. He says we have a "duty of justice with regard to future generations (but not owed to them). Brian Barry (1978) argues that the relevant

---

* In contrast, an argument that seems to severely limit obligations to the future is based on Derek Parfit's work. Policy choice we make now determine which people will be born in the future. Since their very existence is involved, these future people will accept our policy choices as the best they could have been for them. See, for example, Parfit (1983).

concept of justice is equal opportunity, "the overall range of opportunities open to successor generations should not be narrowed." David Richards agrees with the equal opportunity concept, and bases his argument on John Rawls's *Theory of Justice* and contractarian theory of law.

So there is little consensus in the literature on the basis of our obligations to future generations. Bryan Norton argues that, "Despite remaining disagreements regarding the exact foundations of our obligations to the future, there is emerging a broad moral and political consensus in favor of a sustainability ethic. Practically, this ethic will, for the foreseeable future, consist in the pursuit of goals that are supportable on a number of different theories" (Norton, in press). While this may well be true, the "broad political consensus" has yet to be tested where serious sacrifices by powerful interests may be required. Moreover, the theoretical disagreement leaves us in a weaker position to argue for whatever sacrifices might be required from the current generation.

The Motivation Problem

Whether people are willing to accept additional risks/burdens for themselves instead of shifting them to future generations depends on a variety of factors, including the sense of obligation to the future and its importance. The greater the perceived importance of the future, the higher one's motivation to sacrifice for the benefit of future generations.

However, as Norman Care (1982) observes: "A motivation problem may arise when morally principled public policy calls for serious sacrifice, relative to ways of life and levels of well-being, on the part of the members of a free society ... What will, could or should move people to make the sacrifices required by morality?" Care identifies three kinds of motivation that could provide a basis for people's willingness to sacrifice for the future:

- Love or concern, grounded in particularity about persons;
- Community bonding, grounded in reciprocation between persons; and
- Extended shared-fate motivation.

He argues that these three motivations are usually unavailable or unreliable. In their absence, coercive public policy, problematic in a free society, may be required (Care, 1982).

Uncertainty

Uncertainty dominates attempts to address problems that will or might affect far-future generations — 1000 or 10,000 years from now. There are different types of uncertainty; some are similar to that in near-term decision making, some are complicated by the span of time, and others are unique to intergenerational problems. One type is associated with predicting the magnitude and probability of outcomes of large, complex, dynamic physical systems.

In general, the more distant in time, the more uncertain is the predicted event and its consequences. Small changes in the near term are amplified and altered through interaction with other changes over time, increasing uncertainty. Even highly precise and predictable events, such as eclipses, cannot be accurately predicted on very long time scales.

Far greater uncertainty exists about social and political phenomena. No one could have predicted in any detail the events of the past decade or so in the former Soviet Union, South Africa, or the Middle East, to pick just three striking examples. How can we predict anything about societies well into the future? Will political and social institutions remain in existence and maintain their present character? Can we assume they will act reliably and respect earlier public decisions? (Keller and LaPorte, 1994). How will technology and society evolve?

The inherent uncertainties when considering the lives of future generations are especially vexing. Will the outcome of an event, good or bad, even matter to future persons? Will future generations value, monetarily or personally, the same objects and qualities we value today? The more distant the generation from the present, the more likely that generation will view the world differently.

Is it proper to impose the current generation's values on a future society, which may have no choice in accepting, declining, or adjusting to what occurs in the present? Recognizing this problem, Kristin Shrader-Frechette (1988) states that:

> ...although we do have limited knowledge regarding the social ideal of posterity, our partial ignorance does not free us completely from an ethical obligation to our remote descendants. Rather, our ignorance in this respect necessitates our acting on the assumption that future social ideals are not radically different from our own. I conclude therefore that we must ascribe the same basic rights to future generations as those we claim for ourselves.

She argues that future people are likely to want an ethical code based on equity very much like ours, "in which one does not permit treating humans as means, leaving one's debts for others to pay [e.g., acid rain], distributing resources inequitably [e.g., consuming nonrenewable resources], ignoring due process [e.g., future people cannot collect damages], or failing to protect the utterly helpless members of society [e.g., future people cannot speak for themselves]" (Shrader-Frechette, 1988). Although decisions involving intergenerational consequences are necessarily addressed in the context of the present generation's vision of the future, relying on principles such as these in the present would provide considerable protection for the future.

The future is often conceptualized through the use of models, which clarify and add coherence to our perceptions of events that affect distant generations. Models also help frame decision-making processes and can be used as organizational, computational and conceptual aids in problem solving and policy making (Dowalatabadi, 1994). Even if models cannot provide definitive information or reliable predictions, they can be enormously useful in identifying contingencies that can yield important insights and affect future action.

How reliable are computer models that predict events thousands of years into the future? Given large uncertainties in data, assumptions, and the models themselves, how much credence should be given predictions based on these models? Arguably, such models should only be used heuristically over these very long time periods. They should be used in a circumscribed fashion — to help illuminate possible consequences of present action, but not to drive decision making.

What appears to be needed is an acceptable method of making trade-offs between present and future generations — one that can be used even in the absence of definitive knowledge of the far future. This is addressed in the next section.

COMPARING PRESENT AND FUTURE: DISCOUNTING

Discounting is the technique designed by economists to make trade-offs between the present and the future. To choose between two investments with different net benefits occurring at different times, comparisons across time are standardized by calculating the "present value" of future benefits.* Economists disagree how to choose discount rates, and what type of discount to use (e.g., social discount rate, shadow price of capital). However, when intergenerational comparisons are made involving the lives and health of future, unborn people, there is general agreement that, for various reasons, traditional discounting is problematic.

One reason the technique does not work well over long time periods is an artifact of the mathematics involved: since the present value of future net benefits declines exponentially with time, a large benefit enjoyed 100 years (let alone 10,000 years) from now can have a negligible present value. To illustrate the point dramatically, "a complete loss of the world's GNP a hundred years from now would be worth about one million dollars today if discounted by the present prime rate (D'Arge et al., 1991, cited in Plater, 1992). Differences in the rate used also make a great difference: employing the usual formula, $1 billion received 200 years in the future discounted with a 1% discount rate has a present value of $137 million; at a 10% discount rate, the present value is only $5.27!

There are also several arguments that discounting is ethically inappropriate for decisions that affect future generations. Mishan (1975) argues against discounting benefits to any future generations at all, since they accrue to

---

* "Present value" can be calculated using a simple formula, $PV = NB/(1 + i)^t$, where PV is the present value of a future net benefit, NB is the value of the net benefit, i is the discount rate, and t is the amount of time in the future the benefit is received. See, for example, Tietenberg, 1992.

different individuals. Someone in the present has "...no business in evaluating the future worth of 100 by discounting it for 50 years at 10 percent when he himself is not, in any case, going to receive it ... Whenever intergenerational comparisons are involved ... it is well to recognize that there is no satisfactory way of determining social worth at different points of time. In such cases a zero rate of time preference, though arbitrary, is probably more acceptable ..." than other alternatives.

The U.S. Environmental Protection Agency (EPA) (EPA, 1987) has enunciated three arguments against discounting future health risks in particular. First, some argue that no discount rate should be used because there is no actual life-saving market mechanism that can value society's benefits from future vs. present lives saved.

Second, discounting can lead to inequitable distribution of health benefits: "When using a 10 percent discount rate, for example, we value 100 lives saved 30 years in the future the same as 6 lives saved in the present. Thus, when a high discount rate is used, expenditures made to save lives in the future appear to be much less effective than expenditures that will save lives today." As Howarth argues, "...it is difficult to argue that future generations are as a group less deserving than the present. To argue otherwise would be to discriminate against future generations based on the arbitrary happenstance of their birth dates" (Howarth, 1993; see also Cullen, 1991; Wallace, 1992).

Third, it may be inappropriate to use the same discount factor for money and for human life. The argument here is that, while the value of money varies with time, the value of human life does not. Raiffa, Schwartz, and Weinstein strongly disagree, arguing that discounting is

> ...merely an accounting device to place the dollars spent and the lives saved at the same point in time. In effect, we discount future lives precisely because dollars invested today should be expected to yield more life-saving in the future than in the present. It is because of our concern that resources be applied at the point in time where they save the most lives that we "discount" lives. It is, emphatically, not because we wish to value future lives less than we value present lives in any absolute or utilitarian sense. It is because we do not want to be wasteful of scarce resources in saving lives, either present or future (Raiffa, Schwartz, and Weinstein, 1977).

According to a different argument, discounting "...is designed to help assess only whether an action is efficient, not whether it is equitable....Discounting for environmental regulations that span several generations may obscure intergenerational inequities." (EPA, 1987). Similarly, Norgaard and Howarth argue that "discounting is appropriate with respect to the efficient use of this generation's resources but is inappropriate when this generation is primarily concerned with redistributing resource rights to future generations" (Norgaard and Howarth, 1991).

Others believe that intergenerational discounting is acceptable under some circumstances. Farber and Hemmersbaugh (1993) believe that society's concern should focus on the well-being of future persons, being careful not to expose them to serious deprivation. With respect to the next generation, a low discount rate, perhaps the social discount rate, should be used. Nijkamp and Rouwendal (1988) propose adjusting the social rate of discount when intergenerational issues are involved.

Peter Burton's intergenerational discounting technique incorporates a personal discount factor for present generation concerns and a generational discount factor for matters affecting future generations. Both factors are incorporated in calculations and interact, producing an lower overall discount factor (Burton, 1993).

Without widely accepted methods of handling intergenerational benefits and costs, long-term projects cannot be effectively defended or compared precisely to present ones. For instance, Richard Howarth states that "we can reasonably speculate that society would be willing to spend extra resources to mitigate the threat of potentially catastrophic risks. However, the appropriate sum to pay is beyond the reach of economic analysis and thus depends on the exercise of raw value judgements regarding what is acceptable and what is not" (Howarth, 1993). D'Arge and Spash (1991) argue that: "Because of classical and new problems in valuing public goods, it is currently impossible to quantitatively estimate the amount of optimal compensation [to future generations for environmental damage caused by the present generation]."

Given all these arguments, we are left with a need to make decisions that involve balancing risks and benefits between present and future generations without adequate, accepted analytic tools. In the absence of general agreement on how future generations should be treated, no comprehensive treatment of long-term discounting can be developed.*

However, despite the lack of an adequate *technique,* decisions affect future generations are made — and increasingly so, as current actions have greater long-term impacts. And we do frequently discount the future *in fact,* even when no technique or rate is explicitly used. For example, in those many situations when current decisions have long-range implications that are not accounted for, the long-term future is effectively discounted at an infinite rate. Whenever a discount rate is used for, say, a 30-year period, longer-term effects are effectively ignored entirely. In these instances, the result is a bias favoring the present generation.

On the other hand, it seems peculiar not to discount the future at all. For example, if federal agencies like the U.S. Department of Energy or the Environmental Protection Agency are not permitted to discount when setting priorities

---

* An anonymous reviewer of an earlier draft suggested that doing risk–cost–benefit analysis without discounting is like playing *Hamlet* without the Prince of Denmark. On the other hand, perhaps we have unwittingly elevated Guildenstern, like discounting, to the leading role. If discounting is inappropriate, how can economic analysis of any sort remain applicable? Some suggestions are made later in this chapter.

for waste clean-up, they must treat hypothetical risks to unknown people 10,000 years in the future as if they were exactly the same as current risks to clean-up workers. The people, as well as the risks, 10,000 years from now seem quite hypothetical. In addition, treating the present and the long-term future as if they were the same makes no allowance for technological improvements in our ability to handle risks.

A stronger philosophical foundation regarding the equitable treatment of future generations seems to be necessary before discounting or some other more adequate technique for making these trade-offs can be developed. In an overview of a conference on discounting issues held in 1988, Charles W. Howe, then President of the Association of Environmental and Resource Economists, concluded that "a defensible philosophical basis for long-term, intergenerational discounting has yet to be found" (Howe, 1990). At the same conference, Robert Lind (1990) stated that, "for long-term policies, the benefit-cost rationale for discounting breaks down and must be reestablished on principles incorporating intergenerational equity." We believe that such principles can be helpful in making resource allocation decisions across generations even though precise trade-offs do not seem to be possible.

SUGGESTED PRINCIPLES
FOR INTERGENERATIONAL EQUITY

At a recent NAPA/DOE/Battelle workshop, participants developed a preliminary statement of intergenerational equity principles contained in Table 1.* Used as a set, and with further development and public discussion and debate, these principles can provide guidance for incorporating intergenerational equity into environmental risk decisions and setting priorities.

The fundamental principle may be recognized as a version of the "sustainability" ethic, a concept that has received widespread attention in the past decade since the World Commission on Environmental and Development, known as the Brundtland Commission, was established by the United Nations. While more than 60 definitions of "sustainable development" have been enumerated, the best known is that of the Brundtland Commission (1987): "Development that meets the needs of the present without compromising the ability of future generations to meet their own needs." This definition also probably enjoys wider acceptance than any other; for example, it was recently adopted by the President's Council on Sustainable Development.

Note that the fundamental principle stated in Table 1 is phrased as an equal opportunity principle in a way that the Brundtland Commission version is not. Moreover, the notion of an opportunity for an "equivalent" quality of life is not part of the Brundtland Commission definition. In both of these ways, the statement of the principle here appears to be more stringent than the Brundtland

* See *DesignShop Synthesis,* July 1994. Note that NAPA has not yet taken any action with respect to these principles; they have not been endorsed or recommended to DOE.

Table 1   Proposed Intergenerational Equity Principles

**No generation should [needlessly] deprive its successors
of the opportunity to enjoy a quality of life equivalent to its own**

1. Every generation is the trustee for generations that follow.
2. There is an obligation to protect future generations, provided the interests of the present generation and its immediate offspring are not unduly jeopardized.
3. Near-term, concrete hazards have priority over long-term hypothetical hazards.
4. However, this preference for the present and the near future is reduced where questions of irreversible harm exist.
5. When an action poses a plausible threat of catastrophic effects, then that action should not be pursued, absent some significant countervailing need.
6. The reduction of resource stocks entails a duty to develop substitutes.

statement. This version also seems to avoid one problem the Brundtland version is subject to: the present generation is relieved of any responsibility of determining what the needs of future generations are, and to what extent they will be similar to ours. Note also that the principle is stated negatively; it does not say, as Talbot Page argues, that intergenerational justice will be served if we pass on to future generations what we ourselves have inherited. Page (1983) says that "…if the present generation provides a resource base essentially the same as it inherited (including roughly the same level of contamination), it has satisfied intergenerational justice."

The word "needlessly", bracketed in the basic principle, reflects a fundamental point of disagreement: are living generations ethically permitted under any circumstances to knowingly deprive those yet unborn of equal opportunity? Some believe that no circumstances warrant intentional, serious degradation of the future quality of life, while others believe that, under some circumstances, the quality of current life takes priority. This principle acknowledges a strong obligation to the future while the word needlessly maintains some undefined latitude to favor the present. Similarly, there is underlying disagreement about the implications of the sustainability ethic for economic growth and limitations on the use of private property. In the U.S. at least, there has not yet been a willingness to confront these issues in the public discourse.

Principle 1 identifies the nature of the relationship of the present toward future generations as that of a "trustee." There are various examples of the trustee concept in U.S. public policy. In some cases, it is written into law (e.g., the National Environmental Policy Act of 1969), while in many others it is implicitly present. But it is perhaps most apparent in U.S. history during the debate over setting aside public lands and establishing the National Park system. The analogy here with the legal concept of "trustee" as an instrument for preserving something of value for others is not perfect, and some people prefer the use of "stewardship". However, the purpose of the principle is to fix responsibility in the present for the consequences of present actions on future generations.

The literature on intergenerational equity, which overwhelmingly supports the notion that we have ethical obligations to the future, also strongly opposes making trade-offs favoring the future that fail to meet crucial obligations to present generations or that impose an injustice on the present. Principle 2 recognizes both these points, while emphasizing the interests of the present. As stated, this principle arguably provides too convenient a rationalization for the current generation to pursue its own narrow self-interest. A great deal hinges on the definition of "interests" here. If interpreted broadly, the principle could be invoked to justify much greedy and wasteful behavior; if defined narrowly, only "vital interests" or "basic needs" of the present could be used to justify not satisfying the obligations toward future generations. Given that self-interest is a very strong and pervasive motive, a narrower interpretation seems desirable to provide ethical protection against rationalization. The NAPA/DOE/Battelle workshop also suggested some more operational guidelines to aid in making decisions and setting priorities:

- Emphasize protecting present and near-future generations by: (1) addressing the highest near-term risk first; (2) giving additional priority if high long-term risks are *also* involved; and (3) seeking to minimize long-term risk consistent with the principle for intergenerational equity.
- Recognize and respond to the obligation to protect distant future generations, but do not do so at the expense of current and near-future generations. [If this guideline is to provide any protection, "expense" cannot just mean "monetary costs," but something like "significant sacrifice."]

Principles 3, 4, and 5, as a set, can also provide some guidance to decision makers. Principle 3 indicates that explicit current risks, like the risk to clean-up workers, should be given greater weight than hypothetical risks many generations in the future, such as the possible exposure of people to a hazard through some plausible scenario. But notice that the principle is compound, which makes it ambiguous. It gives priority to near-term risks over long-term risks, as well as concrete over hypothetical hazards, but it does not give us any guidance in comparing a near-term hypothetical risk to a longer-term concrete one.

The notion that we have greater obligations to the near term has some support in the literature. For example, Baier (1984) claims that we have extra obligations to the next few generations beyond those to all future generations because "...they are close enough in time to us for their *particular* needs and abilities to be foreseeable, and for us to have control over how many of them there will be, what opportunities they will have, what supply problems they have." Golding (1980) agrees, arguing that our conception of the good life is more likely to be relevant to nearer generations.

Principles 4 and 5 identify exceptions to Principle 3 based on projected irreversible harm and catastrophic effects. These are aspects of what is called the "precautionary principle". As articulated by Richard Howarth (1993), the principle holds that "inhabitants of today's world are morally obligated to take steps to reduce catastrophic risks to members of future generations if doing so would not noticeably diminish their own quality of life." Catastrophic risk or damage can be defined using such notions as increased risk, irreversibility, the scale of human activity, and the planetary impact of a project. If a significant irreversible decision can be deferred at low cost, it should be, thus preserving options for later generations. Howarth (1993) claims that the precautionary principle can be made operational by reducing it to a two-part test: "Does a particular environmental insult impose catastrophic risks on members of future generations? Can we take steps to reduce those risks without substantively compromising our own well-being?" Although he acknowledges that the principle depends on an explicit value judgment, Howarth claims that the principle yields a policy criterion that is operationally decisive under a wide array of circumstances. This point is arguably, but even when the principle is not "operationally decisive", it can be useful as part of a larger decision strategy, as discussed below.

Principle 6 addresses resource depletion; while not as relevant to environment safety and health, the underlying notion of an obligation to provide compensation is quite relevant. Barry (1983), for instance, argues that justice requires compensation for "loss of productive potential", not for resource depletion itself.

A SUGGESTED DECISION MODEL FOR INCORPORATING INTERGENERATIONAL EQUITY CONCERNS INTO RISK DECISION MAKING

Many of the ideas expressed in the literature by economists, philosophers, and others can be synthesized into a set of suggestions for an intra/intergenerational decision-making process. First, regardless of whether there are intergenerational concerns, some suggestions can be made. Following the suggestion of Batemen (1991), all benefits, including ecological and aesthetic/existence values as well as the costs of waste management, should be included in cost-benefit analysis. Until recently, nonmarket goods, such as aesthetic values, and the costs of the disposal of waste products were not included in most studies. However, with the growing acceptance of nonmarket valuation techniques, for instance the contingency valuation method (for example, see Mitchell and Carson, 1989), these benefits and costs can and generally should be included in any cost–benefit analysis.

As Cline (1992) suggests, benefits and costs should be separated into categories. One possible grouping of benefits and costs might be: commercial/material, ecological, and aesthetic/existence values. Intergenerational equity issues aside, each of these categories can be discounted using different rates. The literature suggests various ways to determine rates, reflecting relationships with time, markets, and individual preferences. Portney (1990) wonders whether nonmonetary or nonuse benefits, such as existence and aesthetic values, should be discounted at all. For example, the "psychic benefit" of enjoying the view of the Grand Canyon or knowing that a wilderness area is being preserved might not be different for an individual today than for the same individual years from now or for another individual in a future generation. Then, using an agreed-on definition of the duration of a generation, say 30 years, projects that affect the next generation may be systematically analyzed.

Figure 1 is a simplified decision model which portrays one way (among many) to address the intergenerational issues outlined earlier. For projects that affect the current generation, commercial/material benefits and costs can be discounted, with the choice of rate depending on the project. If the project is undertaken by the government, a nonmarket based rate should be used, for instance the social rate of discount. Similarly, ecological damage that affects a single generation may be discounted, though the choice of rate is again problematic. It might be inappropriate to discount aesthetic/existence values since their value might not be time dependent and substitute goods might not exist.

Decision making for projects that affect future generations is more complicated, and may take several different forms. The simplest is that costs and benefits affecting future generations should not be discounted since we do not and cannot know their preferences. Another possibility would be to apply the same discounting techniques as mentioned for projects that affect a single generation.

Perhaps a better approach is to combine the use of the precautionary principle with Burton's method for intergenerational discounting, as portrayed in Figure 1. According to Howarth (1993), all projects that affect future generations should be examined under the conditions of the precautionary principle before discounting occurs.* Following Howarth's advice, the first question to be addressed is "Will the project impose catastrophic risks or damages on another generation?" If the answer is that there is no catastrophic risk, then Burton's method would be applied to material/commercial and ecological benefits and costs, each using a different intergenerational and intertemporal discount rate (Burton, 1993). However, if the answer is "yes, there is catastrophic risk", then another question must be asked: "Can we take steps to substantially reduce risk without compromising our well being?" If the answer is yes, we should proceed as above. If the answer is no, then serious consideration should be given as to whether the project should be undertaken and whether discounting or cost–benefit procedures should be used at all.

It must be stressed that, while this suggested model takes future generations' welfare into account, it does not solve all the problems of intergenerational

* Perrings (1991) also makes some suggestions for applying the precautionary principle, although he acknowledges there is no consensus on how to use it for decision making under uncertainty. He advocates using the precautionary principle when both the level of fundamental uncertainty and the potential cost or stakes are high — where science is inadequate and ethical judgments are ubiquitous.

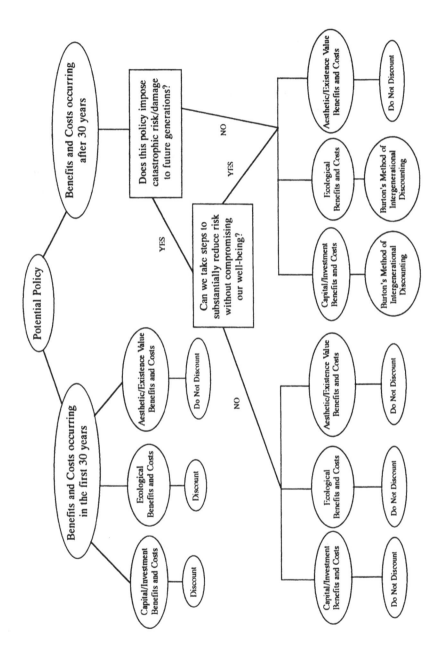

Figure 1   Possible intergenerational decision model.

fairness. For example, in all cases, the choice of a nonzero discount rate is still problematic. In addition, using Burton's technique only slows the exponential decrease over time of net benefits. The suggested decision process will not necessarily lead to equity between generations, especially when benefits and costs are distributed asymmetrically over time.

CONCLUSION

While some aspects of discounting can be preserved in making comparisons across generations, the discounting technique is inadequate to handle the concerns of intergenerational equity. Further development of strategies to make trade-offs between present and future generations seems to depend on a deeper understanding of the requirements of intergenerational equity. Specifically, we need to understand more fully the nature, basis, and extent of obligations of the current generation to future generations; and we need operational principles and guidelines to improve decision making when these trade-offs are necessary.

Uncertainty, especially about the far future, severely limits our ability to make these trade-offs intelligently, reliably, confidently. Therefore, long-term modeling should not be used to drive decision making; it should be used as an aid to judgment, not as a substitute for it. When dealing with the far future, precise calculations and optimizing strategies are generally inappropriate.

Our limitations in making trade-offs systematically mean that we will continue to make these decisions incrementally, using such notions as the "rolling present". According to this concept, the current generation has a responsibility to provide the next succeeding generation the skills, resources, and opportunities to cope with any problems the current generation bequeaths. Likewise, the next generation is obliged to do the same for the generation that follows it, and so forth. In this way, future generations are given consideration and compensated for any harms passed on by the previous one. The rolling present involves an iterative decision process — succeeding generations re-evaluate the policies of the past, using new information and technical capacity together with their own values and priorities, and make appropriate policy changes.

This rolling present process has the advantage of being familiar, incremental, and easy to implement. However, it also has some limitations and deficiencies. One deficiency is that it tends to ignore "time bombs" — that is, risks that do not threaten immediate generations, but will affect later generations, such as nuclear wastes that remain isolated for several generations and then contaminate groundwater. More generally, this decision process can be criticized for making it too easy for the current generation to ignore the long-term implications of its actions.

It is important to acknowledge the pervasiveness of value judgments in intergenerational decisions and view decision making as a social process in which public participation is especially important. Even as we use incremental processes to make current decisions, we should be focusing on educating the next generation about intergenerational concerns. By doing so, over time we might expand the time horizons of our political culture and improve our collective capacity to make decisions that treat future generations fairly.

REFERENCES

Baier, A. "For The Sake of Future Generations," in *Earthbound,* T. Regan, Ed. (Prospect Heights, IL: Waveland Press, 1984).

Baier, A. "The Rights of Past and Future Persons," in *Responsibilities to Future Generations: Environmental Ethics,* E. D. Partridge, Ed. (Buffalo, NY: Prometheus Books, 1980).

Barry, B. "Circumstances of Justice and Future Generations," in *Obligations to Future Generations,* R. I. Sikora and B. Barry, Eds. (Philadelphia: Temple University Press, 1978).

Barry, B. "Intergenerational Justice in Energy Policy," in *Energy and the Future,* D. MacLean and P. G. Brown, Eds. (Totowa, NJ: Rowman and Littlefield, 1983).

Bateman, I. "Social Discounting, Monetary Evaluation and Practical Sustainability," *Town and Country Planning* 60(6):174 (1991).

Boyer, L. and B. Catron. "Modeling the Future, Making Current Decisions," in *Deciding For the Future: Issue Papers,* B. L. Catron, Ed. (Washington, D.C.: National Academy of Public Administration, 1994).

Brown-Weiss, E. "Our Rights and Obligations to Future Generations for the Environment," in "Agora: What Obligations Does Our Generation Owe to the Next? An Approach to Global Environmental Responsibility," *Am. J. Int. Law* 84:198–207 (1990).

Burton, P. S. "Intertemporal Preferences and Intergenerational Equity Considerations in Optimal Resource Harvesting," *J. Environ. Econ. Manage.* 24(2):119–132 (1993).

Callahan, D. "What Obligations Do We Have to Future Generations?" in *Responsibilities to Future Generations: Environmental Ethics,* E. D. Partridge, Ed. (Buffalo, NY: Prometheus Books, 1980).

Care, N. S. "Future Generations, Public Policy, and the Motivation Problem," *Environ. Ethics* 4:195–214 (1982).

Catron, B., L. Boyer, J. Grund, and J. Hartung. *Ethical Dimensions of Environmental Policy and Decision Making: Risk Management, Intergenerational Equity, and Discounting* (Washington, D.C.: National Academy of Public Administration, 1993).

Cline, W. R. *The Economics of Global Warming* (Washington, D.C.: Institute for International Economics, 1992).

Cullen, R. "Discounting the Economic Costs of Conservation and Consumption." *Environ. Planning* A 23: 1121–1132 (1991).

D'Arge, R. C. and C. L. Spash. "Economic Strategies for Mitigating the Impacts of Climate Change on Future Generations," in *Ecological Economics: The Science and Management of Sustainability,* R. Costanza, Ed. (New York: Columbia University Press, 1991), p. 367.

D'Arge et al. in Brown, *Greenhouse Economics: Think Before You Count.* Institute for Philosophy and Public Policy, p. 10–11 (1991); also cited in Plater, 1992, p. 62.

Deciding For the Future: Balancing Risks and Benefits Fairly Across Generations: NAPA/DOE/Battelle DesignShop™ Synthesis, Washington, D.C.: National Academy of Public Administration, 1994).

De George, R. T. "Do We Owe the Future Anything?" in *Values in Conflict: Life Liberty and the Role of Law,* B. Leiser, Ed. (New York: Macmillan, 1981).

DesignShop Synthesis, National Academy of Public Administration, July 1994.

Dowalatabadi, H. "Uncertainty, Ignorance, & Chaos in Models," Working Paper, Department of Engineering & Public Policy, Carnegie Mellon University, Pittsburgh, April 7, 1994.

"EPA's Use of Benefit-Cost Analysis: 1981 1986," U S Environmental Protection Agency, U.S. EPA Report EPA-230–05–87–028 (1987).

Farber, D. A. and P. A. Hemmersbaugh. "The Shadow of the Future: Discount Rates, Later Generations, and the Environment," *Vanderbilt Law Rev.* 46(2):267–304 (1993).

Golding, M. P. "Obligations to Future Generations," in *Responsibilities to Future Generations: Environmental Ethics,* E. D. Partridge, Ed. (Buffalo, NY: Prometheus Books, 1980).

Goldman, B. *Not Just Posterity: Achieving Sustainability with Environmental Justice* (Washington, D.C.: National Wildlife Federation, 1994).

Howarth, R. R. "Environmental Risks and Future Generations: Criteria for Public Policy," in U.S. EPA, *Clean Water and the American Economy Proceedings: Groundwater,* Vol. 2, U.S. EPA 800-R-93–001b, 1993: GW4–31–GW4–43.

Howe, C. W. "Introduction: The Social Discount Rate," *J. Environ. Econ. Manage.* 18(2):S1–2 (1990).

Hubin, D. C. "Justice and Future Generations," *Philos. Public Affairs* 1(1):70–83 (1976).

Keller, A. and T. R. LaPorte. "Assuring Institutional Constancy: A Crucial Element of Public Trust and Confidence in Managing Hazards of the 21st Century," in *Deciding for the Future: Issue Papers,* B. L. Catron, Ed. (Washington, D.C.: National Academy of Public Administration, 1994).

Lind, R. C. "Reassessing the Government's Discount Rate Policy in Light of New Theory and Data in a World Economy with a High Degree of Capital Mobility," *J. Environ. Econ. Manage.* 18(2/2):S8–S28 (1990).

Macklin, R. "Can Future Generations Correctly Be Said to Have Rights?" in *Responsibilities to Future Generations: Environmental Ethics,* E. D. Partridge, Ed. (Buffalo, NY: Prometheus Books, 1980).

Mishan, E. J. *Cost Benefit Analysis: An Informal Introduction.* (London: Allen and Unwin, 1975).

Mitchell, R. and R. Carson. *Using Surveys to Value Public Goods: The Contingent Valuation Method* (Washington, D.C.: Resources For the Future, 1989).

Nijkamp, P. and J. Rouwendal. "Intergenerational Discount Rates in Long-Term Plan Evaluation," *Public Finance* 43(2):195–211 (1988).

Norgaard, R. and R. Howarth. "Sustainability and Discounting the Future," in *Ecological Economics: The Science and Management of Sustainability,* R. Costanza, Ed. (New York: Columbia University Press, 1991), p. 88.

Norton, B. "Future Generations, Obligations to," *Encyclopedia of Bioethics* (New York: Macmillan, in press).

Page, T. "Intergenerational Justice as Opportunity," in *Energy and the Future,* D. MacLean and P. G. Brown, Eds. (Totowa, NJ: Rowman and Littlefield, 1983).

Parfit, D. "Energy Policy and the Further Future: The Identity Problem," in *Energy and the Future,* D. MacLean and P. G. Brown, Eds., (Totowa, NJ: Rowman and Littlefield, 1983).

Perrings, C. "Reserved Rationality and the Precautionary Principle: Technological Change, Time and Uncertainty in Environmental Decision Making," in *Ecological Economics: The Science and Management of Sustainability,* R. Costanza, Ed. (New York: Columbia University Press, 1991), p. 160.

Plater, Z. J. B., R. H. Abrams, and W. Goldfarb, *Environmental Law and Policy: A Coursebook on Nature, Law, and Society* (St. Paul, MN: West Publishing Company, 1992).

Portney, P. R. "Comments on `Discounting' Session," *J. Environ. Econ. Manage.* 18(2):S63–4 (1990).

Raiffa, H., W. Schwartz, and M. Weinstein, "Evaluating Health Effects of Societal Decisions and Programs", *Decision Making in the Environmental Protection Agency, Selected Working Papers,* Volume IIb, Washington, D.C.: National Academy of Sciences, 1977.

Rawls, J. *A Theory of Justice* (Cambridge: Harvard University Press, 1971).

Richards, D. "Contractarian Theory, Intergenerational Justice, and Energy Policy," in *Energy and the Future,* D. MacLean and P. G. Brown, Eds. (Totowa, NJ: Rowman and Littlefield, 1983).

Shrader-Frechette, K. "Technology, the Environment, and Intergenerational Equity," in *Environmental Ethics,* K. Shrader-Frechette, Ed. (Pacific Grove, CA: Boxwood Press, 1991).

Tietenberg, T. *Environmental and Natural Resource Economics* (New York: HarperCollins Publishers, 1992).

Wallace, L. "Discounting Our Descendants," *Finance and Development,* March 1993.

The World Commission on Environment and Development, *Our Common Future* (New York: Oxford University Press, 1987).

SECTION III

Values and Value Judgments

11 INTRODUCTION TO QUANTITATIVE ISSUES

David W. Schnare

As Doug MacLean explained in his keynote presentation at the symposium from which this text arose (Chapter 13), "the undeniable truth is that we must make trade-offs between risks and between methods of risk reduction." To do so, we have to describe the risks around us and make choices on how to deal with them. As the previous section of this volume makes clear, assessments and judgments are inevitably value laden. As well, they will always reflect the moral suasion of those involved. The challenge is whether to expose these values and morals, and if so, how to do so.

In Chapter 14, James Nash announces that risk assessment and risk management are considerably more than a scientific enterprise. They are value laden and not morally neutral. He argues that to ignore implicit values and moral assumptions is to miss a considerable perspective on decisions. In fact, it is to ignore the question of ethics in decision making.

Nash suggests that to incorporate ethics it will be necessary not only to expose the value base implicit in assessments and decisions, but to expand them beyond those typically applied using traditional assessment and management tools like animal studies and benefit–cost analysis. He highlights four value considerations that are often bypassed or underemphasized in risk analysis. Three of these reflect the issue of distributive justice, going well beyond justice among people to the question of "biotic justice".

While rich with insights on the need for consideration of ethics in risk assessment and management, Nash's contribution also offers a practical direction to ensure this gets done. He suggests that evaluations cannot be restricted solely to scientific peer review. Some form of public involvement and evaluation, particularly by ethicists, is essential to expand the value base and ensure consideration of ethics.

In Chapter 15, Christopher Paterson and Richard Andrews offer a review on efforts to implement the Nash dictum to broaden the value base when making decisions on risk. Tracing the use of risk assessment under real world conditions, this chapter revolves around the fact that choices must be made. Thus, risk assessment is translated into "Comparative Risk Assessment", to allow consideration of alternative risks and alternative solutions.

Paterson and Andrews describe the mechanisms for incorporation of public values into assessments and decisions. These methods significantly altered the assessment and decision-making processes in cases where public involvement was used. For example, the list of risky concerns deserving public attention often grew. Threats not typically considered by the scientific risk assessment community, but added by the public, were uncontrolled population growth, urbanization, consumptive lifestyles, and lack of environmental awareness. In like measure, public participants expanded the list of potential risk reduction options.

To the risk assessor, the challenge of an expanded value base is the challenge of how to present and display values, morals, and ethics in terms that effectively characterize risks, options, and the meaning of choices. Resha Putzrath's chapter (Chapter 17) describes the most common challenge — the selection, comparison, and combination of information on value judgments where data are numeric and must be combined or directly compared.

This chapter discusses the breadth of the presentation problem, giving clear announcement of the complexity and size of the challenge. There are cases where data cannot be combined and other cases where values are difficult to present numerically. Putzrath offers suggestions on how to attack these problems using nonquantitative means that maintain distinctions between options, yet characterize the diversity of values involved.

Putzrath and Nash both highlight the case where there is significant uncertainty inherent in risk assessment and risk management. Nash eschews the use of benefit–cost analysis because it does not provide for an exposition of the values associated with uncertainty. Putzrath recognizes that there are cases where uncertainty bars the use of quantitative data. The chapter offered by David Schnare assembles a series of analytical cases that addresses both of these problems.

Schnare (Chapter 16) begins from the point that often more is known about the uncertainty surrounding data than the actual condition data are intended to represent. Demonstrating the use of Monte-Carlo analysis, it is possible to identify cases where uncertainty is clearly too great to permit differentiation between risks or options. In other cases, however, differences between risks are so large that even quite uncertain data are useful in describing risks.

This chapter also presents examples of how to ensure that common values are not ignored or forgotten in assessments with highly uncertain and complex data. Schnare suggests asking, for example, what ethics must be applied when seeking to impose a mandate requiring the investment of billions of dollars so as to delay a single death by one year.

Schnare and Nash recognize, however, that the calculus of risk management does not adequately calculate certain types of outcomes — especially intergenerational effects. In a chapter that is as much philosophical as quantitative (Chapter 12), Bryan Norton carefully examines the differences between the question of human risks and the more complex issue of ecological health.

Like Nash, Norton suggests that economic analysis fails to reflect concerns that stretch over long time periods and affect several generations. Norton argues for a multivalued multitiered assessment approach that better captures the contextual pluralism of the public. His special concerns are for values associated with long periods of time (generations) or large areas (global).

Norton's chapter raises the need to distinguish between having many tools and having useful tools. When a tool like economics ignores some values, new tools are needed. When a tool such as a public referendum captures values, but does so implicitly rather than explicitly, new tools may also be needed. Norton provides the outlines of a new way to clearly display risk information about time and space in a manner that can significantly inform the public and the decision maker.

The limitations of any text preclude a definitive treatment of how to incorporate ethics, morals and values into risk assessment. Nevertheless, this section provides a strong introduction to the mechanics of how values can be incorporated into risk assessment and risk management. The serious student will find this chapter is a doorway to more honest, fair, and sensible analysis.

12

ECOLOGICAL RISK ASSESSMENT: TOWARD A BROADER ANALYTIC FRAMEWORK

Bryan G. Norton

CONTENTS

Introduction: Risk and Risk Models 155
Multitiered Systems of Analysis 156
Scaling Social Values.. 161
Risk Decision Squares as a Multiscalar Method of Valuation 164
A Lexicographic, Scale-Sensitive System of Policy Analysis 171
Conclusion 173
References 174

INTRODUCTION: RISK AND RISK MODELS

The term "risk" has two lives. On the one hand it is, in ordinary language, a useful, catchall term, somewhat vague in its application, but directing our attention to a class of situations in which uncertainty and danger are both well represented. However, "risk" is also a term in environmental and health policy discussions, and in this context the term has taken on a quite different life as a technical term that is defined precisely within carefully calibrated scientific models for risk assessment. This duality is not a problem, as long as we remember that no technically defined concept of risk can ever capture all of the richness of the ordinary concept — some models will capture some aspects of risk, some will capture others, but the ordinary concept involves complexities that cannot be fully comprehended in any single model. It is therefore not possible to identify and support a single risk model as the exclusively correct one. The best we can hope for is to construct a variety of risk analysis models, to be as precise as possible in developing them, and to learn what we can from a variety of models under diverse applications. It must always be remembered

1-56670-131-7/96/$0.00+$.50
© 1996 by CRC Press, Inc.

155

that we are not looking for the "right" risk model, but rather the model that will illuminate risk as it is addressed in particular contexts, and for the insight that can be gained by looking at a complex thing through multiple lenses.

In this paper I will contrast two broad types of risk assessment models, single-tiered and multitiered models, noting that most models hitherto have been single tiered in an important sense. All values countenanced in currently used systems of analysis keep accounts only of "present values" — preferences which are insensitive to changes in temporal scale. They have analyzed risk within a single spatiotemporal dimension and these systems therefore have a characteristic I will refer to as "nonscalar". Nonscalar systems do not necessarily deny the importance of impacts on future generations, but they require that these impacts be valued in a metric that makes them commensurate with present values. Concern for future generations is therefore expressed as what the present is willing to pay to protect the interests of future persons. I will argue that nonscalar approaches to risk assessment, while they have been useful in understanding and measuring risk to human life and health, will not prove adequate to characterize or analyze longer-term risks to ecological systems. There have been a few promising attempts to develop two- or multi-tiered systems, and this paper continues the work of Talbot Page and others who have developed such systems (Page, 1977; 1991; Norton, 1990, 1991; Toman, 1994). Discussions of ecosystem health and integrity often assume, implicitly, a multitiered system in that these management criteria have not been (and probably cannot be) stated in a single-scaled system of value. (See, for example, Costanza, Norton, and Haskell, 1992; Edwards and Regier, 1990.) My approach differs, however, from integrity theorists such as Laura Westra (Westra, 1994), who believe that the concept of integrity must be given a nonanthropocentric interpretation. By contrast to Westra, I will offer a multi-tiered system of analysis which takes account of human values and risks to them in multiple scales of space and time. I will characterize and evaluate changes in ecosystem states from a human, but multigenerational, viewpoint. This alternative approach is based on a more pluralistic conception of human values and of risks and attempts to develop alternative risk-decision criteria that are applicable in different situations. While the approach proposed is admittedly pluralistic, it is *not* relativistic or nihilistic; I believe that there are good reasons to guide choices as to which criterion is applicable in particular situations. This form of pluralism is best described as "contextual" or "inte-grated" pluralism; it applies different criteria according to a rational assessment of the relevant characteristics of a risk encountered in specific contexts.

MULTITIERED SYSTEMS OF ANALYSIS

Risk analysis as practiced thus far has mainly employed a conception of value based in individual welfare, which is not surprising, given the usual focus on human health effects of exposures to pollutants. According to this conception, an increment of risk of a negative outcome is always and by definition a decrement in the expected welfare of some human individual or individuals. Conversely, a decrement in such a risk is an increment in expected welfare of individuals. This definitional connection characterizes the nonscalar nature of current risk decision making. This connection is possible because the whole system assumes a utilitarian, welfare-based definition of value. This type of analysis has the advantage that information about risks can be aggregated with other forms of information about welfare, providing a single accounting system for risks and other types of costs and benefits. Those who favor the economic conception of decision making can thereby achieve a further simplification — the degree of perceived risk can be measured as the willingness of an informed "consumer" to pay for decrements of risk. This simply elegant theoretical framework provides a definition of risk that makes risk measurable in ways that encourage integration of risk calculations into welfare economics. It also has the unquestioned advantage that representations of risk can be registered and aggregated within a monolithic system of values which are all commensurate.

Theoretical elegance can mask important complexities, and I question whether a nonscalar value analysis can be adequate to the task of ecological risk assessment. I begin by establishing that there is an important disanalogy between tools available to analyze risk to human health and those available to analyze ecological risk: human health risks can be understood as directly related to human welfare, whereas ecosystem risks are related only indirectly to the welfare of individuals (at least given currently available analytic techniques). It is possible, in principle at least, to gather scientific evidence to establish links in a causal chain that connects a discharge of a chemical into the environment, for example, to an exposure of a population to the chemical and eventually to an increase in human illness. Since nobody questions that human illness or death is a bad thing, changes in the physical world such as increased concentrations of a toxic chemical in a city's water supply can thus be directly linked to an unquestioned value — individual human welfare. This is not, of course, to say that it is *easy* to trace such a causal chain or that it can be accomplished without important assumptions, but only that *if* the chain is established, nobody will question that reducing the risk has value measurable in human welfare.

Attempts to state descriptors for ecological risk, however, apparently break this direct link because there are no techniques for linking changes in the states of ecological and physical systems to individual welfare. This is true for two reasons. First, while ecologists are often able to foresee impacts of various stress regimes on ecological systems, the pace of these changes is uncertain. And pace of change is crucial because, as Aldo Leopold realized in the 1920s, the most difficult problem in environmental policy is to separate changes in ecological systems caused by anthropogenic impacts from normal changes — what he called "natural cycles" (Meine, 1988). Second, it is impossible to predict the specific needs and preferences of future people, because the emergence of new technologies and uses for resources are notoriously difficult to predict (Faber, Proops, and Manstetten, 1992). It is therefore impossible to establish correlations between changes in ecosystem states and changes in the welfare states of individuals.

This is not to say there would be no changes in the welfare states of future people as a result of degradation of ecosystems. If, for example, the Chesapeake Bay is choked with plankton and its waters become largely eutrophic, economic and other opportunities will be lost, even if it is impossible to quantify these losses as decrements of individual welfare. The point is simply that these changes in what might be called a "keystone resource" in a whole region play themselves out on different scales. Even if ecological models can predict changes over decades, there is no method of representing changes on the ecosystem scale as welfare effects on the individual, economic scale. The systems involved are so complex and unpredictable that it would be impossible to build a model relating changes in ecosystem states to changes in individual welfare states. It is not obvious, I think, whether it is impossible in principle to make such a connection, or whether it is a matter of lack of scientific data and models. Practically, this difference is immaterial: at least for the foreseeable future, given analytic techniques currently available, it will be impossible to establish a causal link between projected changes in ecosystem function and welfare states of human individuals.

This important disanalogy creates a painful dilemma for advocates of single-scale systems of valuation as the basis of risk analysis. In some cases, ecological impacts are simply ignored and left out because there is no way to measure their impacts on human welfare (see, for example, NAS, 1992). It would, of course, be possible to use contingent valuation questionnaires to determine what consumers are willing to pay to reduce nutrient loading into the Chesapeake Bay by a given amount. This approach simply shifts the burden of calculating the likely impacts of a unit of nutrients going into the Bay on consumers from scientific experts (who in this case have no methods by which to make the necessary connections) to lay persons (who, of course, have even less chance of constructing a model to connect impacts of human today's actions on their future welfare). Because I see no escape from this dilemma for models that have only present welfare values as measuring sticks of risk, I believe we must explore approaches to risk analysis that are more pluralistic in the values they recognize and in the measuring sticks they employ.

A pioneer in this field, Talbot Page, introduced a two-tier system of analysis as part of his examination of intergenerational aspects of the problem of "materials policy" in his 1977 book, *Conservation and Economic Efficiency*. Page introduced two criteria for judging policies. The *efficiency* criterion as usually employed embodies a "present value criterion". "Everything is done from the point of view of the present; it is their time preference, and everything is discounted back to them" (Page, 1977: 170). A *conservation* criterion would be more time sensitive, and would require that current usage of natural resources protects the resource base. Success in such protection will be indicated if the "real" price of materials can be projected as constant or nondeclining into the future (Page, 1977: 185). Page argues persuasively that these two criteria might diverge significantly in their policy recommendations and that the efficiency criterion is inadequate to protect the legitimate interests of the future. He concludes that the two criteria must apply independently, and that they really exist on different levels, and proceeds to introduce a Rawlsian contractual obligation to protect the resource base. This obligation, in effect, limits the range of efficient outcomes that can be considered as morally acceptable policy. The intergenerational moral constraint excludes decisions based on the choices of one generation. Page argues persuasively that neither rule can describe the limits of its own application, so we must look to the particulars of situations to decide which criterion, each of which is useful in some applications, is most useful in a given situation.

Page's argument is also applicable to risk assessment, which might employ one criterion in cases of risk to human health and welfare, and another, intergenerationally sensitive criterion to cases of risk to ecological systems. Since individuals apparently discount future risks (Lind, 1982), exhibiting willingness to incur risk in the future in order to enjoy consumption in the present, we can pose the question: what is a fair allocation of risks across an entire society and across time? Page argues convincingly with respect to materials policy that, provided we assume resources are limited, the discounted present value test cannot be defended as "fair" across generations. Similarly, if the present generation makes choices that predictably distribute risk according to their own time preference — by delaying the expenditures necessary to safely store long-lasting hazardous wastes of current production and consumption, for example — it could be argued that they have ignored morally important obligations based in intergenerational equity.

Page's multitiered analyses are deserving of close study because he provides a complex evaluative structure that allows application of different criteria in different situations; in particular, he distinguishes the situations by identifying some decisions that have important intergenerational implications. He applies, we might say, multiple criteria depending on the temporal scale on which the risk situation will play itself out. Page employs standard welfare evaluation to apply his "efficiency" criterion (which works for decisions without significant intergenerational impacts) and he introduces the idea of an "intergenerational contract", derived from the work of John Rawls, to express obligations that are applicable when today's decisions can have strongly negative impacts on the future. Page reasons that we should follow rules of intergenerational equity that would be adopted by a rational individual who designed the rules for intergenerational equity from behind a veil of ingnorance — the individual is understood to be choosing the rules without knowing in which generation that individual would actually live, which encourages the "filtering out" of self-regarding advantages (Rawls, 1971; Page, 1977; also see Norton, 1989).

Page's insights, however, have not resulted in policy applications, because establishment of intertemporal obligations implies important constraints on current practices only if there is a significant danger that those practices will actually harm the future. Page's work has been largely ignored by mainstream economists because most of them are technological and resource "optimists" — believing that every resource has a suitable substitute and that every risk is compensable, they conclude that the best way to fulfill our obligation to the future is to maximize economic growth in the present. Provided some of this wealth is reinvested in productive processes, the future cannot fault the present, because these investments will ensure that the future has the same opportunity to fulfill their needs and desires as the present has (Solow, 1993). The problem with Page's intertemporally sensitive analysis, then, is that its application will be quite different, depending on one's optimism regarding technology's ability to provide substitutes for resources and solutions to pollution problems. If one emphasizes caution, one will be a conservationist and oppose many economic developments; if one accepts technological optimism and a growth strategy, one will oppose conservation efforts and try to produce to compensate environmental degradation with productive technologies. However, that choice between two variant "rationalities" is not a scientifically decidable question.

My approach, which will have much the same structure as Page's model in that it employs multiple criteria in tandem, differs in that the first-order criteria that allow decisions about what to do are applied according to a second-order criterion that sorts risks (problems) according to the temporal *and spatial* scale of the impacts of a risky activity. While there apparently exists considerable substitutability across resources at smaller scales, there exists less substitutability among resources at the larger scales of the ecological and physical systems that provide the context for economic systems (Norton and Toman, 1995).

The difficulty that the rules cannot decide their own range of application need not bother us, provided we keep firmly in mind the caution that the concept of risk is too rich to be entirely captured by any simple risk model. Even though systems of risk analysis and aggregation are inherently underdetermined by a commitment to a single value, it is still possible to use the welfare model as one useful model among others. Recognizing that every model will be an incomplete representation of some aspects of risk encourages us to use several models, comparing and contrasting their results, and to seek a more integrated understanding of risk in all its complexity by creating models of varied aspects of it.

An important outcome of our decision to limit the applicability of first-order criteria is that information relevant to particular criteria need not be expressed in terms that are aggregatable with information relevant to other criteria. Once these applicability decisions are made on the second level, the specific criteria used in particular situations might be calibrated in different terms. Since we do not conceptualize our problem as one of maximizing welfare aggregated across generations, we could use physical measures, for example, to indicate how well our policies are succeeding in situations where risk to ecological systems are involved, and welfare measures to quantify resulting changes in economic status. Conceptually, in our models, we trade off the advantage of aggregatability of goods across generations for the advantage of using particular criteria that are appropriate in various situations. I will argue that this trade-off is useful in thinking about long-term risks because it makes our models sensitive to scale and to values that emerge only in intergenerational time. Further, I believe this change is essential if ecological risk assessment and ecosystem management are to become important contributors to environmental policy formation.

In the remainder of this chapter, I will explore one particular approach to a pluralistic, but integrated, system of value, an approach which uses scalar determinations to locate the range of application of the multiple decision criteria. These criteria are applied in different contexts, and the context is understood according to the the scale at which threatened values are manifest.

SCALING SOCIAL VALUES

The method proposed here is based in hierarchy theory, an application of general systems theory to ecological organization. The method, however, is applied with both descriptive and evaluative goals in mind, leading to an evaluative framework that embodies scalar aspects of human valuation within the models used to state the goals of environmental managment and to measure success in achieving them (Norton, 1995a). Hierarchy theory — really a method for organizing information regarding complex systems into multiple scales, rather than a theory — rests on two key assumptions/principles: (1) that all observation must be from some perspectival point inside the hierarchically organized system that is being measured and interpreted, and (2) that smaller subsystems within the hierarchy change at a slower pace that represents a quantum difference from the pace of change in the larger system — its environment — of which it is a part (Allen and Starr, 1982; O'Neil, et al., 1986; Norton, 1990).

I propose an alternative approach to policy formation which recognizes the primary role of values in determining what aspects of nature we describe and model. I therefore add Principle (3): all human *valuation* must be from some point inside the hierarchical organizational system that is being measured and evaluated.* Human social values are on this approach as important as descriptive adequacy in determining which models are used to measure progress toward our goals of environmental protection. Therefore, I begin to sketch a multiscalar approach by positing that humans tend to experience and value nature in three differing scales, or contexts. I will call these three "horizons of valuation", corresponding to three different policy contexts, and yet recognizing that any structure we impose will in some ways simplify the complex system we describe and evaluate. My claim is only that looking at the valuation

---

* See Norton and Hannon (under review) for a more detailed discussion of this procedure of evaluating place-relative values.

problem as embodying these three scales helps us to see a more comprehensive picture of human valuation of nature than does the single-scaled view of welfare economics.

Consider, first, the most common context in which most persons make evaluative decisions, the context in which we choose a means to make a living and in which we choose one product rather than another. These are the ordinary choices we make with regard to our (broadly) economic well-being, and in making these decisions we assume as background to the decision the current market conditions and rules — as well as a relatively constant climate and availability of ecosystem services — and we generally act with a relatively short time horizon, say up to 5 years. However, this is not the only context in which we make decisions and choices. Consider the example, discussed by Page (1977) and by Toman (1994), of a constitutional convention. In this context, participants and citizens are asked to set aside their personal likes and dislikes and to consider how they might design institutions that will create stable and useful political institutions over multiple generations. Many conditions and rules that are normally taken for granted as background to decision making are now called into question. And, most pertinently for our topic, participants and citizens are asked specifically to think in the expanded frame of multigenerational time; the concern in this expanded frame of time is with the type of society we want to be; attention is turned away from the goal of efficiency in the fulfillment of individual desires and toward concerns about the development of the community and its cultural and political base (Daly and Cobb, 1989).

Today, with growing emphasis on a "global" culture and concern for changes in global ecosystems, it appears that a third scale of concern is evolving, and it can be associated with the indefinite time scale on which the human species evolves. On this scale, more and more environmentalists and policy makers are examining our impacts on ecological and physical systems over indefinite time. Since these very long-term impacts will affect global systems, we can once again apply hierarchy theory, which correlates very long temporal scales with very large extent in space. This third, emerging horizon of concern is associated with the emergence of a species-wide consciousness — a concern for the indefinite future and well-being of the human species.

Hierarchy theory provides a framework for integrating these multiscalar horizons of concern; more important, it allows a correlation between temporal horizons of concern and dynamics of differing temporal and spatial scales, as shown in Figure 1 (Norton, *The Varieties of Sustainability,* forthcoming). Short-term concerns of humans are associated with changes in economic conditions and behaviors. These changes express themselves in the individual-scale decisions of the farmer to plow and plant wheat on a given field or to purchase a new tractor, or of a waterfront landowner to divide and sell parcels for vacation homes.

The second level is especially important because it is the level at which humans shape their own culture and also make decisions that affect the landscape.

Temporal Horizon of Human Concern	Time Scales	Temporal Dynamics in Nature
Individual/Economic	0-5 years	Human economies
Community, intergenerational bequests	up to 200 years	Ecological dynamics/ Interaction of species in communities
Species survival and our genetic successors	indefinite time	Global physical systems

Figure 1 Three horizons of human concern. Human "concerns" (and associated values) have differing temporal horizons. This figure shows three such temporal scales, one corresponding to individual, economic concerns; one corresponding to the community time scale; and one corresponding to indefinite time. It is suggested that these scales of concern, which will embody human values, can be associated with different physical dynamics.

It is on this level that a human cultural unit (a population or deme) interacts with the other species that form with it a larger ecological community. We might say that it is on this level that a culture and society articulates "aspirations" as opposed to the "preferences" they express as consumers (Norton and Hannon, under review; Norton, *Environmental Values*, 1994). It is this second, expanded level of perception that Aldo Leopold referred to as "thinking like a mountain" (Leopold, 1949; Norton, 1990; Norton, 1991). Leopold, who regretted his participation in the intentional eradication of wolves and mountain lions and the resulting overgrazing of wilderness areas by deer, used this simile to stand for an expanding consciousness of impacts that unfold in the frame of time typical of ecological and interspecific interactions. We can therefore think of Aldo Leopold as the first "ecological economist", in that he advocated a dual accounting system, one for keeping track of the short-term, economic impacts of our actions and a second accounting system for keeping track of the larger and longer-scale impacts of human decisions and actions. It was these longer-scale impacts that prompted Leopold to emphasize ecological integrity as a central concern of environmental management: "A thing is right when it tends to preserve the integrity, stability, and beauty of the biotic community. It is wrong when it tends otherwise" (Leopold, 1949: 224–225).*

* While some interpreters of Leopold take this criterion to attribute "intrinsic" value to ecosystems that is commensurate with anthropocentric value and sometimes overrides human value (Callicott, 1988; Westra, 1994), I interpret Leopold as recognizing multiple values and seeking an integration of pluralistic values at multiple levels. See Norton, 1988; Norton, *The Varieties of Sustainability*, forthcoming; *Environmental Pragmatism*, 1995c.

RISK DECISION SQUARES AS A MULTISCALAR METHOD OF VALUATION

I will now sketch a method of valuation and decision analysis that is, as suggested in Leopold's simile, sensitive to the multiple scales on which humans perceive and value their natural surroundings. The mainstream economic viewpoint succeeds in abstracting from the multiscalar aspects of human impacts by reducing all impacts to present costs and benefits. However, this abstraction/simplification also deprives us of two types of information that would be important on a scale-sensitive analysis. One of the problems with the extirpation of wolves over the entire Southwest Territories was that, whereas hunters might control populations of deer in accessible areas, they will not reach the most remote areas. What Leopold learned was that the spatial extent of a human-caused impact is important to evaluating its full ramifications. A highly aggregated accounting system may ignore significant differences in impacts of an activity in different areas, impacts that differ because of local conditions (Norton and Hannon, under review).

Second, the eradication of the predators was a virtually irreversible action because of biological and political barriers to reintroduction of predators (Beck, 1995). The policy of eradication did not allow reconsideration in the face of new evidence that human hunters would not control populations of deer in remote areas. Our scalar approach allows us to take these two important variables into account by calibrating irreversibility as the period of time required for degradation of the system to be corrected by natural processes or rehabilitation efforts and by correlating the physical scale of impacts of a policy with system characteristics at the appropriate physical scale. The resulting decision framework can be represented by a device I call "risk decision squares", which reflect the magnitude of an impact in the vertical axis and the degree of reversibility of a decision/impact in the horizontal axis. Figure 2 locates decisions and their possible outcomes according to severity on the vertical axis and according to degree of reversibility of possible impacts on the horizontal axis (Norton, 1992; Norton, 1995b).

We can represent what we said, above — that economists and ecologists employ a different "paradigm" — by recognizing that ecological economists would, at the expense of being unable to aggregate fully across the whole space, follow Leopold in drawing distinctions in *types of risks*, breaking the risk decision space into distinct regions where different considerations and criteria apply, according to differences in the temporal and spatial scale of impacts of human activities.

For mainstream economists, options for preference satisfaction and, accordingly, for future welfare, are not prejudiced by irreversible changes in ecological systems or other alterations of the physical world (Solow, 1993). For the mainstream economist, then, the decision square need not specify a time scale for reversibility; the horizontal axis can be understood simply as the

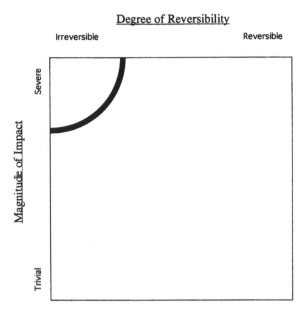

Figure 2 Risk decision squares. Risk decision squares locate actions and policies
according to the impacts involved in a worst-case outcome of the action/policy;
reversibility of impacts is in this case plotted against the magnitude of impacts.
Decisions of high possible impact and low reversibility deserve special treatment.

degree of substitutability of a new resource for any damaged resource, with the
goal being to ascertain the correct dollar figures a consumer would be willing
to accept for the destruction of a resource that will require development of a
suitable substitute. This simplified version of the risk decision square can be
represented as Figure 3 and represents the economists' belief in unlimited
substitutability of resources for each other, regardless of the scale of impacts
on larger systems of resource production (from Norton, 1995b).

   In the context of an economic analysis, all risks remain of the fungible and
compensable type. The risk decision square also exhibits the differences between ecologists/ecological economists and mainstream economists regarding
substitutability of resources. Since the only measure of value for mainstream
economists is consumer welfare, they are comfortable in assuming that access
to any lost resource will be compensable, provided the future has sufficient
wealth. They therefore see the decision space as unitary, without conceptual
divisions among types of decisions that must be addressed in policy analysis.
For an ecological economist, the type of risk involved will differ depending on
physical characteristics of the area/system under risk, and thus spatial scale
becomes an important determinant of the decision process. In particular, the
ecologically minded economist will prefer alterations of ecological systems
that are partial and leave the essential structure of the natural system unchanged.

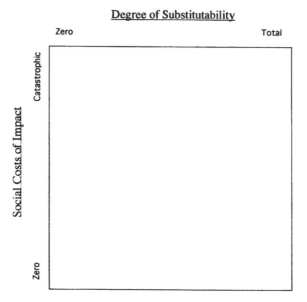

Figure 3   Risk decision square: economists' version. Mainstream economists, many of whom believe all resources have adequate substitutes, do not emphasize irreversibility or scale because their evaluations are stated in terms of human welfare. Irreversible impacts can be compensated by providing the future with alternative capital to support production and opportunities to achieve welfare.

Within a scale-sensitive system of analysis, on the other hand, it is possible to argue that ecological systems change discontinuously, and that some changes — such as the extinction of a species or the destruction of a rainforest — introduce crucial thresholds into the decision process. Questioning that there are "suitable substitutes" for functioning rainforests or for grizzly bears in wilderness areas, many ecologists urge in strongly moral terms that extraordinary efforts to protect such resources are mandatory. Time of reversibility and scale of possible impacts can therefore function as a meta-level criterion to separate decisions with short-term and/or reversible impacts from decisions with long-term and practically irreversible impacts. Decisions in the former category can be decided according to economic criteria; decisions in the latter category are judged by more stringent criteria of intergenerational equity and fairness.

We can elaborate Figure 2 in a more ecological direction by incorporating the hierarchical organization, creating a more complex decision space which is represented here as Figure 4. First we can note that ecological economists, purely by virtue of their emphasis on ecological systems and processes, will focus on the temporal aspects of change; for them, irreversibility is not just interpreted as an abstract concept of substitutability of one resource for another — as measured against units of welfare available to consumers — it will include physical parameters such as the possibility of reversal of impacts in

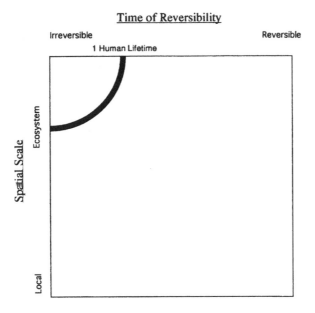

Figure 4 Risk decision square: ecologists' version. Ecologists' concerns regarding temporal and spatial scale of impacts can be incorporated into the analysis by superimposing the assumptions of hierarchy theory on the risk decision square: temporal scale can be calibrated as time of recovery from the possible impact, while spatial scale can be associated with the physical level of the system affected.

ecological time and space. The horizontal axis can therefore be calibrated as a measure of how long, given a particular impact or disturbance, would be required for natural processes to reverse that impact. The horizontal axis therefore locates decisions and policies which incur certain risks according to the restoration time necessary to repair damage if negative impacts occur as a result of that decision or policy. The risk decision space therefore represents the differences between mainstream economists and ecological economists, with Figure 2 representing a neutral space, and Figures 3 and 4 representing the risk decision space as it would be interpreted, respectively, by economists confident of substitutability among resources and by ecological ecologists who are not.

Applying reasoning such as this, Norton and Ulanowicz (1992), for example, have argued that, since a commitment to sustain biodiversity is consensually understood to be a commitment to do so for many generations of humans (at least 150 years), we can conclude that the focus of biodiversity policy should be landscape-level ecosystems which normally change on a different temporal scale, an ecological scale that is slow relative to changes in economic behavior. If we can isolate the dynamic driving local economic opportunities from the dynamic supporting biodiversity, it may be possible to encourage both, and avoid policy gridlock such as has occurred over the spotted owl in the ancient forests of the Pacific Northwest. If it is possible to identify multiple dynamics that unfold at different scales, and if there are identifiable social values asssociated with those various scales, it may be possible to devise policies that have positive effects on all scales. For example, programs that combine economic development with ecological restoration — such as small-scale tree planting programs in deforested areas — can have positive impacts on both economics (by providing thinned trees for firewood) and on the ecological scale by improving hydrology and reducing erosion. A scalar, pluralistic approach takes good environmental policy to be a problem of designing policies that are scale-sensitive. In the best case, a scale sensitive policy would have positive spillover consequences from one scale to the other, as in the case of the tree-planting programs. Alternatively, one should choose a policy that has positive impacts on one scale and predictably inconsequential effects on other levels. If value pluralism can be linked to a scalar hierarchy so that human values are "scaled" in correspondence to important economic and ecological processes, it may be possible to measure progress toward multiple goals simultaneously and seek policies on one level that have positive spillovers on other valued processes at other levels.

Human economics, which is paced to individual decision making, would on this perspective be understood as describing values that emerge in a subsystem of the larger physical environment. Individual economic values are expressed on short scales and, provided they do not, collectively, add up to a trend that is significant on the larger, ecosystem scale, these can be governed by individual free choice. In this sphere, economic analysis (perhaps constrained by considerations of interpersonal equity) models individual decision making. When individual choices cumulatively impact large systems in an irreversible way, morally based considerations of equity across generations come into play. Good management, therefore, involves identifying and protecting processes crucial to the complex structure of the ecological system, which is to say that good management allows economic freedom, provided choices of individuals are damped out at a larger scale and do not threaten to introduce ecosystem-level change that is irreversible. We can therefore define decisions that have impacts within two distinct spatiotemporal dynamics — individual choices affecting small subcomponents of a system that are reversible in one human lifetime, and decisions which threaten to create change that affects whole ecosystems. If these large-scale changes cannot be reversed within a single human lifetime, then they must be treated as decisions affecting intergenerational equity. Sustainability of whole-system processes and the structure necessary to continue them is therefore one of those social decision areas that are a matter of intergenerational equity.

Given these contrasts, we can begin to characterize a multitiered approach to environmental decision making as an alternative to the unidimensional decision processes of mainstream economics. This position would see the decision space faced by environmental managers as split into regions, with the corner of the square that is characterized by risks of major negative outcomes that are irreversible representing an area of risk where values do not vary continuously with consumptive values and where moral strictures apply. This area of the decision space differs in kind from the regions where reversibility is high, cost is low, or both. In these latter regions, we will be inclined to accept the usefulness of economic methods on the assumption that, in these decisions, the future cannot fault us if we compensate them for our use of resources and the impacts of our activities on natural systems with increased technological know-how, monetary capital, etc. In these regions, in other words, impacts of decisions are regarded as fungible across time.

The central issue facing the risk analytic community can now be formulated quite simply. Are there any risk decisions that are located in the upper left corner, and thereby governed by rules of intergenerational equity, rather than economic criteria? If there are any, the goal should be to determine how many decisions and precisely which decisions comprise these two broad categories. It must of course be recognized that introduction of constraints is always under a burden of proof because these imply limitations on the freedom of individuals to pursue their welfare as they see fit. Human, individual freedom is highly valued in Western societies; constraints must in this context always be justified against a background preference for freedom. The values proposed here are adaptive values, self-consciously recognized as responses to environmental constraints. The point is that conscious adaptation, including conscious concept building, can increase our ability to remain adaptive if we avoid irreversible negative changes in the health and integrity of ecological systems.

I advocate, then, "adaptive management", (Walters, 1986; Lee, 1993) which emphasizes policy innovations that will increase knowledge of the functioning of ecosystems and also of the impacts of changes in the system on social values. Adaptive management treats proposals for action as "hypotheses" which have both factual and evaluative elements. A program of adaptive management would emphasize incremental improvements and pilot projects and it would emphasize "social learning" through a managed approach to uncertainty and social conflict (Lee, 1993), including exercises in dispute resolution. Most important, it would pursue policy options that increase knowledge and exchange of views, while avoiding polarization of the community. The goal of good management will be the development of management plans that have positive results on all three levels of social concern; the means to this goal will be adaptive management.

Adaptive management can be given a scalar dimension by noting that we undertake "experiments" in adaptation on multiple system levels. We have outlined a tri-scalar system of analysis that posits three relatively distinct scalar dynamics. Risk assessors who turn their attention to ecological risk will immediately face the difficult problem of categorizing the type of risk according to its magnitude and its irreversibility. One aspect of this process is determining "ecological significance" (see Harwell, Gentile, Norton, and Cooper, 1994). However, determining ecological significance cannot be a value-neutral process (Norton, 1995b). If the hierarchical view of nature is correct, it should be

possible through dimensional analysis (Norton and Ulanowicz, 1992) to identify driving variables at various system levels. It is up to decision makers which
levels of multileveled nature are of interest to us. Risk assessors cannot, it
follows, act as pure scientists — we can only model "ecological risk problems"
once we have identified some ecological (such as eutrophication of a bay) or
physical processes (destruction of stratospheric ozone) that are associated with
important social values. Environmental goals, it follows, must be indexed
according to scale. The overall goal is scale-sensitive management that applies
different criteria in different contexts. These contexts are characterized according to the different cycle times of ecological and physical processes, processes
that are associated with the "production function" for important social goods.

One important characteristic of this two-tiered system is that it uses scalar
principles at the meta-level to provide guidance regarding which of several
decision criteria are appropriate in a given case. Different criteria, based in
different social values, emerge on different levels in the complex, multileveled
interactions of economies and the systems that form their ecological and
physical context. The human economy represents a subsystem within larger
and normally slower-changing ecological and physical systems — the environment. In many cases — especially when populations are low and technology
is small-scaled — individual activities have few ecological impacts. Tropical
forest ecosystems, for example, can absorb the impacts of shifting cultivation,
provided it is done on a small scale compared to overall forest size. As clearing
increases in scope, recovery times of the system increase until an unsustainable
situation arises. The key to environmental management, then, is to determine
which human, economic activities are likely to result in irreversible, system-
level damage. These activities should be monitored closely, and special attention should be paid to gathering baseline data and ascertaining pace of ecological change. Problems that require this special attention are those problems
which are located in the upper left corner of risk decision squares.

For decisions with possible impacts only at the level of field or farm, our
choices usually have high reversibility and low likelihood of serious, large-
scale impacts. Here, the economic, benefit–cost criterion is favored. As long as
there are good reasons to believe that economic activities are either reversible
or pervasive, we assume that any damages to the future will be compensated
by our advances in knowledge, technology, and accumulated wealth. For these
decisions, we can assume a high degree of substitutability of resources because
impacts are narrow in scope and are compensable over time. In decisions on
this scale — choices to cut one tree or a small woodlot — we can therefore use
standard benefit–cost analysis to evaluate risks and impacts in terms of human
welfare. When decisions have possible impacts that are irreversible, we may
apply the precautionary principle, even if the scale is relatively small. When
decisions have possible consequences that are both irreversible and large-
scale — such as clear-cutting most of a watershed which is likely to have
system-level impacts — then we have a clear case where intergenerational

concerns come to the forefront. Good judgment on the part of all members of the risk analysis/risk-management team is required if problems are to be properly sorted and indexed according to the possible scale of their impacts and the likelihood that impacts will threaten important social values.

In a scale-sensitve, hierarchical analysis such as the one outlined here, there is no attempt to aggregate valuations across levels of the system. In modern, especially urban and developed contexts, almost every action has potential impacts on larger levels. For example, permitting septic tanks in a rapidly developing exurban area will eventually lead to problems as growth overtakes the assimilative capacity of the ecological and physical systems in the area. Increasing density and technological intensity results in spillover effects from one level to the next — the integrity of hydrological systems is threatened by the cumulative impacts of many individual actions. Similarly, large deficits in the uptake of carbon in many areas that use fossil fuels regularly can threaten global-scale physical systems and introduce risk on the global, many-generational scale. My system of analysis attempts no reduction of all values and associated environmental goals to a single, aggregatable level. The system includes both physical and monetary accounting systems and information from these accounting systems contributes to measuring management performance at different levels and using different management criteria.

A LEXICOGRAPHIC, SCALE-SENSITIVE SYSTEM OF POLICY ANALYSIS

To apply the system, the risk analyst should perform the following, lexicographically ordered steps in order to evaluate an action or policy. First, it is necessary to determine the possible scale of the impacts of the action or policy and the likelihood of irreversible change. Second, it must be ascertained whether processes unfolding on that scale are likely to affect social values; if so, specific categories of social values should be articulated and associated with processes unfolding on a scale that helps to model natural and anthropogenic changes affecting those social values. Once one has constructed multiple models and goals/criteria associated with particular dynamics, the third step is to attempt integration according to a meta-level application of multiple management criteria. If economic-level impacts are implied, and no others, then a benefit–cost criterion, based in individual welfare is dominant. If there is also a likelihood of cumulative impacts with irreversible impacts on larger systems, then a criterion of ecological integrity and health, stated in physical descriptors of ecosystem states (such as energy flows) must also be applied. Similarly, if there are impacts on global-scale physical systems, impacts of the action or policy on these scales must be monitored and categorized. These are the considerations that allow us to decide which criteria are dominant within a scale-sensitive system of adaptive management. The advantage of this system is that it at least holds out the possibility of integration of policy across several

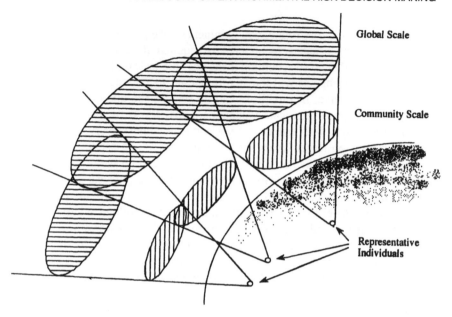

Figure 5   Scalar Pareto Optimality. Decisions based on multiple criteria can be integrated within a scalar approach by applying multiple criteria at different scales,
relying on the Scalar Pareto Optimality principle, which is applied from each
local position. Good decisions made from such positions will have positive or
neutral outcomes for all three scales.

levels or scales. In the ideal case, policies will be chosen that have positive
impacts on all scales or, failing that, the choice will be policies that have
positive impacts on one level, and neutral impacts on other levels, according
to the decision criteria applicable on those levels.

   This approach can be stated as fulfilling the meta-level criterion of "Scalar
Pareto Optimality". According to this criterion, favored policies are ones that
have a positive impact on all three levels in our system of values. (1) They
contribute to economic activity (measured in human welfare) — they improve
the lot of a representative, individual, economic actor; (2) they contribute to
increasing integrity and health on the ecosystem level; and (3) they reduce
impacts on the large-scale physical systems that provide the global context of
adaptation. If no such policies exist, choose those policies that have a positive
impact on the local level, and neutral impacts on goals formulated at higher
levels. Figure 5 represents a three-level application of Scalar Pareto Optimality.*

   While I have provided general guidance regarding which rules should
dominate in which contexts, there remains an inevitable question what is to be

---

* See Norton, *The Varieties of Sustainability,* forthcoming for a more detailed explanation of the
scalar Pareto optimality meta-criterion; See Norton and Hannon (under revision) and Norton and
Hannon (under review) for an explanation of why local values must take precedence in a democratic society.

done when all opportunities to pursue goals on one level have negative impacts on other levels. Good overall management should reduce these situations to a minimum, but in situations with rapidly growing populations and concentrated technological means to alter natural systems and landscapes, there will inevitably exist some tradeoffs. The best rule, here, is to choose that policy path that has positive economic impacts and reduces negative spillovers affecting goals at higher levels of the system to a minimum.

CONCLUSION

I have argued for two basic, though negatively stated, principles to guide our understanding of environmental risk. First, I have denied that there can be a single, unidimensional accounting system that will capture all of the aspects of real-life risk in a single dimension. I have not, from this principle, concluded that risks cannot be rationally analyzed, but rather that they should be analyzed within a variety of systems. I have tried to avoid leaving the decisions as to which system of analysis to use in particular situations as a subjective or nonrational decision, because I believe careful attention to the scale at which a human value is manifest in space and time can tell us which criteria should be applied in given situations. Because there exist no methods for predicting changes in individual welfare as a function of changing states of ecological and physical systems, a second principle was forced upon us: the values protected when we reduce the risk of damage to ecological functioning must be expressed in a more complex and scale-sensitive system of value analysis than nonscalar welfare measures expressed as aggregations of present values.

If these principles stand, it follows that the expansion of risk analysis to include ecological risk will necessarily require a major rethinking of the moral foundations of risk analysis itself. If ecological risk analysis is to be implemented in the foreseeable future, it will have to introduce some nonwelfare and nonutilitarian values and principles into the analytic framework of risk assessment. Besides calling into question the dichotomy between risk assessment and risk management, this conclusion requires a complex and pluralistic value theory to inform a broadened, ecologically sensitive, conception of risk.

Building on these conclusions, a multitiered decision model and a corresponding scalar classification of human values was proposed. This model applies multiple first-order criteria according to meta-criteria based upon contextual information regarding the temporal and spatial scale at which a risk unfolds. These risks to ecological processes are thus risks to human values which occur on an individual-value scale, on a community-value scale, and on a species-value scale. Although this tripartite classification will undoubtedly require refinement, we can say that an ecological risk is a risk that is manifest at the ecosystem level as unusually rapid changes in the states of the system. Further, these changes threaten the community-level values, the aspirations and "cultural capital", which emerges when a human culture evolves within a specific ecological context. The solution suggested is to model system changes at multiple levels, associate social values with various system scales, and pursue policies that have positive or nonnegative impacts on goals formulated at each scale.

REFERENCES

Allen, T. F. H. and Starr, A., *Hierarchy: Perspectives for Ecological Complexity*. University of Chicago Press, Chicago, 1982.

Beck, B., "Reintroduction, Zoos, and Conservation," in *Ethics on the Ark: Zoos, Animal Welfare, and Wildlife Conservation*, Norton, B.G., Hutchins, M., Stevens, E., and Maple, T., Eds. Smithsonian Institution Press, Washington, D.C. (1995).

Callicott, J. B., *In Defense of the Land Ethic*. State University of New York Press, Albany, 1988.

Costanza, R., Norton, B.G., and Haskell, B., *Ecosystem Health: New Goals for Environmental Management*. Island Press, Covelo, CA, 1992.

Daly, H. and Cobb, J., *For the Common Good*. Beacon Press, Boston, 1989.

Edwards, C. J., and Regier, H., Eds., *An Ecosystem Approach to the Integrity of the Great Lakes in Turbulent Times*, Great Lakes Fisheries Commissions Special Publication, Ann Arbor, 1990.

Faber, M., Proops, J., and Manstetten, R. "Toward an Open Future," in *Ecosystem Health: New Goals for Environmental Management*, Costanza, R., Norton, B.G., and Haskell, B., Eds., Island Press, Covelo, CA, 1992.

Harwell, M., Gentile, J., Norton, B.G., and Cooper, W., Ecological significance, in *Ecological Risk Assessment Issue Papers*, United States Environmental Protection Agency, Risk Assessment Forum, Washington, D.C., 1994.

Lee, K. N. *Compass and Gyroscope: Integrating Science and Politics for the Environment*. Island Press, Covelo, CA, 1993.

Leopold, A. *A Sand County Almanac*. Oxford University Press, Oxford, 1949.

Lind, R. C., *Discounting for Time and Risk in Energy Policy*, Resources for the Future, Washington, D.C., 1982

Meine, C. *Aldo Leopold: His Life and Work*, University of Wisconsin Press, Madison, 1988.

National Academy of Sciences, Panel on Policy Implications of Greenhouse Warming, *Policy Implications of Global Warming*. Washington, D.C., 1992.

Norton, B. G. Intergenerational equity and environmental decisions: a model using Rawls' veil of igorance, *Ecol. Econ*. 1, 137, 1989.

Norton, B. G., Context and hierarchy in Aldo Leopold's theory of environmental management, *Ecol. Econ*. 2, 119, 1990.

Norton, B. G., *Toward Unity Among Environmentalists*, Oxford University Press, New York, 1991.

Norton, B. G. Economists' preferences and the preferences of economists, *Environmental Values*, 1994.

Norton, B. G. Ascertaining public values affecting ecological risk assessment, in *Ecological Risk Assessment Issue Papers*, United States Environmental Protection Agency, Risk Assessment Forum, Washington, D.C., 1994.

Norton, B. G. Ecological integrity and social values: at what scale? *Ecosystem Health,* 1995a.

Norton, B. G. Evaluating ecosystem states: two competing paradigms, *Ecol. Econ.,* 1995b.

Norton, B. G. Reduction or integration: two approaches to environmental values, in *Environmental Pragmatism,* Light, A. and Katz, E., Eds. Routledge and Kegan Paul, 1995c.

Norton, B. G. Sustainability, ecosystem health, and sense of place values, in *The Varieties of Sustainability,* Thompson, P. and Dixson, B., Eds., forthcoming.

Norton, B. G. and Hannon, B. Toward a biogeographical theory of environmental values, under revision.

Norton B. G. and Hannon, B. Democracy and sense of place values, under review.

Norton, B. G. and Toman, M., Sustainability: ecological and economic perspectives in the United States, Washington D.C., Jan. 10, 1994.

Norton, B. G. and Ulanowicz, R. E. Scale and biodiversity policy: a hierarchical approach, *Ambio,* 21, 244, 1992.

O'Neill, R. V., DeAngelis, D. L., Waide, J. B., and Allen, T. F. H., *A Hierarchical Concept of Ecosystems.* Princeton University Press, Princeton, NJ, 1986.

Page, T., *Conservation and Economic Efficiency.* Johns Hopkins University Press, Baltimore, 1977.

Page, T., "Sustainability and the Problem of Valuation," in *Ecol. Econ.: The Science and Measurement of Sustainability.* Columbia University Press, New York, 1991.

Rawls, J., *A Theory of Justice.* Harvard University Press, Cambridge, MA, 1971.

Solow, R. M., "Sustainability: an economist's perspective," in *Economics of the Environment: Selected Readings,* Dorfman, R. and Dorfman, N., Eds., W.W. Norton and Company, New York, 1993.

Toman, M., "Economics and 'sustainability:' balancing tradeoffs and imperatives," *Land Econ.* 70, 399, 1994.

Walters, C. J., *Adaptive Management of Renewable Resources.* Macmillan, New York, 1986.

Westra, L., *An Environmental Proposal for Ethics: The Principle of Integrity.* Rowman and Littlefield, Savage, MD, 1994.

13   ENVIRONMENTAL ETHICS AND HUMAN VALUES

Douglas MacLean

CONTENTS

Introduction 177
A Role for Philosophy 178
Public Participation and Moral Inquiry 178
Three Views About Environmental Values 180
     Anthropocentrism 181
     Biocentrism 182
     Ecocentrism 184
Intrinsic Value vs. Valuing as an End 188
Nature and Morality 191
References 191

INTRODUCTION

     People have strong environmental preferences, which often conflict and seldom can fully be satisfied, but that is not the crucial point. It is not the strength of environmental preferences, but the nature of people's attitudes that matters. People tend to care about the environment in special ways. They regard nature as sacred, or they think of wilderness and species as moral entities, which have rights and impose certain kinds of duties on us. The problem for the theorists, analysts, and administrators of environmental policy is to understand and figure out how to respond appropriately to these attitudes.

     Why are we so selective in our concerns for environmental risk? Why do we care so much about certain small environmental risks and ignore many large ones? Why do we care at all about preserving some remote area of wilderness or saving a species whose existence few of us knew about until we learned that it was threatened with extinction? I suspect that there are some very deep-seated

1-56670-131-7/96/$0.00+$.50

needs and fears involved, which we have hardly begun to comprehend, and I am sure it would be beneficial to devote more effort to understanding these matters. At the heart of what people say about these things, however, are claims about values, and we must also try to understand and assess these claims. This is the subject matter of environmental ethics.

A ROLE FOR PHILOSOPHY

I am suggesting that philosophy has a role to play in the design of environmental policies, for the nature and justification of values is an important issue here, and this is a philosophical subject. However, it is not at all clear what the role of philosophy might be, for surely progress in moral theory or environmental ethics alone cannot be expected to solve many environmental problems. Philosophy has no privileged position for determining how to deal with environmental uncertainties or to make the relevant trade-offs, nor can it tell us what weights to give competing interests, especially at the margins. Philosophical training may help to clarify some issues and bring a degree of objectivity and rigor to our discussion of values, but these advantages should not be overemphasized. Other disciplines have as much to teach as philosophy about how to think clearly about these subjects.

Philosophy examines the nature of values and how they connect to reasons for action. These sorts of issues are often somewhat distant from the urgent decisions that dominate the political process. Some of the most important contributions philosophy can make in our thinking about the environment may not be directly related to any practical decisions, and it would be a mistake to insist that they must be. They are, as Bernard Williams puts it, "reflective or explanatory considerations, which may help us to understand our feelings on these questions, rather than telling us how to answer them."[1] Williams also says, correctly I think, that philosophical considerations should be seen as joining political discussion in no special way, but, rather, "in various of the ways in which other forms of writing or talking may do: ways that include not only marshalling arguments, but also changing people's perceptions a little, perhaps, or catching their imagination."[2]

PUBLIC PARTICIPATION AND MORAL INQUIRY

Everyone will agree that environmental values are at the heart of many policy conflicts and are crucial to environmental risk management. A consensus exists on this point. A second claim for which consensus seems recently to have emerged is that we must have greater public participation in the decision-making process. These claims are often linked. Recognition of the importance of values in this area is often cited to support greater public participation in the policy-making process. These claims are sometimes taken to be equivalent, or at least it is frequently thought that supporting greater public participation is a sufficient way of taking environmental values seriously.

I want to call attention first to a possible inconsistency in this way of thinking. It seems to me that policy analysts are frequently skeptical about the objectivity of environmental values, and support for greater public participation in the policy making process flows from this skepticism, because for the skeptic there is no rational way an analyst or policy maker can decide which values are right.

The skeptic thinks that claims about moral duties or environmental rights reduce to expressions of individual preferences, and it is therefore a mistake to believe that value claims may be true or false in the way that empirical claims are. So when the skeptic agrees that values are important in making and winning acceptance for environmental policies, he means that the environmental preferences of different individuals and groups are strongly held and are important to the people who hold them. If there is no rational or objective way to determine which claims are right or true and which are wrong or false, then perhaps the only way to take these preferences seriously is to create a forum in which they can be expressed and weighed according to the strength with which they are held. This is how skepticism about values can lead to combining the recognition of the importance of value claims with support for greater public participation.

Of course one does not have to be a moral skeptic to support greater public participation in the policy-making process. Someone who is not skeptical might believe that a more inclusive decision-making process will reduce the adversarial nature of many disputes and break stalemates that prevent some environmental improvements from being made. This would be an important goal, even for someone who believes in the intrinsic moral value of mountains and forests. A more open process of decision making might also generate new ideas and new approaches for solving difficult problems, which would lead to reducing pollution, improving health, and protecting the environment. One might thus support greater public participation for these kinds of pragmatic or instrumental reasons.

One might also support greater public participation in environmental policy making by appealing to democratic principles. If the moral skeptic appeals to democratic principles to support public participation, then he has to worry about the consistency of his position, for he would seem to be appealing to the kind of objective values whose existence he wants to deny. This inconsistency seems to me to be common in discussions of environmental risk management. Many policy makers and analysts seem to be skeptical about the objectivity of moral values, but they support greater public participation on grounds of democratic principles. However, then we need to know why political values may be objective, if moral or environmental values are not. The person who tries to combine moral skepticism with democratic principles cannot appeal to the idea that most people would prefer participatory or democratic procedures as a method of decision making, because that merely shifts the appeal to democratic principles to another level. Why should we do what most people prefer?

In any event, neither pragmatic concerns nor democratic principles address a more basic concern about combining moral skepticism with support for greater public participation in environmental decision making. How should the people who participate in the process think and decide? What environmental values should they support? The skeptic thinks we have no rational grounds for answering these questions, but despite the many disagreements about values that seem to arise in our society and to be resistant to resolution, skepticism about the objectivity of values is not very plausible. It is not a position that many people sincerely believe in a consistent manner. I think very few people believe, for example, that it makes no rational difference what a person wants, or that no preferences are subject to rational criticism. How many people would be willing to take that attitude toward *any* preferences whatsoever, regardless of how disgusting or repellent the object of preference strikes us as being? And how many people sincerely believe that if someone you cared about — one of your children, for example — were perplexed, troubled, and in need of advice, all you could honestly tell her is that it makes no rational difference what she wants or does, or that although you might have your own preferences and opinions about the matter, they are nothing more than that and cannot be supported by reasons that she should share?

This is not to say that if we reject moral skepticism, we must believe that objective or rational considerations can settle all questions about values, or that we should be able to rank all values objectively on a single scale. The objectivity of values is compatible with incommensurabilities of values and irreducible pluralities of ends. The rational ineliminability of some conflicts is not a threat to objectivity. We might not be able to tell our children, for example, whether pursuing a career in commerce is objectively better or worse than devoting a life to artistic or intellectual interests, but this does not mean that there are no lives that, without any hesitation, we would declare to be bad and try to steer them away from, not merely because we do not want our children to lead certain kinds of lives (as perhaps you would not want your son to go off to sea or your daughter to join a rock band), but because we think some lives are bad or evil, such as lives that encourage dishonesty, viciousness, or the degradation of oneself or others.

Keeping these facts in mind, we may proceed to examine our environmental values and the moral claims made on behalf of nature, without worrying about moral skepticism or assuming that public participation is the end, rather than the first step, in the process of taking environmental values seriously.

THREE VIEWS ABOUT ENVIRONMENTAL VALUES

Our concern here is with what it is rational to think about nature. How should we understand environmental values? I will approach this question by considering three different views that have dominated the discussion of environmental ethics. My description of these views will necessarily be sketchy, and it will be further skewed in order to illustrate what I believe is most unsatisfactory about each one.

Anthropocentrism

The first is an anthropocentric view about moral values that is sometimes associated with a simplistic economic cast of mind. We can characterize this view by saying that only human interests have intrinsic moral value. Interests are usually interpreted in terms of preferences to be satisfied, and (at least in older versions of this view) we measure satisfaction in terms of welfare or happiness and the absence of suffering. Anything other than the satisfaction of interests or preferences has at most only instrumental value, or is valued only as a means, to the extent that it contributes to human happiness and the reduction of suffering. The first head of the U.S. Forest Service, Gifford Pinchot, was a bold and colorful exponent of this view. He said, "There are just two things on this material earth, people and natural resources." And he declared that the object of our policies should be "not to preserve the forests because they are beautiful ... or because the are the refuges for the wild creatures of the wilderness ... but ... [for] the making of prosperous homes."

William Baxter, a contemporary defender of this kind of anthropocentrism, traces the implications of this view for environmental policies. Baxter interprets what we value most highly as that which yields the greatest human satisfaction, so, he says, "damage to penguins, or sugar pines, or geological marvels is, without saying more, simply irrelevant."[3] He rejects the idea that we ought to respect nature or preserve the environment "unless the reason for doing so, express or implied, is the benefit of man."[4]

Critics of this kind of anthropocentrism are quick to claim that our environmental problems today, which also pose threats to human health, result in part from a long tradition dominated by this way of thinking about the environment as a commodity to be consumed, whose only value is to promote human welfare. We have in the past carelessly regarded nature as a vast set of resources to be exploited for our pleasure, but we are now too numerous, and our technologies too powerful, to survive much longer with this attitude. This kind of anthropocentrism, according to its critics, contributes to our problems. We need to think about nature in a radically different way.

Whether or not it is appropriate to blame anthropocentrism for our environmental problems, there are serious philosophical objections to the kind of view that Pinchot, Baxter, and others defend. It simply fails to do justice to the way we live our lives and experience our values. We value as an end not only our own happiness but many other things as well. This is how we value friends and loved ones, objects of beauty or historical and symbolic importance, and this is how many people value parts of nature. The way we value these things may have little or nothing to do with a desire for our own welfare or happiness, and to the extent that these things do connect to happiness, it is because we value them, rather than the other way around.

I can illustrate this point with a simple thought experiment. Think of something you value, for any reason whatsoever, and ask whether you would be willing to give it up or substitute for it something that would give you the same or more happiness. The things that we are most willing to give up for an increase in happiness are things we value instrumentally or merely as means. They are the things we value not in themselves but for the happiness or the reduction of suffering they bring us. However, there are a lot of things that we would not be willing to substitute for an increase in happiness, things that we would be reluctant to give up, exchange, or replace. This reluctance is often due to the fact that we value these objects for what they are or represent, and not because of the happiness they produce, either in ourselves or in others. This is what it means to value something as an end and not merely as a means.

Now, it is important to emphasize here that I am not making a metaphysical claim about the existence of values *in the world*. It is not a claim about what has intrinsic moral value, but rather about the nature of our moral experience. It is a claim about the different ways we value things. The anthropocentric view I have described cannot give an adequate account of these different ways of valuing. This is the problem with any view that sees human welfare as the only valued end and the only reason for acting. This view cannot explain large parts of our moral life.

Biocentrism

The second view I will consider rejects anthropocentrism and attributes moral value not only to humans but to a wider range of entities. It calls for "biotic justice", which requires a form of moral reasoning that will give equal or due weight to the interests of all living things.

It should be mentioned that some of the important traditional moral theories also reject the kind of narrow anthropocentrism described above. Even Kant, who is usually singled out as one of the primary defenders of anthropocentrism, would not identify human welfare as the only thing valuable as an end. Kant locates moral worth, not in human nature per se, and even less in human welfare, but in reason. He does at one point state the categorical imperative as, "Act in such a way that you treat humanity, whether in your own person or in the person of another, always at the same time as an end and never simply as a means."[5] But he explains this by insisting that "man, and in general every rational being, exists as an end in himself."[6] So any priority for humanity on Kant's view is due to the contingent fact that on this planet human beings are the only rational beings we know to exist. That is, if it turned out that computers, Martians, whales, or apes were capable of rational thought, then for Kant they would have to be treated as ends too, and not merely as means.

The classical utilitarians, for whom happiness was the highest moral good, would also have rejected the anthropocentric view described above. They locate the grounds for moral worth in sentience, the ability to experience pleasure and pain, and sentience is surely shared by at least the higher animals. Even if these animals are not rational and cannot act as moral agents, the classical utilitarians would insist that their experiences deserve moral consideration, just as do the experiences of infants and other humans who lack reason but are capable of suffering.

However, neither Kantian nor utilitarian reasons for expanding the class of beings who must be treated as ends or whose interests deserve moral consideration will help the cause of environmentalism, because these reasons are too selective as criteria of moral worth. They do not allow us, for example, to include plants, lower animals, species, or ecosystems as entities whose interests we are morally required to take into account, and a concern for these kinds of things is fundamental to environmental ethics.

Thus, the defenders of biocentrism reject both Kantian and utilitarian theories. They regard both the ability to reason and the capacity to experience pleasure or pain as arbitrary criteria for moral considerability, and they appeal instead to an interest principle, which has a far broader scope. The argument for this principle is clearly stated by Kenneth Goodpaster, an early and influential proponent of this way of thinking. He claimed that "nothing short of the condition of *being alive* seems to me to be a plausible and nonarbitrary criterion" of moral considerability.[7] For something to have a claim to moral treatment, according to Goodpaster, it is sufficient for it to have interests and be capable of being benefited or harmed. Thus, he concludes, it is arbitrary to exclude anything that can be said to have interests from the realm of things with intrinsic moral value.

Now plants have interests, clearly enough, and we can treat them in ways that cause them harm or help them to survive and thrive. They have needs, even if they are not subjects of experience. So, according to Goodpaster, plants deserve moral consideration. He concludes that "the interest principle either grows to fit what we might call a 'life principle' or requires an arbitrary stipulation of psychological capacities (for desires, wants, etc.) which are neither warranted [by the argument] ... nor independently plausible."[8]

I believe that biocentrism is not a reasonable foundation for morality, however, and the argument supporting it is flawed. The problem can be seen in the way Goodpaster assumes that the "interest principle", which is surely a necessary condition for a being to have intrinsic moral worth, is also a sufficient condition. Notice first that even Goodpaster's expansive approach to moral worth might not allow us to assign noninstrumental value to rivers, canyons, mountains, deserts, or other nonliving things, so it will fail to capture the environmental concerns of many of us. Why can't we also include these things as being morally considerable by appealing to *their* interests? Surely a desert or a river can thrive or be harmed; it has needs that we can understand and interests that can be represented when we deliberate about our actions or policies; it can be a beneficiary of our actions; and so on. So it seems that biocentrics like Goodpaster are themselves being arbitrary if they restrict moral considerability by a criterion of being alive.

This would appear to be a *reductio ad absurdum* of biocentrism, but it also points to other unacceptable implications of this view. For if it is arbitrary not to extend the interest principle to plants and even rivers, then it would also seem arbitrary not to extend it to other kinds of entities, including collective entities like nations or firms. They too have interests and the capacity to thrive or be harmed, and even, in a sense, "lives" of their own. Are we unjust if we do not balance their interests against our own? To what extent should the welfare of individual humans be sacrificed for the interests of a nation or a corporation? Some philosophers who have noticed that questions like these are naturally raised by biocentrism have responded by calling these views "environmental fascism."[9] Whether or not that is a fair label, the reason for being alarmed is justified.

Biocentrism is implausible for other reasons as well. Consider a gardener who grows vegetables — organically, let us say, and with great care — but who caters to contemporary culinary tastes. Among his specialties are baby carrots and baby zucchini squash, the latter picked carefully with the flowers still attached. Now this gardener surely thwarts the interests of the living vegetables in his garden. If he continued to care for them, they would grow and eventually reproduce. He cuts them off, so to speak, even before their prime. Their interests, according to biocentric views of ethics, must be determined by their "natural" life, and although it is not entirely clear what that means, it must mean at least a life that extends to full maturity. Are the gardener's practices morally bad? Is he perhaps committing only a very trivial moral wrong which is more than balanced by the pleasure he brings to his customers? Surely it is absurd to think that he is doing anything wrong at all by picking his vegetables young. Having a life cannot be a sufficient condition for giving carrots and squash claims of justice against the gardener.

Ecocentrism

The third popular view, which is also nonanthropocentric, is based on Aldo Leopold's idea of a "land ethic". We might distinguish it from the biocentric view just described by calling it an ecocentric approach to ethics, but this may be somewhat misleading. The purpose of the land ethic is not so much to cast the net of intrinsic value or moral considerability still further, so that it brings rivers and mountains, as well as animals and trees, within the scope of individuals with interests to consider and claims to make. The point is rather to suggest a basis for a more communitarian and holistic perspective on ethics from the start. The land ethic takes its methodological lesson from ecology, with its holistic approach to understanding how biotic systems survive and change, and it suggests that our ethics and perhaps our values generally should be guided by a similar perspective. Leopold says that along with animals and plants, and the earth, water, and air around us, we are "all interlocked in one humming community of cooperations and competitions, one biota."[10] Defenders of the land ethic also tend to emphasize what we can learn from Darwin about our natural origins, the complex interactions among species and between species and their environments, and the evolutionary determinants of norms and ethics themselves. The holistic perspective of the land ethic is suggested as a complement or antidote to the individualistic methodologies (from economics?) that have dominated much of moral philosophy in the past.

The guiding principle of the land ethic is Leopold's claim that "a thing is right when it tends to preserve the integrity, stability, and beauty of the biotic community. It is wrong when it tends otherwise."[11] Many people are attracted to a view like this, and not unreasonably. It can be salutary, as well as awe-inspiring, to contemplate one's self as simply one small part of nature, in which, like other living entities, we struggle as individuals or species to carve our own niche, exploiting our environment, but remaining utterly dependent on it as well. The question I want to press about the land ethic, however, is whether it is a perspective that can provide us with a solid foundation for moral reasoning. What can it tell us about justice, or about trade-offs and priorities? How, for that matter, are we guided by Leopold's basic principle itself? How, that is to say, should we understand notions like the integrity and the beauty of the biotic community? Is integrity greater when the marks of humanity and civilization are strong and dominating, or when they are almost invisible? How do we understand beauty from this perspective, rather than in more familiar anthropocentric terms?

Why should we care about this community as such? The environmental ethics literature is full of warnings about the damage we are doing to the environment through pollution and overexploitation. These arguments appeal to our long-term interests and our concerns for our descendants and future generations. It is good political strategy to argue in this way, for appeals to self-interest remain the most effective arguments for bringing about changes in behavior; but what these arguments show, if they are sound, is that we have instrumental reasons for protecting the environment. Defenders of the land ethic, and many others who discuss environmental values, are concerned to say more than that.

Leopold writes about a distinctive kind of value at the heart of the land ethic.

> It is inconceivable to me that an ethical relation to land can exist without love, respect, and admiration for land, and a high regard for its value. By value, I of course mean something far broader than mere economic value; I mean value in the philosophical sense.[12]

Defenders of the land ethic, along with many other environmental philosophers, contrast their views with a kind of crude economic understanding of
rationality, similar to the anthropocentric view described above. This crude
view is characterized as an instrumentalist conception of rationality, in which
reasons are restricted to maximizing the satisfaction of one's preferences and
desires. The preferences and desires themselves remain unexamined. The idea
of "consumers' sovereignty", which is popular among many economists, calls
for taking individual preferences and desires as they are given, and it justifies
market mechanisms (as well as intervention and regulation in order to correct
market failures) for determining the allocation of resources that most effi-
ciently and equitably satisfies individual preferences. The objection to this
instrumentalist view of rationality, as we have noticed, is an objection to
treating nature and the environment strictly as means for satisfying these
preferences.

One way to interpret Leopold is to read him as urging us to think differently about these issues, to adopt values that would give greater emphasis to
love and appreciation of what surrounds us and less emphasis to our consumer-
driven desires to satisfy our own preferences. Bryan Norton similarly defends
a concern for endangered species by appealing to the "transformative" effect
such a concern can have in changing our values and guiding us to richer (and
ultimately more satisfying?) ways of living.[13] I am concerned here not with
whether this claim is correct as a matter of moral psychology, but rather to
point out that this line of argument is congenial to anthropocentrism and not a
repudiation of it. It appeals not to the intrinsic value of the land but to the
beneficial effects that certain kinds of attitudes toward the environment have
on people who accept them. These enlightened attitudes, so it is claimed, will
help lead us away from shallow consumerism and toward more satisfying ways
of living.

Many of Leopold's followers would nevertheless reject this anthropocentric interpretation of his views and insist that the land ethic must be read
differently. J. Baird Callicott, for example, sees Leopold as offering an alternative to traditional moral theories. The problem with traditional theories, as
Callicott sees it, is that

> ...[the] standard modern model of ethical theories provides no possibility
> whatsoever for the moral consideration of wholes — of threatened population
> of animals and plants, or of endemic, rare, or endangered species, or of biotic
> communities, or, most expansively, of the biosphere in its totality — since
> wholes per se have no psychological experience of any kind.[14]

He interprets Leopold in the following way: "By 'value in the philosophical
sense', Leopold can only mean what philosophers more technically call 'intrinsic value' or 'inherent worth'."[15] Having intrinsic value, Callicott explains,
means being "valuable in and of itself, not because of what it can do for us."[16]

If we accept this nonanthropocentric interpretation of the land ethic, then we are forced back to wondering how it provides a basis for moral reasoning. Like the biocentric view we considered above, the land ethic also seems susceptible to some inhumane and morally repugnant implications. Of course these implications are not intended or accepted by those who embrace such views. As I have indicated, some writers consider biocentrism or ecocentrism as part of a process of human moral development. In earlier stages of civilization, they would say, we progressed from narrow tribal and family loyalties to the recognition that morality requires treating all humans with equal concern and respect. The idea of moving a step further, from the "family of man" to the biotic community, as Callicott puts it, should be seen as a further stage in our development. Using an image he borrows from other environmental philosophers, Callicott says that, "The biosocial development of morality does not grow in extent like an expanding balloon, leaving no trace of its previous boundaries, so much as like a circumference of a tree. Each emergent, and larger, social unit is layered over the more primitive, and intimate, ones."[17]

Does this image help alleviate our concerns about the inhumane implications of nonanthropocentric views of morality? Even now, the way we are supposed to combine our natural loyalties to family and friends with an impartial concern for the claims of all humans remains a vexed philosophical issue. The demands of impartiality threaten to swallow up the natural concerns we have (which all humans presumably share) for those who are closer to us. At least we have some bases for carrying on a discussion of this issue. We can talk about the role of natural loyalties in a normal person's life and about the alienation that would be involved in giving them up. We can talk about the demands of respect that other people can reasonably make on us, the connection between respect and self-esteem, the suffering involved when respect is denied or needs ignored, and so on. Perhaps philosophers are making some progress in sorting out these kinds of conflicts between the demands of impartial morality and our natural, partial concerns for people closer to us, but even these inner rings on the tree of moral development are not so clearly defined, the one merely enlarging and adding to the other. And what about the newer layers? How are we supposed to balance our concerns for family and friends, or even our concerns for humanity or the suffering of all sentient creatures, against the interests of the larger social or biotic unit, which is not a subject of experiences at all? Should we be willing to sacrifice our loyalties to family and friends, or even our impartial concerns for justice for all, in order to advance the interests of larger units, like a nation or an ecosystem? What reason is there for thinking that an entity that is not a subject of experiences can have intrinsic moral worth? Why should we give up anything at all for the interests of these entities? Why deny to a single individual the lumber he needs for a home — or even for a yacht — for nature's sake, if indeed it makes sense to attribute a "sake" to nature at all? The land ethic appeals to some intuitions that most of us share, I think, but it offers very little help in understanding our values in a useful way.

INTRINSIC VALUE VS. VALUING AS AN END

I have described and criticized one anthropocentric view of environmental values and two alternative nonanthropocentric views. The fundamental problem with each of these views, in my judgment, lies in a confusion they share, which is a failure to distinguish between the different ways we value things and the different kinds of value a thing may have.[18] If we become aware of this confusion and learn to avoid it, then I believe we will be on the road to a more satisfactory way of understanding our feelings about nature and environmental values.

Leopold says that the land ethic requires that we love, respect, and admire the land. This is to value the land in a certain way. No doubt this means valuing the land as an end, not merely as a means for satisfying our preferences. But we can value something as an end without insisting that it has a certain kind of value, whether we call it intrinsic value, philosophical value, or something else. We can value things as ends — that is, we can love, respect, or admire them — without insisting that they possess some moral value property, so to speak, as part of their nature.

Consider an example. Most people value automobiles in an instrumental way, as an efficient and comfortable means for getting around. They would be willing to trade their automobiles for better ones or for better ways of getting around, thus passing the exchangeability test for instrumental values that I described above. However, some people come to love automobiles and even collect them. For these people, it would be wrong to say that they value automobiles merely as a means to something else. The automobile collector might work and save in order to buy a 1956 Corvette. That is the thing he values as an end, and he orients other parts of his life as a means for getting it. Once he has the desired car, he will not be willing to give it up or trade it for something else.

People come to love art, antiques, wine, baseball cards, historical relics, and many other things, in similar ways. It is wrong to say that these things must be valued merely as a means for getting something else. Think of the joy of a child who uncovers an arrowhead or an Indian head penny. If it is tempting to say that these things have only instrumental value, this is because we seem to be reluctant to acknowledge that we might reasonably value something as an end that does not itself have intrinsic moral worth, but this reluctance may not itself be reasonable.

We can use a different thought experiment to test for intrinsic moral worth, which follows an idea of G.E. Moore.[19] We can ask of any object whether a world containing that object (and nothing else) has greater moral value than a world without it. We would not say that of baseball cards in a world without baseball; we would not say that of automobiles in a world without human communities built in ways that require movement over long distances; we would not say that even of Beethoven symphonies in a world with no atmosphere for transmitting sound or without any creatures who have a sense of hearing — but of course we know that in our world people value each of these things as ends and can love and admire them.

Both of the nonanthropocentric views we have examined begin by insisting that we do or should value nature as an end, not merely as a means — and then, because we should value nature in this way, they insist that we must regard nature as having intrinsic value or moral properties that are not dependent on humans or human nature. The anthropocentric view we examined also makes this same connection, but in the other direction. It begins by insisting that the only things that can have intrinsic moral value are beings that are the subjects of experience. In fact, the view we described locates the source of this value in the experiences of happiness or suffering themselves. The anthropocentric view described above insists that if intrinsic value is found only in these experiences or in beings capable of having them, then these experiences or beings are the only things that can be valued as ends in themselves, and everything else must therefore be valued only instrumentally as a means.

This is a mistake. We can of course distinguish between two ways of valuing a thing. We can value it in itself, as an end, or we can value it instrumentally, as a means toward something else (which must of course be part of a process which ends up with something valued as an end). To say that something has intrinsic value, however, is to talk about the kind of value a thing may have. Of course the kind of value an object has may have implications for the way it should be valued. We might say that it is wrong to treat other people merely as means because people have intrinsic moral worth; but it does not follow that if something does not have intrinsic value, then it can be valued only as a means. The contrast to intrinsic value is extrinsic value, which can mean conferred or "borrowed" value. Thus, a religious relic might be an otherwise ordinary material object whose value is conferred on it by God or by a religion's history and teachings. The value is not intrinsic to the wooden or cloth object, but the relic is an objectively valuable object and should not be valued merely as a means.

Most of the things we value "in themselves" are valuable in this way. Their value is conditioned on facts about their context: features of the world, the object's history, and especially facts about us, that is, about the kinds of beings we are, the things we are able to appreciate, and the reasons we have for appreciating them. It does not help us very much to insist that the source of value must be in the objects or in us. In most cases, the account we give of an object's value will necessarily have a lot to do with us. This does not make values subjective or a matter of choice, nor does it suggest that an object whose value is conditioned or nonintrinsic in this way can be valued only as a means. The struggle with anthropocentrism and its denial, in the end, is not very illuminating. The real issue is about how different objects should be valued, or what is a reasonable and appropriate way to respond to different valued objects.

Bernard Williams correctly observes that "conservation and related matters are uncontestably human issues, because, on this planet at least, only human beings can discuss them and adopt policies that will affect them. That is to say, these are inescapably human questions in the sense that they are questions for humans."[20] Surely nobody would deny that. Williams also notices that this implies something further, "that the answers must be human answers: they must be based on human values, values that human beings can make part of their lives and understand themselves as pursuing and respecting."[21] In the debates I have been considering here, this further claim seems to tip the scales in favor of anthropocentrism — and it does, in a sense, but not in a way that has any important consequences for these debates. For the content of our values includes many things that go beyond humans and their experiences, and it leaves open the question of whether and in what ways the satisfaction of our own preferences should be favored over other values in our policies.

Sorting out the confusion between the ways we value things and the kinds of value a thing may have will not by itself resolve all our questions about environmental values, but it will help to clear the way for more useful attempts to understand them. We can talk about respecting nature and seeing it as something more than resources that provide the means to our satisfaction, without worrying about how we know that nature has some property of intrinsic value and whether this value is an extension of or is incompatible with other human interests. We can better understand the skepticism and dissatisfaction with analytic methods for establishing priorities that are based on measuring the strength of different preferences to be satisfied. We can insist, that is, that measuring the strength of preferences gives a distorted account of values, without implying that some values have their source outside human interests; and we can better understand our duties to future generations if we do not have to ask whether these duties have their source in our own preferences or in the claims that members of future generations can make against us. The first alternative makes these concerns merely a subjective matter that is contingent on our having some specific kinds of altruistic concerns, while the second is not coherent if our actions and policies determine the identities of the individuals in future generations.[22] We would do better to understand how the value of future generations is conditioned by our other values. We care about many things in ways that go beyond the satisfaction they bring us, and we care about some of these things in ways that show a commitment to their surviving our own lives and even the lives of our children. So it is in understanding the way we value some things that we can hope to understand why other things have any value at all.

Perhaps nothing has intrinsic or unconditioned value. Perhaps all values are human constructions, just as all of science and mathematics is a human construction. This does not mean that values are not objective or that some reasons for caring about different things and valuing them in different ways are not better than others. The fact remains that we value some things as ends, and not merely as means, and we appeal to reasons to explain and justify these attitudes. These reasons in turn refer to cultural and natural facts, and many of these facts will be contingent. What matters, however, is not that these facts are contingent or that values are conditioned by them, but that we can share and be moved by the reasons for valuing something as an end. To appreciate environmental values is not to learn to perceive some morally compelling property of animals, plants, or ecosystems, but simply to feel the force of the reasons for admiring and caring about them.

NATURE AND MORALITY

As a final thought, I would suggest that we might also question whether the value of nature is best seen as a moral value at all. Nature has traditionally been viewed as an alien and threatening force. It has been valued as something different and apart from our moral community. It has been spoken of as sublime, awe-inspiring, and a challenge that tests our own human strength and courage. I believe that the concern to reject anthropocentrism and find intrinsic value in the environment is in part a concern to "moralize" our relationship to nature. It is an effort to move us away from seeing nature as alien and valuing it for that reason, to seeing it instead as an appropriate object of our love and concern. Do not see the environment as threatening us; see instead how we now threaten the environment. Do not see nature as an alien force, testing our courage but rewarding our efforts to tame it; see it instead as a fragile and delicate object, upon which we utterly depend, but which also needs our care and support. It is no longer enough to respect nature; we must learn to love and nurture it as well. Perhaps this is a rational response to a shift in the balance of power, as our population increases and our technologies become more sophisticated and powerful. However, something important will have been lost if we must give up the image of nature as something to be tamed. The development of environmental ethics, leading to biocentrism and ecocentrism, insists on the need for a change in our moral outlook. It defends environmental values and poses a challenge to our traditional moral outlooks — but it has mistakenly focused the issue on anthropocentrism and its denial. The real issue is about how we form and comprehend attitudes and feelings that we find appropriate to our current situation.[23]

REFERENCES

1. Williams, B., Must a concern for the environment be centred on human rights?, in *Ethics and the Environment*, C.C.W. Taylor, Ed., Corpus Christi College, Oxford, 1992, 60.
2. Williams, B., Must a concern for the environment be centred on human rights?, in *Ethics and the Environment*, C.C.W. Taylor, Ed., Corpus Christi College, Oxford, 1992, 60.
3. Baxter, W., *People or Penguins: The Case for Optimal Pollution,* Columbia University Press, New York, 1974, 5.

4. Baxter, W., *People or Penguins: The Case for Optimal Pollution,* Columbia University Press, New York, 1974, 7.

5. Kant I., *Grounding for the Metaphysics of Morals,* J.W. Ellington, trans., Hackett, Indianapolis, 1981, 36.

6. Kant I., *Grounding for the Metaphysics of Morals,* J.W. Ellington, trans., Hackett, Indianapolis, 1981, 36.

7. Goodpaster, K., On being morally considerable, *J. Phil.,* 75, 310, 1978.

8. Goodpaster, K., On being morally considerable, *J. Phil.,* 75, 321, 1978.

9. See, for example, Aiken, W., Ethical issues in agriculture, in *Earthbound: New Introductory Essays in Environmental Ethics,* T. Regan, Ed., Random House, New York, 1984, 269; and Regan, T., *The Case for Animal Rights,* University of California Press, Berkeley, 1983, 262. For discussion, see Callicott, J.B., The conceptual foundations of the land ethic, in *Companion to A Sand County Almanac: Interpretive and Critical Essays,* J.B. Callicott, Ed., University of Wisconsin Press, Madison, 1987, 186.

10. Leopold, A., *Round River,* Oxford University Press, New York, 1953, 148.

11. Leopold, A., *A Sand County Almanac,* Ballantine, New York, 1970, 224–225.

12. Leopold, A., *A Sand County Almanac,* Ballantine, New York, 1970, 223.

13. Norton, B., *Why Preserve Natural Variety?,* Princeton University Press, Princeton, 1987, chap. 1.

14. Callicott, J.B., The conceptual foundations of the land ethic, in *Companion to A Sand County Almanac: Interpretive and Critical Essays,* J.B. Callicott, Ed., University of Wisconsin Press, Madison, 1987, 85.

15. Callicott, J.B., The conceptual foundations of the land ethic, in *Companion to A Sand County Almanac: Interpretive and Critical Essays,* J.B. Callicott, Ed., University of Wisconsin Press, Madison, 1987, 212.

16. Callicott, J.B., The conceptual foundations of the land ethic, in *Companion to A Sand County Almanac: Interpretive and Critical Essays,* J.B. Callicott, Ed., University of Wisconsin Press, Madison, 1987, 212.

17. Callicott, J.B., The conceptual foundations of the land ethic, in *Companion to A Sand County Almanac: Interpretive and Critical Essays,* J.B. Callicott, Ed., University of Wisconsin Press, Madison, 1987, 212. Callicott attributes the image to Sylvan, R. and Plumwood, V. (formerly Routley, R. and V.), Human chauvinism and environmental ethics, in Mannison, D., et al., Eds., *Environmental Philosophy,* Department of Philosophy, R.S.S.S., Australian National University, Canberra, 1980, 96.

18. My discussion in the paragraphs that follow is greatly influenced by Korsgaard, C., Two distinctions in goodness, *Philo. Rev.,* 92, 169, 1983.

19. Moore, G.E., *Principia Ethica,* Cambridge University Press, Cambridge, 1903; also Moore, G.E., The conception of intrinsic value, in *Philosophical Studies* Routledge & Kegan Paul, London, 1922.

20. Williams, B., Must a concern for the environment be centered on human rights?, in *Ethics and The Environment,* C.C.C.W. Taylor, Ed., Corpus Christi College, Oxford, 1992, 61.

21. Williams, B., Must a concern for the environment be centered on human rights?, in *Ethics and The Environment,* C.C.C.W. Taylor, Ed., Corpus Christi College, Oxford, 1992, 61.

22. See Parfit, D., *Reasons and Persons*, Oxford University Press, Oxford, 1984, part 4.

23. I have benefited from the comments and suggestions of friends and colleagues. I am especially grateful to Talbot Page for many discussions of environmental values over a period of years, and especially for a memorable conversation we had while climbing Mt. Jefferson in New Hampshire. I am pleased also to acknowledge the particularly helpful comments I have received from Jennifer Welchman and Susan Wolf.

14   MORAL VALUES IN RISK DECISIONS

James A. Nash

CONTENTS

Introduction 195
The Moral Foundations of Science 196
Moral Values in Risk Decisions 198
Expanding the Value Base 201
Distributive Justice 202
Responsibilities to Future Generations 203
Biotic Justice 205
Uncertainty and Moral Decisions 207
A Final Brief for Ethical Inclusiveness 209
Notes 210

INTRODUCTION

This essay is addressed to all my fellow moralists–especially, in this case, to both risk managers *and* risk assessors. Risk assessment and risk management[1] are not exclusively scientific enterprises, neither in present nor potential practice. They are also, inescapably to some degree, moral enterprises. They are not and cannot be value-free or morally neutral. Moral values — and I am concerned only about moral, not aesthetic and other nonmoral values, in this chapter — enter scientific processes of risk analysis at every stage and shape the processes themselves in various degrees. Indeed, moral values ineradicably permeate human life and all its projects; they are not simply accidental and removable additives.

This permeation — along with political and economic considerations, even manipulations — is generally acknowledged in risk evaluation and management. For instance, when scientists counsel caution or calmness about, say, a possible carcinogen, or, indeed, make any recommendations on public policy, they are no longer functioning strictly as scientists (even if done in the guise of "objectivity"), but also as moralists — scientifically informed ones, no doubt, but still ones who have exceeded the bounds of their formal competency. This moral conditioning of science suggests that the public, especially ethicists, cannot leave the subject of risk to technical elites. All of us need scientific data and analyses for sound decision making, but we dare not confuse technical expertise with ethical discernment, or relinquish our obligations to participate in moral decisions to an assortment of scientists, let alone politicians or any other specialists, including ethicists.

As I shall argue, moreover, this moral permeation is evident not only in risk management but also in risk assessment — generally far more subtly but nonetheless inescapably. Moral values are present in all phases of risk assessment — including its motives, purposes, definitions, methods, and assumptions. The danger to scientific integrity and credibility, however, is not in the embodiment of moral values, but rather in invisible values — values which are not made clear and explicit, and in nonviable values — values which are truncated or otherwise ethically dubious, and which may corrupt the scientific process itself. If so, the moral conditioning of risk assessment also suggests some form of public involvement and evaluation, including by ethicists. Evaluations cannot be restricted solely to scientific peer review, though that process is obviously important.

On these assumptions, my purpose in this essay is, first, to offer an ethical evaluation of current risk procedures, focusing on some of the generally hidden and deficient values. Second, I intend to outline a few major value considerations that are usually ignored or underemphasized but are necessary for an ethically adequate approach to risk.

Risk estimation and evaluation have great social and ecological significance. The process is an immensely prominent and important analytical tool for public regulatory policy. Moreover, it is ethically indispensable, since sound ethical decisions depend on a solid empirical grounding. That is why the ethical enhancement of risk procedures is so important.

THE MORAL FOUNDATIONS OF SCIENCE

The prevailing assumption, in both popular and some scientific circles, is that scientific tools can be insulated from ethical concerns. Authentic science is value-free or morally neutral. A sharp fact–value dichotomy is often assumed; the pursuit of "facts" can proceed untainted by value judgments. Science is objective; ethics is subjective — even arbitrary and relative. Science is quantitative; ethics is qualitative. Science is rational and impartial; ethics is emotional or preferential. Science is empirical and experimental; ethics is intuitive and existential. These frequently encountered dichotomies seriously distort both science and ethics.

These dichotomies distort ethics in several ways. For example, they identify ethics in general with particular and often vigorously rejected ethical theories, notably emotivism, subjectivism, and relativism. They often are also expressions of scientism or positivism, which limits relevant data to sensory perceptions — what can be seen or felt, weighed or measured, and which argues that only the particular methods of empirical verification in science can yield valid understandings of reality. They overlook the fact that ethics itself is a rational process. It too is concerned with internal consistency, external coherence, and adequacy or comprehensiveness of interpretation. Ethics in many interpretations is also, broadly speaking, empirical in character, reflecting on the whole of human experience — the existential and scientific, sociological and psychological, interpersonal and political, cultural and ecological — in search of the universal and discoverable norms of behavior which are built into our being and are essential conditions for our optimal well-being as relational or interdependent creatures. As James Q. Wilson argues, "Our moral nature grows out of our social nature."[2] We are ethical or evaluating animals precisely because we are internally social and ecological animals with unique rational capacities. We are forced by our relational nature to reflect and act on questions of the good and rightness in relationships as means of guiding and limiting behavior. In cognitivist theories of ethics, authentic moral values are real, objective, factual, or true — transcending our inevitable mental and social reconstructions of reality. Moral values are not simply subjective, emotional, habitual, cultural, customary, or otherwise arbitrary preferences; they are structured in the nature of human relationships. They are not merely the projections of cultures; they are the preconditions of cultures, enabling cultures to come into being and to thrive. They are the objective terms for living together effectively.

Equally, the aforementioned dichotomies distort scientific enterprises, overlooking their inherent valuational dimensions. Whatever an adequate interpretation of the relationship between science and ethics might be, it seems clear that science cannot be quarantined from ethics. Moral values are the very foundations of science, the preconditions of doing science. The practice of science is impossible without certain moral commitments and truncated without others. Science seems dependent on certain "scientific values" which are really moral values, including: honesty in the selection and interpretation of data, respect for the rules of rationality, trustworthiness and fairness in the community of peers, commitment to the quest of truth, tolerance of interpretive diversity, freedom of inquiry, corrective dissent from prevailing paradigms, cooperation in the search for knowledge, and open communications. Indeed, even the much-celebrated scientific "objectivity" is, as Langdon Gilkey argues, "a moral and spiritual achievement, a triumph of an intense and impassioned subjectivity, an existential commitment of the self."[3] Thus, science rejects strongly, in theory if not consistently in practice, a variety of moral corruptions of the process: personal biases and ambitions, institutional policy prejudices and objectives, cultural assumptions, plagiarism, institutional secrecy, and particularly political and economic manipulation of science — culminating, for example, in "Republican and Democratic theories of genetic chemistry."[4] The concern for scientific integrity and autonomy reflects the moral values of science. Scientists are moral subjects in a moral guild, or else science itself is impossible or corrupted.

Science not only rests on moral foundations, but scientific methods cannot confirm or explain these foundations on which the possibilities of science depend. Any effort to do so simply begs the question. The moral values which are essential for scientific inquiry, such as honesty or trustworthiness, cannot be validated by scientific inquiry, since even the process of validation must assume the validity of these values! The moral values of science are not scientifically verifiable, because they are the presuppositions of all procedures of verification. This moral dependency points to the limits of the scientific method. This method is not epistemologically exhaustive or exclusive, contrary to positivism. There are other modes of experiencing or knowing reality, which are not competitors but complements to scientific inquiry. The validation of the moral grounds of science must rest finally on philosophical arguments — including that discipline dreaded or denied by some scientists: metaphysics. Moreover, this moral dependency of science points to the objectivity of certain moral values (though hardly all). If all moral values are subjective, relative, or arbitrary, then science itself rests on very flimsy foundations. Indeed, these assumptions undermine the integrity of science. If, however, certain moral values are necessary conditions for vital human projects and relationships, then these conditions suggest, at least, that these moral values are objective — factual and true. The reality of these values is validated in part by their "fruitfulness": they are coherent with our experience of human relational needs and aspirations, and pragmatically they enhance our potential for realizing these needs and aspirations.[5]

MORAL VALUES IN RISK DECISIONS

Particular moral values are also reflected in the specific scientific enterprises of risk assessment and evaluation. The very idea of "risk" implies a value judgment. In fact, risk assessments would not be undertaken without assumptions about what is good and bad. The definitions of "costs" and "benefits" are not only economic and other empirical assessments, but also value judgments that one state of being is better or worse than another. Even the argument that risk assessment should be a strictly scientific activity reflects a moral concern for procedural "integrity" — the capacity for full honesty in the whole process. The typical definition or expressed purpose of risk assessment — to find and evaluate "adverse" effects — also expresses a value judgment in favor of human or ecosystemic well-being. Again, the selection of problems to be examined is often a value judgment. For example, a decision to focus research on carcinogens rather than on the effects of toxics on immune or reproductive systems involves not only a political but also a moral judgment about comparative importance for human well-being.[6] And certainly, moral evaluations are part of setting standards for "acceptable" risks or "safe" dosages, deciding whether or not to regulate, comparing and ranking risks, setting budgetary and other priorities for different hazards, and even designing strategies for reducing risks ("reducing" itself reflecting a moral concern). All of these depend on moral assumptions about relative goods and bads for humans and/or lifeforms.

Thus, the well-known and celebrated document, *Reducing Risk*, by the Science Advisory Board of the Environmental Protection Agency, is on solid ground in acknowledging "inevitable value judgments" as part of the process[7] — though, incidentally, I think it is wrong in describing these judgments as "subjective" rather than moral. This document is filled with moral words and evaluations. Indeed, some of its ten recommendations are indisputably moral judgments. For example, the second recommendation states: "EPA should attach as much importance to reducing environmental risk as it does to reducing human health risk," and its tenth states: "EPA should develop improved analytical methods to value natural resources and to account for long-term environmental effects in its economic analyses."[8]

Moral judgments are no less a part of the methods and the choice thereof in risk analysis. For instance, testings on animals — indeed, multiple millions of rats and mice — rather than on humans represents a rejection of biotic egalitarianism and a comparatively higher evaluation of humans over rodents.

Methodological moral assumptions are perhaps most clearly, and deficiently, evident in cost-benefit analysis (CBA). Some form of cost–benefit analysis — the balancing of values gained and lost to realize the "best possible" — seems to be an inherent, although frequently implicit, characteristic of any ethical approach concerned about the consequences of actions; but this approach does not mean that all values can appropriately be quantified, let alone monetarized. In fact, the opposite is often the case. CBA, however, generally involves monetary estimations of costs and benefits in a problem or project on the basis of persons' "willingness to pay" (in fact, however, the "ability to pay," thus unjustly giving preference to higher income parties). As such, it makes a variety of at least dubious value assumptions:

1. Even moral costs and benefits can be converted into monetary amounts,
2. Moral values can be reduced to market values,
3. Moral values are simply subjective preferences,
4. Financial allocations reliably indicate the acceptability of risks,
5. Costs and benefits can be simply aggregated without regard for just distribution,
6. The interests of future generations can be justly discounted or even ignored,
7. Only anthropocentric interests and not the intrinsic values of other lifeforms count, and

8. A society is only an aggregate of individuals, or really consumers — indeed "an aggregate of preferences,"[9] and not also a corporate entity, a common good, in which the values created by interactions are more than the sum of the parts.

Generally, the systemic value assumption of CBA is a simplistic version of utilitarian ethics, but perhaps Mark Sagoff describes it most fittingly as "antinomianism."[10] Sagoff consummates his argument against CBA with a compelling comment:

The things we cherish, admire or respect are not always the things we are willing to pay for. Indeed, they may be cheapened by being associated with money. It is fair to say that the worth of things we love is better measured by our *unwillingness* to pay for them. . . . These things [like love itself, he adds] have a *dignity* rather than a price.[11]

CBA, however, can be a useful method under appropriate circumstances when the issues at stake are largely economic — for example, comparing the gains and losses or the cost-effectiveness of alternative proposals for a given project.[12] Yet, it is inappropriate and insufficient in most cases. At best, it can be used as a supplement to, but not a substitute for, a morally comprehensive risk analysis. Perhaps, however, CBA can be reformed, as K. S. Shrader-Frechette argues, while noting the serious flaws in such quantitative analyses. Perhaps it can be "ethically weighted", for example, by treating inequities as costs, and taking account of different ethical assumptions and systems.[13] That would certainly be an improvement, but not in my view a solution. That still leaves intact some of the basic, troubling moral assumptions of CBA, notably the incommensurable quantification of qualities and the translation of moral values into market terms.

CBA is an effort to avoid arbitrariness, but it is not a successful one. It assumes, rather than justifies, moral values, and thus remains ethically arbitrary. It also adds conceptual confusion to arbitrariness, by measuring qualities, such as the value of a life, otter or even human, in quantitative terms. Beyond the scandalous ways that particular CBAs sometimes have been abused (unscientifically and unethically) by government and industry to misrepresent the full costs of a project in order to protect the interests of an institution intent on promoting it, the moral deficiencies of the method are more than sufficient to explain and justify the widespread suspicions and rejections of CBA.

CBA, of course, is only one method for examining risk. Whatever strengths or weaknesses other methods may have, they share with CBA one fundamental characteristic: they depend on and reflect moral values. The same is true of risk assessment and evaluation in general — indeed, of all activities of the evaluating animal. A strictly quantitative, morally disinterested, or otherwise scientifically "pure" approach to risk is an illusion. Indeed, a major danger to scientific integrity is the pretense of value neutrality. This pretense hides values that must be explicated and themselves evaluated for their consistency and comprehensive coherence. It opens scientific procedures to abuses by political and economic powers intent on using an allegedly objective and impartial science to promote their institutional values.[14] It is a peculiar form of what Christians and Jews have traditionally called the sin of pride, for it assumes that a certain brand of humans can totally transcend the human condition of finitude and the human propensity to egoistic bias — as well as political, economic, cultural, class, ethnocentric, anthropocentric, and similar biases or values. Walter Rosenbaum soundly argues that this situation is "a strong argument for keeping scientific and technical determinations open to examination and challenge by other experts and laypersons...The myth of a socially neutral science should not be unchallenged within the scientific professions."[15]

Moral evaluations are inevitable — even in all phases of scientific risk decisions, and especially when these evaluations are denied.

EXPANDING THE VALUE BASE

If moral valuing is inescapable in the risk enterprise, then it makes a great deal of difference what moral values are assumed, for not all values are ethically equal. Some so-called moral values really are subjective and cultural preferences. Some are ethically indefensible in whole or in part; in some cases, they might more accurately be called immoral or unethical values. Moral values are "moral" descriptively, in the sense that individuals and institutions accept and express them, but not necessarily prescriptively, in the sense that they can be ethically justified and *ought* to be accepted and expressed by individuals and institutions. Indeed, the discipline of ethics can be described in part as the systematic, rational process of evaluating and justifying (or not) moral values. If so, the moral values embodied in the risk process must be ethically evaluated to determine their consistency and coherence — to see whether or to what degree they can be ethically justified. That task, of course, would be a monumental undertaking, far beyond the purposes and possibilities in these few pages. Thus, I can offer only a small contribution to that task here.

My primary complaint against current risk assessments and evaluations is that the values they assume are in general insufficiently inclusive and comprehensive — that is, they fail to give adequate moral consideration to all parties with stakes in an outcome, and they fail to incorporate all relevant moral values. Consequently, in the remainder of this essay, I intend to highlight four value considerations that are often bypassed or underemphasized in risk analysis. Three, in fact, deal with different dimensions of distributive justice. Obviously, I cannot offer an adequate ethical justification of these values here. Instead, I propose them as moral values which seem to a significant minority of us to be ethically justifiable and which warrant consideration in ethically adequate risk decisions. It is not necessary for risk assessors to be committed to these values in order to conduct a scientific inquiry. Yet, since this scientific inquiry is value-laden, it may be necessary for a scientific enterprise that wants to avoid ethical truncation to give full and fair consideration to all defensibly relevant values, or make a case to the contrary.

DISTRIBUTIVE JUSTICE

Much risk assessment/evaluation seems to suffer from an aggregative prejudice. Its focus is on risk as a whole and not how that risk is distributed within a society or among societies. Cost-benefit analysis generally has been aggregative, balancing costs and benefits without giving adequate consideration to who gets what share of the benefits and burdens. Matters of distributive justice, however, are critically important moral matters, and ought to be structured into risk assessment/evaluation, insofar as is possible and relevant in particular cases.

Distributive justice can be defined as the ethical process of comparing, differentiating, and apportioning benefits and burdens, on the basis of morally relevant similarities and differences, and in the midst of conflicts of claims, in order to ensure that all parties with stakes in the outcome receive their due or proper share. Distributive justice is concerned not with "the greatest good of the greatest number", but with a fair apportionment of "goods" and "bads".

If distributive justice is truly just, however, it must include all human beings as relevant claimants in the process. Ethically, the peoples of this planet are an interdependent community of moral equals, bearing moral rights which are grounded in an equal and universal moral status: the inherent dignity or intrinsic value of humans as ends in themselves. These rights are the essential conditions for lives expressive of human dignity. These rights are numerous, but they include basic biophysical needs (for example, adequate nutrition, shelter, health care, etc.), environmental security as an extension of biophysical needs, full political participation in defining and shaping the common good, and social ground rules for fair treatment and equal protection of the laws. These rights are moral claims on others, entailing duties of justice by others in proportion to their capacities and resources. In fact, however, because of limitations on our individual capacities and resources, governments, precisely because they are the agents of our interests, are also the prime representatives of our corporate responsibilities.[16] These human rights, moreover, demand not only national but international justice.[17]

This thumbnail sketch of distributive justice, of course, veils the complexities in the concept, and those complexities are even greater in practical applications of the concept. Since it is not possible to give an adequate account of even a single application here, I will simply list some questions that give the flavor of the issues in distributive justice and that are particularly relevant for risk assessments/evaluations.

Whose costs and benefits are morally relevant and deserve what weight in determining levels of toleration for pollutants and other forms of ecological degradation? Do the economic interests of, say, agribusinesses and chemical complexes take precedence over other contenders? Is the prevention of job losses in particular locales more important than the social and ecological costs elsewhere? Do only regional interests count?

How should the full and fair costs of production and distribution of socially and ecologically hazardous products be assessed? Should they not include the "externalities", the social and ecological effects of an industrial process which entail adverse public consequences and expenditures?[18]

What about the interests of the poor and powerless, particularly in African American and Hispanic American communities, near whose residences toxic industries and dumps are disproportionately situated? Is this not environmental racism and classism? Is this inequity not a deprivation of human rights to equal protection from arbitrary discrimination? What are just distributional standards for the production and disposal of toxic materials? What compensations or reparations are just for the losers in the unavoidable battles of NIMBY ("not in my backyard")?[19]

Do only national interests count? What about the interests of Third World nations which are often used as cheap dumps for the industrialized nations' toxic wastes, which are often the sites of hazardous industries and technologies transferred from nations like the U.S., and which are often the importers of pesticides, like DDT, banned for use in the U.S.? What about the rights of nations that are the victims of transboundary pollution, like acid rain, from other nations? Does risk assessment stop at the water's edge? Do not environmental impact assessments need to consider global consequences in an ecologically interdependent world?

Is "free and informed consent" by workers, communities, and nations a relevant moral consideration for the siting of toxic dumps and industries when the conditions of that consent cannot be satisfied — when, for example, the full risks are largely unknown or the relevant population is insufficiently educated, when genuine alternatives are not available because of economic necessity, or when their bargaining powers are no match for confronting economic and political principalities?[20]

These are among the many questions that risk assessments and evaluations concerned about distributional justice must confront. The task is obviously formidable. Evaluations must be local and global, economic and environmental, immediate, and, as I shall next argue, long term.

RESPONSIBILITIES TO FUTURE GENERATIONS

Risk assessment and evaluation tend now to be focused on the present or the near future. In fact, cost–benefit analysis tends routinely to discount future interests, as if calculating financial depreciation. In this context, the Environmental Futures Project, conducted by the Science Advisory Board of the U.S. Environmental Protection Agency, is somewhat unusual. This project is scanning the "future horizon" to forecast social and technological trends, to assess their environmental impacts, and to plan appropriate responses. It is an attempt to be proactive. The point of the project is not only to predict the future, but to prevent an ecologically debilitated future. One can hardly fault this project. It is imperative and laudable in light of the rapidity of change. Yet, its temporal scope is severely, and perhaps necessarily, limited: the long-range future is defined as a maximum of 30 years, or scarcely more than one generation! This project is only a start in assessing impacts on and exercising responsibilities for future generations.

The ecological crisis has brought the problem of responsibilities to future generations or intergenerational justice into prominence. This is the essence of the emerging moral concept of sustainability. Sustainability is living within the bounds of the regenerative, assimilative, and carrying capacities of the planet, continuously and indefinitely. It seeks a just distribution between present and future generations, avoiding public policies that sacrifice one for the other, and following policies that ensure the ecological conditions necessary for thriving in both the present and future. While we cannot know the precise forms of future generations' needs, we can reasonably anticipate, since they are our biological heirs, the general functions of their needs — for instance, sufficient and safe resources. Sustainability recognizes that while we mortals are by definition unsustainable, we have obligations to our successors, because the cultural and ecological heritage that we pass on will shape them and their possibilities profoundly for good or ill. These responsibilities apply not only to our children and grandchildren, but also to the progeny of all peoples and generations. Future interests cannot be arbitrarily discounted; that is a euphemism for stealing from the future. Any preferential treatment for the present must be justified on the moral criteria of distributive justice — and these can be a heavy burden of proof.

The evidence is compelling — as the current stress on "sustainable development" testifies — that a primary characteristic of present patterns of using the planet as source and sink is *un*sustainability. The moral tragedy is not only the damage done in the present, but the harm caused to future generations. Their interests are being sacrificed for present myopic gratifications. A portion of humanity is receiving generous benefits by living beyond planetary means, while future generations, if many, will bear most of the risks and costs — from nuclear wastes and possible climate change to extinctions and overpopulation. Future generations will be major victims of our generation's excessive production, consumption, toxication, and reproduction.[21]

Sustainability entails major changes in risk assessment and evaluation. It forces us, first, to think prospectively, in formidable ways that almost defy imagination. Risk ranking may, in fact, change significantly when relevant time periods change. For instance, low risks defined in decades, such as nuclear energy wastes or irrigation from nonrenewable aquifers, are high risks if the benefits are in the present but the liabilities are postponed to the far-off future, or if the full costs over time are counted fairly. Again, adaptability or flexibility must be built into our present projects, such as protecting humanly threatened species or ecosystems, to allow for unpredictable, long-term shifts in ecological dynamics such as climate changes, natural calamities, or normal population fluctuations.[22] Sustainability compels us to think in centuries, perhaps even millennia. Our responsibilities seem to extend as far into the future as our influences are significant and plausibly foreseeable. Thus, a commitment to sustainability asks: considering the diverse social and ecological harms caused by humans over the brief span of economic modernity, what responsibilities are demanded of us now for the sake of the genuinely long run?

Sustainability also forces us to think holistically and relationally. For the sake of both present and future generations, assessments of environmental risk must abandon the prevailing individualistic and isolationistic approaches. Cumulative, synergistic, and persistent effects of, say, pollution and extinctions now must carry much greater weight. How, for example, should we deal with a potential problem that may be extremely serious for future generations: the prospect of cumulative catastrophes, in which many single, small hazards, each of which is now "allegedly acceptable," combine into a dangerously unacceptable aggregate over time?[23] With tens of thousands of synthetic chemicals interacting in the biosphere, ecosystems depending on sometimes unknown keystone species, deforestation affecting hydrological cycles and climate changes, and a host of other connections, the problem is neither trivial nor simple. Sustainability must reflect the fact that we live in an interdependent biosphere in which the cumulative activities of our time may have severe, even catastrophic, consequences for the future.

The confounding question to all who are in the risk business is: how must the various forms of risk assessment and evaluation be transformed or restructured to give appropriate consideration to the interests of future generations?

BIOTIC JUSTICE

Risk assessments and evaluations today generally reflect the anthropocentric imperialism dominant in our culture, though, as I shall note, some changes may be emerging. Under the prevailing cultural paradigm, now in technocratic rationales which seem far more manipulative of nature than any historical theological rationales, humankind is an ecologically segregated species, designed for managerial domination and possessed with a virtually unrestricted right to exploit nature's bounty for human benefit. The earth is defined exclusively by human purposes and subject entirely to human "improvements". The moral significance of otherkind is reduced to instrumental values — "renewable resources" or "capital assets" — for human needs and wants, without regard for the fact that these others are also an astonishing diversity of life forms struggling for sustenance and space in complex interdependency.

The consequences of this paradigm have reached new and dangerous heights of casual cruelty and wanton destruction in the 20th century. Animal experimentation — a mainstay of some methods of risk assessment — seems often to be excessive, frivolous, and/or brutal. Genetic engineering displays the lustful potential for treating animals and plants as simply Cartesian mechanisms to be redesigned. The reduction of nonhuman populations and the extinctions of species are proceeding at a rapid pace. The causes of these decimations are many and interwoven. They include multiple forms of pollution of the air, water, and soils; human overpopulation which presses for Lebensraum into wildlands; and habitat destruction and fragmentation for the sake of social and economic "development." The potential for human-induced climate change and ozone depletion threatens a mega-enhancement of current trends. Thus, if nonhuman life has a moral status independent of human purposes, this anthropocentric imperialism is a serious moral deficiency.

In my view, the most defensible alternative to anthropocentric imperialism is the extension of concepts of distributive justice beyond human relationships, present and future. This extension is an effort to redefine responsible human relationships with the rest of the planet's biota, and grounding these human responsibilities not only, weakly, in human utility or even generosity, but also, strongly, in the just dues and demands imposed on us by the vital interests of otherkind.

Distributive justice, as previously noted, requires giving every interested party its "fair share" of the goods necessary for their well-being in accord with morally relevant similarities and differences. Otherkind seem to qualify for moral consideration in this process of apportionment. They certainly have heavy stakes in the outcomes of public policies; indeed, they are vulnerable and powerless in the face of human hegemony. Moreover, they embody what I regard as the minimally sufficient condition for the recognition of moral standing: *conation* — that is, a striving to be and do, characterized by aims *or* drives, goals *or* urges, purposes *or* impulses, whether conscious *or* nonconscious, sentient *or* nonsentient. At this point, organisms can be described as having "vital interests" (in the sense of needs) for their own sakes. Whatever their instrumental values for human interests or ecosystems, they are also intrinsic values, ends or goods for themselves. These intrinsic values provide a sufficient moral basis for at least basic claims for justice from human communities.

The moral issues surrounding this extension of justice to the rest of the biota are, of course, mind-numbing in their novelty and complexity. I have interpreted "biotic justice" at length elsewhere.[24] Thus, it is sufficient here to say that if biotic justice is warranted, all lifeforms, individuals and species, have a *prima facie* right to a "fair share" of the goods, including the habitats, necessary for their well-being. Of course, defining a "fair share" is extremely difficult, especially when humans must destroy other lifeforms and their habitats in order to survive and create in a predatorial biosphere. Yet, it is a concept that we must struggle to define in order to stifle human imperialism over the rest of nature. In this process, we will encounter a variety of moral dilemmas inherent in the dual moral status of other species — as instrumental values for human needs and as intrinsic values for themselves. Still, one conclusion is clear: biotic justice imposes obligations on the human community to limit production and consumption in order to prevent the excessive commodification, domestication, and toxication of wildlife and wildlands. Profligate production and consumption are anthropocentric abuses of what exists for fair and frugal use in a universal covenant of justice.

If biotic justice is a moral mandate, then risk assessments and evaluations clearly must give due consideration to the vital interests of otherkind, not simply as utility values for human interests. The prevailing anthropocentric value-set in these disciplines can no longer be casually assumed. This moral shift will require, for example, a reevaluation of animal experimentation, at least in search of procedures to minimize harm. The estimation of "safe" dosages and "acceptable" risks for various toxins will need to consider the effects on nonhuman populations. Cost–benefit analyses can no longer simply calculate the values of wildlife and wildlands by measuring human economic preferences, because then intrinsic values are abandoned in favor of instrumental values for humans. This process only disguises the anthropocentric moral assumptions of CBA.[25]

Fortunately, there are signs of some extension of moral concern to nonhuman life in some risk assessments/evaluations, particularly in ecosystemic studies. In fact, *Reducing Risk,* by EPA's Science Advisory Board, recognizes the "intrinsic moral value" of natural ecosystems "that must be measured in its own terms and protected for its own sake."[26] That is an important even if undeveloped affirmation. The challenge now to risk assessors and evaluators is to draw out the full implications and applications of that statement for their work.

UNCERTAINTY AND MORAL DECISIONS

Responsible risk analysts often express concern, even alarm, about the great degree of uncertainty that is present in their enterprise — and justifiably so. Data are generally incomplete, inadequate, inconclusive, or even nonexistent — and, of course, disputable. Tools are imperfect and methods depend on debatable assumptions. Errors are inevitable. Different estimates of risk for the same event may vary widely, sometimes in several orders of magnitude. Causal relations are often extremely difficult or impossible to establish; inferences of various levels of plausibility may be the only options. On some major problems, the level of "ignorance" is "astounding and potentially catastrophic."[27] Risk deals only with probabilities, and that in radically different degrees, not certainties. The public and its politicians may want precision and simplicity, but risk assessors usually can offer only complexity and ambiguity — and even if the scientific evidence were full and flawless, decisions would remain uncertain, since they entail both scientific and moral judgments.

This condition of uncertainty, which appears to be as much or more a part of the human condition as of the state of the art, raises a critical and complex moral question: how much and what kinds of risk are justified under conditions of high scientific uncertainty? It is not clear that the risk process has dealt adequately with this question.

In the context of high risk, where the consequences of miscalculation can be severe or even catastrophic, even if the alleged probability may be quite low,[28] the worst moral response to uncertainty may be an indecisive tolerance of the risk, while waiting for conclusive evidence — which probably will not be forthcoming. In these cases, it may become too late to take remedial action; risks may become intolerable costs. In fact, while the appeal for more time to study a problem, to gather more evidence, may be legitimate, it may also be a cynical tactic by an interest group — a chemical manufacturer or a nuclear energy industry, for instance — to postpone or prevent action indefinitely, to preserve the status quo. Thus, Holmes Rolston's counsel to businesses seems equally applicable to the risk business: "Do not use complexity to dodge responsibility."[29] Waiting for surety is somewhat like waiting for Godot: it never comes or it comes much too late. In the meantime, great social and ecological harm may result.

In contrast, the best moral strategy for dealing with high risks in high uncertainty seems to be to proceed with high caution, exercising a consistent, "preferential option" for social and ecological security. This approach suggests that in situations of uncertainty where there is doubt but still some reasonable grounds for fear of adverse outcomes, it is morally appropriate to act decisively to prevent or minimize a potential social or ecological harm, even at the risk of being wrong and rejecting thereby a relatively harmless substance or act.[30] This approach clearly shifts the burden of proof in favor of health and environmental interests. Thus, for example, the burden rests not on public agencies to show that new or even existing synthetic chemicals are harmful, but on the producers to give reasonable and reliable evidence that these compounds are tolerably safe and nonpersistent in the environment. "Chemicals, unlike persons, are not innocent until proven guilty, but suspect until proven innocent."[31] The same point, however, can be made about a variety of forms of environmental degradation, from deforestation to ozone depletion. This position is essentially the same as Shrader-Frechette's argument that "the typical scientific norm dictating behavior under uncertainty is wrong:"

> It is wrong to be reluctant to posit effects such as serious environmental consequences in a situation of uncertainty. Therefore, it is wrong in a situation in which we cannot adequately assess effects to place the burden of proof on possible victims of pollution or development. Instead, we should be reluctant *not* to posit effects such as serious harm in situations affecting human and environmental well-being and the burden of proof should be placed on the persons who create potentially adverse effects — that is, on polluters and developers.[32]

In the midst of scientific and moral uncertainty, we have no other option than to abide by the standard canons of rational plausibility. This requires efforts to prevent and reduce hazardous risks by following the best lights we have, which in particular cases may be little more than reasonable fears and suspicions, or scientifically informed guesses. When the prospects of danger are sufficient, however, plausibility is also sufficient for acting decisively.

Uncertainty is also a call for humility, the antidote to arrogance, the counter to hubris, which is needed as desperately by science as by theology. Humility is the constant reminder that we are only human. Despite the extraordinary intellectual and technical achievements in the 20th century, humility reminds us that our knowledge of the intricate interdependencies of nature is minor and fragmented, our technical ingenuity is limited, our risk estimates are often best guesses, our moral character is ambiguous, our capacities for error are extensive, our follies are frequent, and one of our worst enemies is the denial of uncertainty in outbursts of technical overconfidence.[33] We are neither wise enough nor good enough to control all the powers we can create or to fully comprehend the powers that created us. No human plans or techniques are failsafe so long as humans are relatively free and definitely finite. Humility may yet save us from a lot of unnecessary risks and griefs.

A FINAL BRIEF FOR ETHICAL INCLUSIVENESS

I have suggested in these pages that ethicists need to be involved in the formal evaluations of risk decision-making. However, the field of ethics is nothing close to a monolith. One must ask, ethicists of what worldview or school of thought? Which perspectives should be included or excluded? To improve risk decision models in our ethically pluralistic culture, participation in the process must be broadened, in my view, to include a variety of potential contributors — including some who now seem regularly excluded. For instance, religious ethical perspectives are often excluded from the public process on the grounds that they are not "neutral" but sectarian, confessional, or revelational. Yet, they are excluded in preference for ethical views, such as philosophical utilitarianism, which are no less sectarian or confessional, no less value laden, and no more rational on claims about the "good" and the "right." This exclusion is in itself a value judgment. On matters of public ethics, there is no "neutral" ground, though we constantly need to search for common or shared grounds in order to live together. Religious traditions, no less than various philosophical traditions, have relevant emphases to offer to risk decision making — for example, on human moral capacities, the breadth of justice, and the nature of the good society. All ethical perspectives which offer "public", rationally accessible arguments, rather than epistemologically privileged or parochial ones, and which are willing to engage in civil discourse, ought to be included in what is largely a normative, and not merely technical, dialogue about our communal life.[34]

NOTES

1. Risk assessment usually means the formal, analytical process of identifying and estimating the probability and magnitude of an adverse ("bad") outcome from some environmental hazard or practice, such as a toxic substance or a construction project. In some definitions, it includes evaluating, weighing, and/or comparing the outcomes of different substances or events. In other definitions, however, these latter procedures are included under risk management, the process by which regulators decide whether, to what degree, and by what means an outcome should be controlled based on the probability and magnitude of its effects. These procedures are widely used in industry and government, including in environmental impact statements. They are usually quantitative analyses, but also, in various degrees, qualitative judgments. Cf. definitions in, for example, "the red book" or, formally, Committee on the Institutional Means for Assessment of Risks to Public Health, *Risk Assessment in the Federal Government: Managing the Process* (Washington, D.C.: National Academy Press, 1983), pp. 18–19; Policansky, D., "Application of Ecological Knowledge to Environmental Problems: Ecological Risk Assessment," in *Comparative Environmental Risk Assessment*, C.R. Cothern, Ed. (Boca Raton, FL: Lewis Publishers, 1993), p. 39; Schierow, L.J. "The Role of Risk Analysis and Risk Management in Environmental Protection," Environment and Natural Resource Policy Division, Congressional Research Service, The Library of Congress (Code 1B94036), June 1, 1994.

2. Wilson, J.Q. *The Moral Sense* (New York: Free Press, 1993), p. 121.

3. Gilkey, L. *Nature, Reality, and the Sacred: The Nexus of Science and Religion* (Minneapolis: Fortress Press, 1993).

4. Rosenbaum, W.A. *Environmental Politics and Policy*, 2nd ed. (Washington, D.C.: CQ Press, 1991), p. 162. For an excellent account of political influence on and abuses of science, see Rosenbaum, esp. pp. 145–167.

5. I have benefited greatly from the discussion of these issues in Gilkey, L. *Nature, Reality, and the Sacred*, especially chaps. 3–6, 8.

6. Cf. Misch, A. "Assessing Environmental Health Risks," *State of the World, 1994* (New York: Worldwatch Institute/W.W. Norton, 1994), pp. 117–136.

7. *Reducing Risk: Setting Priorities and Strategies for Environmental Protection* (Washington, D.C.: USEPA, Science Advisory Board, 1990), p. 8.

8. *Reducing Risk: Setting Priorities and Strategies for Environmental Protection* (Washington, D.C.: USEPA, Science Advisory Board, 1990), p. 6.

9. Sagoff, M. *The Economy of the Earth: Philosophy, Law, and the Environment* (Cambridge: Cambridge University Press, 1988), p. 118. For his telling critique of CBA, see esp. pp. 27–123. For another helpful critique, see Kelman, S. "Cost-Benefit Analysis: An Ethical Critique," in *Readings in Risk*, T.S. Glickman and M. Gough, Eds. (Washington, D.C.: Resources for the Future, 1990), pp. 129–136.

10. Sagoff, M. *The Economy of the Earth: Philosophy, Law, and the Environment* (Cambridge: Cambridge University Press, 1988), p. 49.

11. Sagoff, M. *The Economy of the Earth: Philosophy, Law, and the Environment* (Cambridge: Cambridge University Press, 1988), p. 68.

12. See the critique of Barbour, I. *Ethics in an Age of Technology* (San Francisco: HarperSan Francisco, 1993), pp. 223–226; also, Risler, P. "The Use of Economic Data and Analysis in Comparative Risk Projects: Questions of Policy and Reliability," in *Comparative Risk Assessment,* Cothern, C.R., Ed. (Boca Raton, FL: Lewis Publishers, 1993), pp. 247–255.

13. Shrader-Frechette, K.S. *Risk and Rationality: Philosophical Foundations for Populist Reforms* (Berkeley/Los Angeles: University of California Press, 1991), pp. 183–186.

14. Milbrath, L.W. *Envisioning a Sustainable Society: Learning Our Way Out* (Albany, NY: State University of New York Press, 1989), pp. 60, 65, 233.

15. Rosenbaum, W.A. *Environmental Politics and Policy*, 2nd ed. (Washington, D.C.: CQ Press, 1991), p. 151.

16. See Shue, H. *Basic Rights: Subsistence, Affluence, and U.S. Foreign Policy* (Princeton, NJ: Princeton University Press, 1991), pp. 150–152.

17. For a fuller exposition of these themes, see Nash, J.A. "Human Rights and the Environment: New Challenge for Ethics," *Theology and Public Policy*, 4(2):42–57 (1992).

18. See, among many examples, Hawken, P. *The Ecology of Commerce* (New York: HarperCollins, 1993).

19. For sensible criteria on the siting of waste dumps, see Peters, T. "Not In My Backyard: The Waste Disposal Crisis," *Christian Century*, 106(5):175–177 (1989); and MacLean, Douglas. "Nuclear Waste Storage: Your Backyard or Mine?" *Philos. Publ. Pol.*, 9(2/3):5–8 (1989).

20. Shrader-Frechette, K.S. *Risk and Rationality*, (Berkeley/Los Angeles: University of Claifornia Press, 1991), pp. 72–74, 153–156, 206–207.

21. For a fuller interpretation of sustainability and a critique of sustainable development, see Nash, J.A. "Ethics and the Economics-Ecology Dilemma: Toward a Just, Sustainable, and Frugal Future," *Theology and Public Policy*, Vol. VI, No. 1 (Summer, 1994), pp. 43–50.

22. Thus, some biologists stress the need for redundancy in protecting biotic populations and ecosystems. See Wilson, E.O. *The Diversity of Life* (New York/London: W. W. Norton, 1992), pp. 309–310, 339.

23. Schrader-Frechette, K.S. *Risk and Rationality*, (Berkeley/Los Angeles: University of California Press, 1991), pp. 70–71.

24. See Nash, J.A. *Loving Nature: Ecological Integrity and Christian Responsibility* (Nashville: Abingdon Press, 1991), chap. 6; Nash, J.A. "Biotic Rights and Human Ecological Responsibility," *Annual of the Society of Christian Ethics, 1993*, Harlan Beckley, Ed. (Boston: SCE, 1993), pp. 137–162; and Nash, J.A. "The Case for Biotic Rights," *Yale J. Int. Law*, 18(1):235–249 (1993).

25. Cf. Rolston, H., III. *Environmental Ethics: Duties to and Values in the Natural World* (Philadelphia: Temple University Press, 1988), pp. 257, 283.

26. *Reducing Risk: Setting Priorities and Strategies For Environmental Protection,* (Washington, D.C.: USEPA, Science Advisory Board, 1990), pp. 9, 17.

27. Shrader-Frechette, K.S. "Environmental Ethics, Uncertainty, and Limited Data," *Ethics and Agenda 21: Moral Implications of a Global Consensus*, N.J. Brown and P. Quibler, Eds. (New York: United Nations Publications, 1994), p. 77.

28. Shrader-Frechette, K.S. *Risk and Rationality*, (Berkeley/Los Angeles: University of California Press, 1991), pp. 89–99.

29. Rolston, H., III. *Environmental Ethics: Duties to and Values in the Natural World* (Philadelphia: Temple University Press, 1988), p. 315.

30. For an application of this general approach to climate change, see Nash, J.A. "Ethical Concerns for the Global-Warming Debate," *Christian Century*, 109(25):775–776 (1992).

31. Rolston, H., III. *Environmental Ethics: Duties to and Values in the Natural World* (Philadelphia: Temple University Press, 1988), p. 319.

32. "Environmental Ethics, Uncertainty, and Limited Data," *Ethics and Agenda 21*, p. 78. Cf. the "precautionary principle" (principle 15) in the Rio Declaration, *Agenda 21: The United Nations Programme of Action from Rio* (New York: United Nations Publications, 1992), p. 11.

33. On overconfidence, see Slovic, P., Fischhof, B., and Lichtenstein, S. "Rating the Risks," in *Readings in Risk*, T.S. Glickman and M. Gough, Eds., (Washington, D.C., Resources for the Future, 1990), pp. 65–66, 73–74.

34. On "public theology," see especially Perry, M.J. *Love and Power: The Role of Religion and Morality in American Politics* (New York/Oxford: Oxford University Press, 1991); *Religion and American Public Life*, R. Lovin, Ed. (New York-Mahweh: Paulist Press, 1986); Stackhouse, M.S. *Public Theology and Political Economy: Christian Stewardship in Modern Society* (Grand Rapids, MI: Wm. B. Eerdmans, 1987).

15

VALUES AND COMPARATIVE RISK ASSESSMENT

Christopher J. Paterson and Richard N.L. Andrews

CONTENTS

Introduction 213
The Rise of Comparative Risk 215
State and Local Comparative Risk 217
    General... 217
    Washington Environment 2010............................. 218
Conclusion 222
References 223

INTRODUCTION[1]

The use of risk assessment as a tool for framing and informing environmental policy debates and decisions has grown extensively over the past 15 years, leading one observer to note that "[b]y the end of the 1980s the idea of risk-based decision making had become the framework for environmental policy in the Environmental Protection Agency (EPA), and to varying degrees in other agencies as well."[2] Even more important, the *conception* of risk assessment and how it is being used has changed significantly over the past decade. Since the 1970s, federal and state regulatory agencies have used risk assessment for its "traditional" purposes: to aid in the design of regulations by assigning predicted health benefits and costs to various control options for particular toxic substances and for approving the selection of sites for either the clean-up of existing hazardous substances (i.e., Superfund sites) or the siting of new potentially hazardous facilities.[3] Over the past 10 years, however, a third use of risk assessment — as a guide for setting regulatory *priorities* — has gained in prominence and notoriety.

1-56670-131-7/96/$0.00+$.50

Usually referred to as relative risk or comparative risk assessment (CRA), this more recent usage of risk assessment involves (generally) characterizing the effects of the universe of different environmental hazards of concern, comparing the effects of these different hazards against one another, generating a relative ranking of risks based upon these comparisons, identifying and evaluating a range of potential risk management options for those risks, and then adjusting agency budget and action priorities based upon these rankings. The U.S. Environmental Protection Agency has been the earliest and the most consistent advocate of the comparative-risk approach. In addition to the two most noted examples of CRA — *Unfinished Business*[4] and *Reducing Risk*[5] — the Agency has promoted and financially supported its use among states, local and tribal governments, and internationally.

As the advocacy of CRA has grown, so too has a corresponding debate over its utility and appropriateness. The bulk of the criticism of comparative risk assessment has focused upon one issue: the inability of a risk assessment framework to capture the full range of values relevant to public concerns about environmental protection. Risk analysts tend to characterize risks in terms of expected losses (e.g., excess cancer deaths). Critics of this approach note, however, that the public's perception and understanding of risks includes a much more complex array of factors, including the distribution of effects (geographically, socioeconomically, temporally, etc.), the ability of an individual to control the activity imposing the risks, the catastrophic potential of the activity causing the risks, how the risk-causing activity conflicts with or supports sociocultural institutions, and others.[6] The result is that there is often a fundamental schism between the risk characterizations prepared by agency experts and the public's conception of these same risks.[7]

The debate about CRA, however, has taken place largely without reference to perhaps the most significant example of CRA in practice: the growing number of state and local comparative risk projects across the U.S. This neglect of state and local projects would be less critical if they were no more than copies of the process previously engaged in by EPA. Indeed, some casual observers perceive state and local comparative risk projects as "essentially replicating" EPA's efforts.[8] If this were the case, the debates about CRA as conducted by EPA could be applied with little or no modification to these smaller projects. In significant ways, however, these subsequent CRA efforts have differed from the process which produced both *Unfinished Business* and *Reducing Risk*.

As noted by one observer, "[a]s near as I can tell from academic discussions of, and published accounts about, this enterprise, no one has a clue to what 'comparative risk analysis in the States' means or is supposed to mean."[9] Given that interest in comparative risk assessment continues both at EPA and in the Congress,[10] it is critical that we gain a better understanding of the different working models of comparative risk. This paper examines the rise of comparative risk as a decision-aiding tool and the general structure and processes used by state and local CRA projects. It then explores one state comparative risk project in more detail — Washington's "Environment 2010" — to provide insights into the possibility of conducting CRA in a way that more fully considers and incorporates public values, concerns, and knowledge.

THE RISE OF COMPARATIVE RISK

Although risk assessment was not a fundamentally new concept to environmental policy, its use and promotion by the EPA increased dramatically in response to a concurrence of trends and events in the late 1970s and early 1980s.[11] First, the number and character of problems the agency was asked to regulate changed significantly between 1976 and 1980 (i.e., Toxic Substances Control Act, Resource Conservation and Recovery Act, and the Comprehensive Environmental Recovery and Compensation Liability Act). Whereas previous statutes had focused upon technology-forcing standards for a few readily identifiable pollutants, these new laws required EPA to regulate a wide range of chemicals on the basis of the ultimate harm they caused. In addition, these new statutes and the programs created to administer them (each with its own set of constituencies and unique politics) exacerbated an inherent coordination problem within EPA.[12]

As the Agency geared up to address these new responsibilities, the new Reagan administration cut the EPA budget, arguing that it could "do more with less". Thus, as noted by a former EPA official, "priorities were established by substantial cuts in most EPA programs to make room for the new hazardous waste programs."[13] In addition, the new administration sought to exercise greater oversight and control over EPA's development of regulations. Executive Order 12291 (1981) required the EPA to prepare regulatory impact assessments for any major rule making and to promulgate only those regulations (to the extent permitted by law) whose benefits exceeded their costs. The Office of Management and Budget (OMB) was authorized to review these assessments and to "suggest" modifications to proposed rules. As a result, the environmental regulatory agenda increasingly came to be framed in terms of economic costs and benefits and less of public and environmental health.

Risk assessment offered a powerful tool to address these problems. The concept of risk provided "a common denominator — human health risk — by which the administrator could rationalize and defend" decisions across programs.[14] While this defense could be used by the administrator internally (i.e., to program managers and staff) to justify decisions about which substances or programs would receive higher priorities for action and funding, perhaps more fundamentally it offered a means to respond to the incursions of OMB — and also to the many interest groups lobbying for and against regulation of particular pollutants. The language of risk assessment provided EPA a means to regain control of the debate about environmental policy in language more favorable to the administrator and agency staff. As Andrews has noted:[15]

> Risk assessment, however, made risk rather than dollars the new focus for policy analysis, and provided a criterion by which to compare programs and proposals on terms relevant to EPA's mission and expertise. To be sure, risk assessment was also being used . . . as a weapon *against* aggressive regulation, by lobbying for less cautious inference guidelines. But at least the debate was defined in terms of scientific issues, about which EPA's scientists could argue from strength, and on which the EPA administrator's decisions were normally accorded greater deference than in the broader domains of economics and politics.

By returning the debate to the language of science and health effects, risk assessment "allowed EPA to recapture control of the environmental regulatory agenda...."[16]

Practical experimentation with *comparing* risks and setting risk-based priorities began in 1980 with EPA's initiation of the Integrated Environmental Management Program (IEMP). A stated objective of IEMP was "to reemphasize the central purpose of the [EPA] as the reduction of risk using quantitative risk analysis as the logical common denominator for establishing risk reduction priorities among EPA's air, water, and hazardous waste programs...."[17] Following a series of unsuccessful attempts at a sectoral approach, the IEMP did complete regional "cross-media" analyses that formed the basis for environmental planning and management strategies for five geographic areas: Philadelphia, Baltimore, Santa Clara Valley, Kanawha Valley, and Denver.[18] Among the results of these studies was "the conclusion that quantitative risk analysis could provide a basis for coordinating a seriously fragmented system of media-oriented controls" making it "likely that any [future] integration effort will use risk as a common denominator."[19]

The agency personnel who administered the IEMP assisted in the planning and implementation of the most noted example of using risk assessment to set regulatory priorities — the internal staff review of EPA programs that led to the publication *Unfinished Business* in 1987.[20] Over the course of a year, 75 senior managers within EPA, using a method described as "systematically generating informed judgments among agency experts," attempted to "develop a broad picture of environmental problems in terms of the relative risks they pose...."[21] The resulting report ranked 31 "problem areas" in 4 categories: cancer health risk, noncancer health risk, ecological risk, and welfare risk. Although it had "limited impact on Agency priorities and internal operations," *Unfinished Business* does appear to have generated further interest in and debate about the concept of risk-based priority setting.[22] In particular, it did "catch the eye" of William Reilly, who would vigorously pursue the topic 2 years later as Administrator of EPA.[23]

Shortly after becoming Administrator of EPA, William Reilly asked the Science Advisory Board (SAB) to review and evaluate the methodology and findings of *Unfinished Business* and to develop strategic risk reduction options. The following year the SAB issued its report — *Reducing Risk* — in which it endorsed, with a number of caveats, the concept of comparative risk and urged EPA to further develop it as a tool for setting budget priorities and strategic planning.[24] Reacting immediately to the report, Administrator William Reilly committed the EPA and its regional offices to using comparative risk rankings as a guide to planning and setting their budget priorities.[25] Although the primary factors influencing decisions about budgets and priorities within EPA continue to be statutory and court-ordered directives, by 1990 risk had indeed become "the primary language of analysis and management," as "virtually every EPA administrative decision was couched in terms of how much risk it would reduce...".[26]

STATE AND LOCAL COMPARATIVE RISK

General

EPA's promotion of the comparative risk tool for strategic planning efforts for environmental protection spread beyond its own efforts in Washington, D.C. In 1987, three EPA regional offices initiated CRA projects which largely followed the process used to produce *Unfinished Business*.[27] By 1991, all remaining EPA regions had completed their own comparative risk assessments and were using them to craft regional strategic plans and as part of their planning and budgeting negotiations with EPA headquarters.[28] At the same time, EPA sought out states interested in experimenting with CRA. Beginning in 1987 with three state pilot projects (Pennsylvania, Colorado, and Washington), the EPA has supported a growing number of state CRA projects. By the end of 1994, 19 states had projects in various stages (including implementation) and an additional 9 states were actively planning CRA projects.[29] This support has recently spread to both local and tribal governments. Eight communities and two tribal governments were engaged in or had completed CRA projects by the end of 1994, and an additional two communities were preparing to initiate efforts.[30]

The process used by states and cities has differed significantly from that of the EPA. Whereas the EPA has relied exclusively upon the judgments of its internal expertise in assessing and ranking risks, states and cities have expanded the role of lay representatives in the process. This has been done most notably through the inclusion of members of the public in the decision-making structure of the project itself. Most of the projects have had similar three-committee structures to accomplish the assessment, ranking, and development of management options: a steering committee (SC) consisting of senior policy makers from within state government, technical advisory committees (TACs) who have the task of developing the scientific information for each of the identified problems, and a public advisory committee (PAC).[31]

PACs have most often been comprised of a wide diversity of persons representing the range of groups within a state. Although some state projects have consciously sought to include "average" citizens on the PAC, the norm has been to seek out representatives from the various stakeholder groups likely to play a role in or be affected by that state's environmental policy. Often, PACs have been given significant roles and authority within state CRA projects. Typically, their primary function has been to work with the TACs in identifying problems and their relevant impacts for assessment, to assimilate the technical information provided by the TACs, to produce a final risk ranking, and often to participate in selecting and ranking risk reduction action priorities. Throughout the process, PAC members have been expected to represent, consider, and incorporate public values and interests into the CRA process.

The inclusion of the public in state and local CRA projects is at least in part a reaction to criticisms of previous federal comparative risk efforts. Proponents and practitioners alike have noted that the state and local projects consciously designed public involvement into the process in an effort to link "analysis with judgment in a way that can help to bridge the differences between expert and lay perceptions of risk."[32] The primary means of achieving this bridge is through a collaborative analytical and decision-making process in which the PAC members are equal partners with other project participants. As noted by Jonathan Lash:[33]

> A key in most state projects has been the process through which the public advisory committee works together with policy makers and a technical committee composed of experts from the private sector, academia, and state agencies with responsibilities related to the environment (including health, agriculture, and economic development agencies). They work with the same data, and a common set of objectives and problem definitions, so the interaction compels explicit discussion of the meaning and significance of the data and the values that each participant brings to the process. It is an interaction from which everyone learns.

Thus, to the extent that this interaction actually occurs, the activities at the state and local level may provide an alternative model of the CRA process — one more closely resembling a "democratic model" of risk analysis in which lay values are incorporated directly into the decision-making process through a process of focused discourse among members of the public, technical experts, and government representatives.[34]

Washington Environment 2010

One specific example of this process is the Washington state comparative risk project.[35] From the very beginning, one of the primary goals of Washington's

"Environment 2010" was to create a public consensus on the need for and direction of changes in the way the state addressed environmental problems. Project organizers sought to do this through two related venues: the formation and utilization of a public advisory committee and by the creation of substantial opportunities for public participation across the state.

The PAC was created after both the project's steering committee and technical work groups (TWGs) had already been formed, and it was viewed explicitly as a group-based body. Members were nominated by individuals from the steering committee and appointed by the Governor, based upon the following criteria: their standing within the group they represented and subsequent ability to commit their organization to the project's results; their previous activity in and knowledge of the issues; their willingness to work with others to reach consensus; and balance in the final composition of the PAC among the various interests within the state. The result was a PAC with 34 members from business, industry, agriculture, recreation groups, environmental groups, local governments, Tribes, the state legislature, and academia.

The PAC was given a broad range of responsibilities and decision-making authority. Members were expected to "review and refine the scope and format" of the reports prepared by the technical committees. They were to be "actively involved in the development [and implementation] of the project's public involvement plan." PAC and Steering Committee members were to work together "to develop a long-term action plan based upon the analytical findings and the results of the public involvement process." In addition, they were expected to "have an in-depth knowledge of [their] constituencies'" values and concerns, to represent those within the project, and to present and advocate the goals and findings of the project to both their constituents and the general public.[36]

A list of problem areas for assessment had already been generated by project staff and modified by the technical work group members before the PAC reviewed them. Although the list was not considered final until approved by the PAC, this appears largely to have been a formality as no modifications were made to the list by PAC members. In addition, the PAC was informed of the methods and criteria being used by TWG members to assess the risks of different environmental hazards. While it was within their purview to review, challenge, and modify these criteria, little of this seems to have occurred.[37]

Following the preparation of their assessments, TWG members presented their findings to the PAC in a series of briefings during which members had the opportunity to question technical personnel directly and to critique the reports. Although these interchanges seem to have resulted in no significant changes in the substance or form of the technical reports,[38] it did allow for a better understanding by the PAC of both the relative strengths and the limits of the technical information. In some cases, the PAC asked TWG members for certain types of additional information, but often were told either that such information did not exist or that there was not enough time or resources to produce the information. Several times the PAC used resources committed to it by the project to contract with outside experts itself, "to review initial draft reports in subject areas in which they had a particular interest."[39]

Following these briefings, the PAC held a 2-day ranking retreat. After final discussions of the technical reports, the PAC spent the remainder of its retreat forging a consensus on a preliminary ranking of risks. Initially, PAC members were asked individually to rank the list of hazards on a "ballot". The 34 ballots were then tallied to give a "group ranking" which formed the basis for subsequent discussions moderated by an outside facilitator. Much of the discussion focused on why different groups within the PAC believed a problem should be ranked higher or lower than others. Opposing groups were given a chance to persuade others of their reasoning, restrained only by the limited time available and the agreement among all parties that all reasons for a position must be linked to the information available on the particular problem. Compromise on different problem rankings was often achieved by including "rationale statements" that further clarified the meaning of the ranking.

Attempts to engage the broader public began in earnest after the technical analyses had been completed and a "State of the Environment Report" had been published with the PAC's provisional risk rankings. Following the release of the 1989 State of the Environment Report in October, the project held a 2-day "environmental summit" in November. The nearly 600 participants were nominated for invitation primarily by PAC members as part of the effort to engage their constituent groups. Participants broke into 17 work groups (each with representatives from the various "constituencies" — environmental groups, agriculture, industry, etc.) to review the vision statement developed by the PAC and the list of resources and threats (i.e., problem areas). Subsequently, threats were ranked by each participant and these votes were then totaled for a group ranking. During the second day, each work group was assigned three threats and asked to suggest and analyze possible risk reduction strategies and action tools for these threats.[40]

While there was a desire to maintain a process of open discussion during the summit, time constraints inhibited the ability of participants both to explore fully the material being presented to them and to engage in sustained dialogue about either the problems or proposed solutions. Despite these barriers, summit participants did identify a series of threats that had not been considered by the PAC or technical work groups in their original list and rankings. These included uncontrolled population growth, urbanization, consumptive lifestyles, and lack of environmental awareness. Although not fitting the "traditional" categories of environmental threats considered within risk or comparative risk analyses, these problems were seen as fundamental underlying causes of many of the individual threats being considered. They were introduced by summit participants, further supported by subsequent town meeting participants (see below), and incorporated by the PAC and staff into subsequent project reports.

A second significant contribution of the environmental summit was the generation of a list of potential risk reduction options. These included not only specific recommendations for individual problem areas, but also general actions needed to achieve the chosen environmental goals (e.g., environmental education). This menu formed the basis of the next stage of the process — the development of specific strategic plans for risk reduction.

Following the summit, 12 town meetings were held across the state over a period of 2 months. Their purpose was to communicate the results of Phase I, to ask citizens what they perceived to be the most important threats facing the state, and, most important, to have attendees suggest solutions to the identified threats. Meeting participants had the opportunity to talk with and ask questions of both PAC and TWG members. Approximately 1200 people attended the meetings, and while it is impossible to accurately report the groups' composition, impressionistic evidence suggests that significant publicity efforts were successful in getting "non-regulars" (i.e., average citizens) to attend the meetings.

The meetings had at least two significant impacts. First, the opinions expressed at these meetings made the PAC members feel more confident that they had accurately represented the public's values in the process. Second, citizens provided approximately 200 ideas for the strategic action plan.[41] Significant among these was a repeated emphasis on environmental education, an emphasis reflected in the project's subsequent recommendations for action.

In January 1990, the project's Action Strategies Analysis Committee (ASAC), a combination of some previous TWG members and agency staff reviewed the over 300 action ideas that had been generated by summit and town meeting participants. The ASAC, working closely with the PAC and SC, had already developed criteria for ranking proposed risk reduction strategies (technical feasibility, political feasibility, risk reduction potential, prevention vs. mitigation orientation, public awareness, etc.). Following a preliminary screening of the ideas during which recurring themes and ideas were combined, 12 categories of actions were developed and proposals within each category were analyzed more rigorously by ASAC "subteams". These analyses were then presented to the PAC and SC during a 2-day retreat in April 1990. Following extensive discussion of the various proposals, the PAC and SC selected a number of strategies to develop further, often with significant changes as a result of PAC-SC review. Based upon these discussions, the ASAC produced a draft action agenda.

The PAC took this draft action agenda to another series of 12 town meetings across the state. A review of the "Summary of Public Comment" suggests two observations. First, project staff were very responsive to the concerns expressed within these meetings. On numerous occasions changes were made to proposed action strategies in response to observations and objections made by meeting participants. Second, while it is apparent that organizations were in attendance and suggested a number of these changes, it is more difficult to discern the influence of individuals upon the development of the final action agenda.[42] Based upon the comments received, the PAC-SC agreed to several changes in the proposed action agenda resulting in the final product in July 1990, *Toward 2010: An Environmental Action Agenda.* While the activities continued toward implementing the strategies identified within *Toward 2010*, its release marked the end of Phase II and the completion of the formal comparative risk project.

CONCLUSION

Many of the criticisms of comparative risk assessment are based upon its inability to characterize or reflect public values within its analytic and decision-making structure. Only rarely, however, do these discussions make reference to the growing experiences at the state and local level. This paper suggests that the process used in most state and local CRA projects explicitly attempts to incorporate public values into a risk-based prioritization process through the direct inclusion of lay members within the project structure.

At this point, it is unclear how well state and local comparative risk projects have actually succeeded in this task. As the Washington case description suggests, there are still significant barriers to integrating public values into the process. Participants are often selected based upon group affiliation, potentially biasing which values will be incorporated. Within a risk framework, public members can be reluctant to challenge technical "experts" in selecting the assessment criteria and choosing analytic assumptions. In addition, they can be overwhelmed by the large amounts of information they are expected to assimilate, potentially making them less likely to uncover what those assumptions are and what they mean for the reported risk characterization. Further, the time constraints of political decision-making cycles create pressures against taking the time necessary to redress these types of problems.

Imperfection, however, does not imply a fatal flaw. State and local CRA projects appear to engage a broader spectrum of the community in meaningful dialogue about the risks facing their jurisdictions than may have occurred in the past. Through these discussions, representatives of a broader range of public constituencies have had the opportunity for input into the selection of problems for assessment, the criteria used to analyze those problems, and a set of consensual public rankings of priorities among them. In addition, as the Washington case suggests, the public can and does set the agenda for finding solutions to these problems. This translation from risk assessment to risk management is a fundamental value-based choice within the selection of action strategies and has often been overlooked within the comparative risk literature. Given that the risk ranking does not dictate what the regulatory priorities should be, the extent to which the policies coming from these projects have been based upon a more careful inclusion of the public and their values is a very positive step.

Further study will be necessary to determine how effectively the state and local comparative risk projects are actually being incorporated into state and local decisions. This issue, however, is not unique to comparative risk. Attempts to consider public values within technical decision processes through the direct inclusion of the public have been advocated and used in other settings.[43] This paper suggests that such attempts need closer inspection as a viable option within risk-based decision frameworks, and also that we need to be aware of the barriers and limitations such processes encounter in practice.

REFERENCES

1. All correspondence should be addressed to Christopher J. Paterson, Northeast Center for Comparative Risk, Vermont Law School, South Royalton, VT 05068. The views in this paper do not necessarily reflect those of the Northeast Center for Comparative Risk.
2. R.N.L. Andrews, "Risk-Based Decision Making," in Environmental Policy in the 1990s, 2nd ed., N. Vig and M. Kraft, Eds. (Washington D.C.: CQ Press, 1994), p. 209.
3. M. Russell and M. Gruber, "Risk Assessment in Environmental Policy-Making," Science 236 (April 17, 1987), pp. 286–290.
4. Unfinished Business: A Comparative Assessment of Environmental Problems (Washington, D.C.: Environmental Protection Agency, 1987).
5. Reducing Risk: Setting Priorities and Strategies for Environmental Protection (Washington, D.C.: EPA, 1990). Technically, the EPA Science Advisory Board review which produced Reducing Risk was not a CRA project but rather a review of EPA's earlier efforts. It is often included, however, in discussions of CRA as another example of the Agency's efforts to promote and use comparative risk.
6. For a review of this literature, see Social Theories of Risk, S. Krimsky and D. Golding, Eds. (Westport, CT: Praeger Press, 1992).
7. C.P. Gillette and V.E. Krier, "Risk, Courts, and Agencies," Univ. PA Law Rev. 138 (1990).
8. D. John, Civic Environmentalism (Washington, D.C.: CQ Press, 1994), p. 34.
9. D. Hornstein, "Prepared Statement," Hearings of the Subcommittee on Technology, Environment, and Aviation, Committee on Science, Space, and Technology, U.S. House of Representatives, February 3, 1994.
10. Evidence of continued interest in Congress are hearings on comparative risk held by the House Subcommittee on Technology, Environment, and Aviation, Committee on Science, Space, and Technology, U.S. House of Representatives (February 3, 1994). In addition, although not precisely "comparative risk," debates continue about proposed amendments to Senate Bill 171 (elevating EPA to Cabinet status), which would require some comparisons of risks for certain regulatory actions; see Comp. Risk Bull. 3, nos. 6 and 12 (June and December, 1993).
11. This discussion is not intended to be an exhaustive history of the rise of risk assessment within EPA. Instead, its intent is to illustrate the general conditions which gave rise to the use of risk assessment. For example, two events not discussed here which likely influenced EPA's promotion and adoption of quantitative risk assessment are the U.S. Supreme Court ruling in Industrial Union Dept., AFL-CIO vs. American Petroleum Institute (1980) and agency efforts to regain credibility following a number of controversies during the tenure of Administrator Gorsuch; see Donald Hornstein, "Reclaiming Environmental Law: A Normative Critique of Comparative Risk Analysis," Columbia Law Rev. 92 (1992), p. 565.

12. Created by an Executive Order rather than legislative action, the agency has no statutory mission with which it can organize, integrate, and guide its activities across program areas.

13. A. Alm, "Why We Didn't Use 'Risk' Before," *EPA Journal* 17 (March/April 1991), p. 15.

14. R.N.L. Andrews, "Risk-Based Decision Making," in *Environmental Policy in the 1990s*, 2nd ed., N. Vig and M. Kraft, Eds. (Westport, CT: Praeger Press, 1994), p. 217.

15. R.N.L. Andrews, "Risk-Based Decision Making," in *Environmental Policy in the 1990s*, 2nd ed., N. Vig and M. Kraft, Eds. (Westport, CT: Praeger Press, 1994), p. 218.

16. R.N.L. Andrews, "Environmental Policy-Making in the United States," paper prepared for the International Roundtable Seminar "Toward a Trans-Atlantic Environmental Policy," Washington, D.C., January 10, 1992, conducted by The European Institute, p. 19.

17. Review of the Office of Policy, Planning, and Evaluation's Integrated Environmental Management Program, Integrated Environmental Management Subcommittee, Science Advisory Board, appendix A at 1,9; quoted in Hornstein, "Reclaiming Environmental Law: A Normative Critique of Comparative Risk Analysis," *Columbia Law Rev.* 92 (1992), p. 583.

18. Office of Policy, Planning and Evaluation, USEPA, "An Overview of Risk-Based Priority Setting at EPA," a background paper prepared for Setting National Environmental Priorities Conference, November 15–17, 1992, sponsored by Resources for the Future, p. 4; D. Hornstein, "Reclaiming Environmental Law: A Normative Critique of Comparative Risk Analysis," *Columbia Law Rev.* 92 (1992), p. 583.

19. J.E. Krier and M. Brownstein, "On Integrated Pollution Control," *Environmental Law* 22 (1991): 130; J.C. Davies, "The United States: Experiment and Fragmentation," in N. Haigh and F. Irwin, Eds., *Integrated Pollution Control in Europe and North America* (1990): 60.

20. D. Hornstein, "Reclaiming Environmental Law: A Normative Critique of Comparative Risk Analysis," *Columbia Law Rev.* 92 (1992), p. 583–584; other connections are discussed in J.E. Krier, and M. Brownstein, "On Integrated Pollution Control," *Environmental Law,* 22(1991), p. 131.

21. R. Morgenstern and S. Sessions, "EPA's Unfinished Business," *Environment* 30 (July/August 1988), p. 34.

22. Alm, "Why We Didn't Use 'Risk' Before," *EPA Journal* 17 (March/April 1991), p. 16. It should be noted that the project and report were not designed to be the sole basis for changing agency priorities and operations, see R. Morgenstern and S. Sessions, "EPA's Unfinished Business," *Environment* 30 (July/August 1988), pp. 37–39.

23. R. Morgenstern and S. Sessions, "EPA's Unfinished Business," *Environment* 30 (July/August 1988), p. 16.

24. Relative Risk Reduction Strategies Committee, Science Advisory Board, *Reducing Risk: Setting Priorities and Strategies for Environmental Protection* (Washington, D.C.: EPA 1990).

25. W. Reilly, "Aiming Before We Shoot: The Quiet Revolution in Environmental Policy," address at the National Press Club, September 26, 1990 (USEPA Doc. No. 20Z-1011).

26. R.N.L. Andrews, "Risk-Based Decision Making," in *Environmental Policy in the 1990s,* 2nd ed., N. Vig and M. Kraft, Eds. (1994), p. 217.

27. The three EPA regions were Region 1 (Boston), Region 3 (Philadelphia), and Region 10 (Seattle); see *Comparing Risks and Setting Environmental Priorities: Overview of Three Regional Projects* (USEPA, August 1989).

28. OPPE-EPA, "An Overview of Risk-Based Priority Setting at EPA," a background paper prepared for Setting National Priorities Conference, November 15–17, 1992, pp. 7–8.

29. *Comparative Risk Bull.,* vol. 4, nos. 11/12 (December 1993), pp. 7–9. Some of the projects identified within the "planning" category have since officially begun their projects.

30. *Comparative Risk Bull.,* vol. 4, nos. 11/12 (December 1993), pp. 7–9.

31. R. Minard, "Structure and Process: An Overview of State Comparative Risk Projects," NCCR Issue Paper no. 2 (November 1991). Those projects that have not adopted this three committee structure have usually had some type of PAC merged with the SC or other government personnel; e.g., see "Environmental Risks in Seattle," Seattle Environmental Priorities Project (October 1991).

32. D.J. Fiorino, "Can Problems Shape Priorities?," *Public Adm. Rev.* 50 (1990): 87.

33. J. Lash, "Integrating Science, Values and Democracy," paper prepared for Setting National Environmental Priorities: The EPA Risk-Based Paradigm and Its Alternatives (November 15–17, 1992), sponsored by Resources for the Future.

34. The notion of a democratic model of risk analysis is described and developed further in D.J. Fiorino, "Technical and Democratic Values in Risk Analysis," *Risk Analysis* 9 (1989): 293–299.

35. A slightly modified version of the following section has been previously presented in "Procedural and Substantive Fairness in Risk Decisions: Comparative Risk Assessment Procedures," *Policy Stud. Rev.* (in press). "Washington Environment 2010" is not presented here as representative of all CRA projects (although it does share a number of characteristics common to many of the state projects). Instead, the brief examination of the Washington experience presented here is primarily to illustrate how PACs have functioned within a state CRA project and the extent to which they have been able to affect some of the key elements of risk-based decision making.

36. "Focus: Washington Environment 2010," Washington Environment 2010, State of Washington (September, 1989).

37. This may have been due, in part, to the inclinations of the PAC simply to defer to the technical personnel and accept the problem list and criteria already selected. In addition, they were not encouraged to challenge the criteria because of concern that the project would get "bogged down" in debates over methods and would not produce analyses in time to allow the eventual results of the project to effect the upcoming budget cycle.

38. In part, this is due to the satisfaction expressed by PAC members in the information being presented and, in part, due to the sheer volume of data they were asked to absorb (over 1200 pages).

39. "The State of the Environment Report, Volume 1: Introduction and Overview," Washington Environment 2010, State of Washington (October, 1989).

40. Washington Environment 2010 Summit: Summary of Results," Washington Environment 2010, State of Washington (no date).

41. "Town Meetings Draw Enthusiastic Crowds Around State," Washington Environment 2010, State of Washington (March, 1990).
42. "Summary of Public Comment on Draft Action Agenda," Washington Environment 2010, State of Washington (no date).
43. e.g., L.G. White, "Introducing Norms into Policy Studies," in *Policy Theory and Policy Evaluation* (New York: Greenwood Press, 1990), S.S. Nagel, Ed. O. Renn and T. Webler, "Anticipating Conflicts," *GAIA* 1 (1992): 82–94.

16

RISK AND RATIONALITY IN DECISION MAKING: EXPOSING THE UNDERLYING VALUES USED WHEN CONFRONTED BY ANALYTICAL UNCERTAINTIES

David W. Schnare[1]

CONTENTS

Introduction 227
Risk Assessment and Risk Management Analysis 229
Incorporating Uncertainty into Analysis 230
    Hazard Analysis... 230
    Exposure Assessment 233
Examples of Monte Carlo Analysis............................ 236
    Example 1 — A Typical Regulation 236
    Example 2 — Nematicide Pollution Control 238
    Example 3 — Arsenic Removal 239
    Example 4 — Alachlor 240
Conclusion 242
References and Notes....................................... 242

INTRODUCTION

Branded by environmental interest groups as corrupt and disingenuous, risk–cost and benefit–cost analysis are at the center of a conflict of values and ethics associated with management of environmental risks. Traditional economic and risk analyses fail to expose underlying values that may be central to risk management decisions. A close examination of the hidden biases and values of scientist-advocates and analysts discloses that underlying values are ripe for replacement or ratification by decision makers. This chapter will present means to expose these values in a manner that permits their rapid and rational consideration.

Some have argued that if this analysis is done, the public and the public's decision makers will be confronted with an unending barrage of information indicating that many, and perhaps most, federal environmental health requirements are very bad buys. In fact, there is some evidence that the Environmental Protection Agency (EPA) investment portfolio may be inverted. One examination of EPA budgets has shown the Agency to spend 20% of its money on the most serious problems and 80% on those that are the least serious.[2] This is not a new finding. As long ago as 1987, the Congress admonished EPA's upside-down budget:

> The [Appropriations] Committee was pleased with the Agency's earnest effort last year to assess comparative risks across programs. While progress is clearly being made, a number of serious mismatches among environmental risks and the resources devoted to them highlight the need for improvements.... While the Committee appreciates that comparative risk is only one of a number of factors that should determine resource allocations, the Agency is expected to make risk considerations more explicit in the development of its 1989 budget.[3]

Five years later, the EPA still did not include in its budget information on the relative risk reduction offered by the various programs. Nor did it publish routine budget documents that make it easy to compare risks and the resources devoted to them. Some believe the lack of easy risk-budget comparisons makes it easier to restrict the use of risk assessment and risk management data throughout Agency decision making. Over the years, however, there have been enough attempts to make these comparisons that there is a growing pressure to engage in routine comparisons. Because of the potential upside-down nature of EPA's environmental protection investment, use of risk management data in the budgetary process might cause large-scale investment growth in some programs with inevitable disinvestment in other programs.

Some who have strong interests in programs with little risk reduction capacity could be expected to oppose expanded use of risk assessment and risk management tools. It is this group that strongly opposes the proposed analysis and risk-based decision making.

The argument opponents make to limit the use of risk management data is to attack risk assessment itself, in the main by highlighting the uncertainties inherent in risk assessment. By claiming that risk assessment is a fatally flawed approach, the opponents can argue against risk-based management of federal, state, and local environmental protection resources and continue their drumbeat for a pristine environment that excludes any vestige of man and mankind as denizens of the global habitat.

This chapter provides examples of how analysts can expose underlying values and deal with uncertainty. It shows that uncertainty analysis can be the anvil upon which risk management information can be hammered into a sturdy decision tool.

RISK ASSESSMENT AND RISK MANAGEMENT ANALYSIS

The calculus of managing risks is straightforward, even if not always easy to employ. Table 1 provides a simplified check sheet for the process.[4] The inherent hazard of a pollutant or impacting physical action is determined. The potential for exposure to that pollutant or the likelihood of the physical impact is examined. Alternative means to prevent or reduce exposures or impacts are assembled. The technological efficiency of pollution prevention or control for each policy alternative is identified. The associated costs of prevention or control are assessed, as are the values of the benefits. The costs and benefits of risk prevention or reduction are compared. The results of these analyses are presented in a useful manner. These presentations are intended to expose the certainty level of the analysis as well as any underlying social, ethical, moral, or professional values the analyst applied to the problem.

Each of the analytical steps above must expose its uncertainties if an honest picture of available knowledge is to be presented. A variety of papers have been written on analytic and philosophic means to deal with the various types of uncertainties.[5] Unfortunately, some disregard these techniques, arguing only that uncertainties are too large to make analysis meaningful. These individuals tend to fall back on polemics rather than science when dealing with this issue, saying, for example, that to rely on risk assessment and risk management principles would open a Pandora's box.[6] This is "lobbyist-speak" for the complaint that some element of the public would not have to comply fully with a federal environmental regulation (for example, in the case where risks are meaninglessly small and costs are unaffordable). These lobbyists generally seek zero risk in the face of uncertainty. To counter this basically political complaint, the federal and state bureaucracies need to routinely identify and incorporate uncertainties through means that strengthen risk assessment, so that the political process can be fully informed.

Table 1   The Risk Management Assessment Process

- Resource assessment (current human and ecosystem health)
- Societal values assessment
- Hazard assessment
- Dose/physical change–response assessment
- Exposure response assessment
- Prevention and control cost assessment
- Benefit valuation assessment
- Risk, cost and benefits characterization (including marginal or incremental benefits–cost comparison of risk management alternatives)
- Presentation of analytical results, uncertainties, and implicit social and professional values and biases

In the sections that follow, examples of means for dealing with uncertainties in each step of the risk assessment and risk management analysis process are identified. Routine use of these kinds of techniques will allow competent, rational, and appropriate use of risk assessment and risk management principles — a critical need during a period marked by the enormous size of the unfunded and unaffordable regulatory mandate.

INCORPORATING UNCERTAINTY INTO ANALYSIS

While some elements of risk assessment are uncertain, analysts often know a great deal about the nature of that uncertainty. This additional information is often forgotten or ignored. When knowledge about the nature of uncertainty is included in analyses in a overt manner, decision makers can often place greater reliance on the outcome of analysis. This is especially true in the case where the expanded analysis documents a clear difference between alternatives, despite uncertainty. Mechanical inclusion of data on uncertainty can be accomplished in key steps of the risk assessment.

Hazard Analysis

Toxicologists are asked to specify the degree of human risk posed by exposure to pollutants. They look for any measurable change in metabolic, enzymatic, neurological, or organ function, as may be found in animals or man. The scientific methods used to estimate hazard are often subjective and routinely based on extremely small sets of data. The resulting uncertainty about the actual effect of an exposure can be enormous.

Interestingly, although this step in the risk assessment/management process usually contributes the greatest degree of uncertainty, it is the step the opponents to risk assessment are least interested in opening to discussion. This is because estimates of hazard are now highly conservative.[7] So too have been the means used to incorporate our understanding of uncertainty in hazard assessment. Thus, any discussion of hazard analysis would bring to light the fact that we routinely select the extreme high end estimate of the hazard of a pollutant. Two cases make this point.

Consider first the hazard of exposure to low levels of lead. Hazard assessment for lead is based on human data and one remarkable simplifying assumption. The assumption is that there in no threshold of effect for lead. All risk assessments conducted by the EPA since the mid-1980s make this assumption. However, the EPA is so uncertain of this assumption that it has stated categorically: "the Agency has not concluded that adverse health effects occur at blood lead levels below 10 µg/dl."[8]

Data on actual health effects is based on levels in lead in the blood (blood-lead) greater than 4 µg/dl. Models are used to estimate the dose-response relationship between lead and humans. Naturally, the only models used are

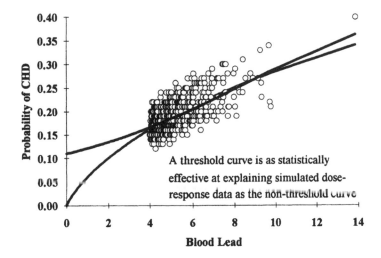

Figure 1 Threshold vs. nonthreshold curves.

those that assume no threshold of effect. EPA scientists have never directly tested whether a threshold assumption is as satisfactory as a no-threshold assumption.

For purposes of this chapter, synthetic data derived from the no-threshold assumption for lead and coronary heart disease are plotted in Figure 1. Also plotted is the no-threshold model used by the EPA and a threshold curve often used in toxicology. Statistical analysis of the two curves indicate they are equal in describing the dose response estimates.[9] A good analyst would ensure that the implications of this dose-response uncertainty are made obvious to the decision maker, perhaps through presentations like Table 2.

To highlight the importance of uncertainty analysis associated with this lead example, the reader may be interested to note that 98.7% of the U.S. population have blood-lead levels below 10 µg/dl and about 40% are below 4 µg/dl. Thus, if uncertainty about a threshold of effect were incorporated in the regulatory analysis, the implications of EPA's lead fittings and fixtures rule, now in preparation, would not be adequately described as avoiding 788 cases of coronary heart disease per 100,000 people. Rather, taking hazard

Table 2 Health Effects Uncertainty for Lead Fittings and Fixtures Rule

	10 µg/dl Threshold Assumption	Nonthreshold Assumption
Population with less than 10 µg/dl	98.7%	
Coronary heart disease avoided per 100,000	12 cases	788 cases

Best CHD estimate: 12 cases. (Based on EPA official position, October 12, 1992.)

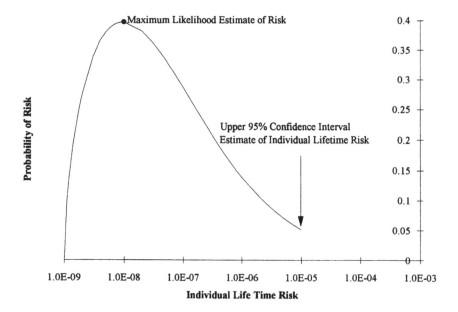

Figure 2   Probability of alternative risk estimates.

uncertainty into account, the benefits would range from 12 to 788 cases per 100,000, with EPA's official position on adverse health effects favoring the estimate of 12 cases.

In like measure, the I.Q. (neurological) effects of lead on children have also been assumed to have no threshold of effect. Rabinowitz et al. report, however, that a threshold of effect does exist.[10] Again, the decision maker should be informed that the hazard analysis on I.Q. effects is uncertain, and that previous estimates of I.Q. benefits associated with this rule would decrease by 96.5% if a threshold assumption were used.

Carcinogens are also assumed to have no threshold of effect (although the universality of this assumption is undergoing serious reconsideration). Figure 2 is a plot of the possible risk estimates for one drinking water contaminant. Note that the most likely estimate of risk is the peak of the distribution curve. This is what a scientist would state is the best estimate of the unit risk (hazard) of a specified carcinogenic dose. EPA, however, does not use this best estimate. Rather, it usually uses the upper 95th percentile estimate. Observe that the point used by EPA implies that the hazard from this pollutant is 1000 times larger than the best estimate would suggest. Instead of burying this assumption in the hazard analysis, the entire distribution can be used in Monte Carlo analysis of benefits and costs, thus giving weight to the best estimate without ignoring the potential for the hazard to be larger or smaller. Although it is standing EPA policy to conduct and present analysis of uncertainty, it is almost never done.

These two examples demonstrate that assumptions about hazard are so-phisticated, explicit, and open to large distortions reflecting underlying politi-cal values. Most important, they show that hazards can be modeled and alternative assumptions underlying hazard assessment are available. In any case, however, uncertainty does not doom risk assessment. Knowledge about the uncertainty in hazard assessment is usually quite good and it can be used to overcome concerns about uncertainty, as will be shown below.

Exposure Assessment

Although hazard assessment may contain the greatest degree of quantifi-able uncertainty, the opponents to risk assessment have most often condemned the inadequacy of exposure data. Again, it can be shown that these concerns can be minimized in the face of appropriate analysis.

The loading of pollutants into the environment is not often measured directly and even then requires assumptions on fate, disposition, and uptake before the actual dose to the target organism and organ (exposure) can be estimated. This is often the most frequently missing data element in risk assessment. What is usually forgotten is that this element is not always needed.

Most risk management decisions appear as the selection between more or less protective alternatives. In many cases, a national estimate cannot be developed due to lack of nationwide exposure data. However, that does not limit examination of localized scenarios that can be easily modeled and which have often already been modeled for cost purposes. Further, as control or prevention methods are based on the degree to which they limit pollution, direct estimates of exposure reduction are usually available for the alternatives of interest. By examining these alternatives for a variety of expected population and locality scenarios, it is possible to determine the relative cost-efficiency of each succeedingly stringent regulatory alternative.

Of course, in cases where exposure data are available, knowledge on uncertainty can be used directly in Monte Carlo analysis, thus incorporating uncertainty directly into risk assessment and risk management presentations. The more interesting example is where no exposure data are available.

Table 3 provides an example of a drinking water contaminant (pentachlo-rophenol) where there was inadequate national exposure data. Instead, a sce-nario approach was used. Specifically, the water supply industry is commonly split into 12 size categories. Although the ambient level of contamination was not well known, the meaning of a decision to control below a 10^{-4} individual lifetime risk was easily calculated. The second and eighth lines show the incremental cost per case of cancer avoided in typical large and small water systems. Also shown is the period of years during which one case of disease is expected in an affected water system of the associated size.

The immediate result of this analysis is that the two most stringent regu-latory alternatives (the two right columns) are extremely cost-ineffective, even

Table 3 Pentachlorophenol Incremental Cost-Effectiveness Analysis

Individual life risk MCL alternative Population served	1.7 × 10⁻⁵ Risk 5 µg/l		3.4 × 10⁻⁶ Risk 1 µg/l		1 × 10⁻⁶ Risk 0.3 µg/l		0.2 × 10⁻⁶ Risk 0.1 µg/l	
	Years/Case	$M/Case	Years/Case	$M/Case	Years/Case	$M/Case	Years/Case	$M/Case
25–100	24,390	$ 190	100,000	$ 711	1,000,000	$ 4,062	10,000,000	$ 14,216
101–500	5,435	73	20,408	272	125,000	1,557	500,000	5,449
501–1,000	1,754	42	6,579	157	38,462	894	142,857	3,130
1,001–3,300	706	32	2,646	121	15,152	694	55,556	2,429
3,301–10,000	236	30	855	112	5,076	643	17,857	2,249
10K–25K	85	19	320	73	1,862	417	6,410	1,459
25K–50K	40	17	149	64	850	367	2,976	1,285
50K–75K	22	11	82	43	471	246	1,650	860
75K–100K	17	11	64	42	365	238	1,277	833
100K–500K	7	8	26	29	150	167	524	584
500K–1 Mil	2	7	7	28	42	160	148	559
>1 Million	1	4	2	16	12	93	44	325

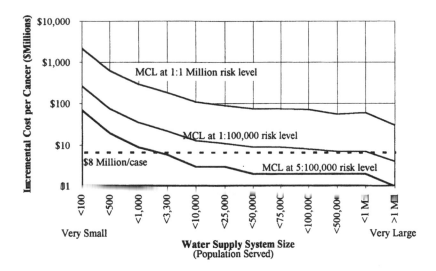

Figure 3   Incremental cost per cancer avoided by different size water systems for three alternative MCLS.

for very large water systems where maximum economies of scale might be expected, with costs near or well above $100 million. The question then arises, which of the remaining two are most appropriate.

Because there is enormous variation in cost-effectiveness, depending on water systems size, the decision maker will look for any useful information on what size systems pollution is most likely to be found. Information from environmental and public health experts was available to indicate the contaminant is a ground water pollutant that is found only in small water systems, if found at all. Thus, without any detailed exposure data it was possible to develop useful risk management data and thereby enable decision makers.

A 1 µg/l standard was selected, although the regulation development specialist had proposed a 0.1 µg/l alternative. The responsible EPA executive recalled and amended the regulatory package at the last minute only because he was advised by his policy staff through the use of this table. Specifically, although the $272 million per case for typically polluted water systems above the standard is very large, for a large water system (one basis used to establish drinking water standards) the cost per case was reduced from $860 million to $43 million, a barely acceptable level for the executive involved.

Hence, despite the absence of exposure data, EPA was able to deal with the uncertainty and apply both risk assessment and risk management principles to its regulatory decision.

Because some managers find tabular information difficult to interpret, graphical representations can also be used. Figure 3 provides another example where there were no exposure data available to EPA. Much like the table, the incremental costs per case of moving to increasingly stringent drinking water standards are shown for each of the 12 model water systems. In this chart, a dashed line indicates the EPA guidance on excessive cost per case of cancer avoided ($8 million). Note that the **cost** axis is logarithmic. For this chemical (hexachlorobenzene) a standard of 5×10^{-5} was promulgated. Note also that this decision, made by a different executive, still reflects a fairly reasonable cost per case for a typical large water system. More important, these risk management data were made available to the decision maker despite the best efforts of a subordinate manager opposed to use of uncertain risk assessment data, who sought to exclude it from both the record and the executive strata briefing materials.

These two examples indicate the power of risk management data, despite missing exposure data. In the section below on Monte Carlo simulation, examples are given that use knowledge on uncertainty of exposure estimates when such estimates do exist.

EXAMPLES OF MONTE CARLO ANALYSIS

The power of desktop (and laptop) computing has reached a point where any regulatory analyst can routinely examine the significance of literally thousands of alternative assumptions. Thus, if the analyst is uncertain about the true value of an assumption, he can examine many alternative values for the assumption. Monte Carlo analysis allows different assumptions to be varied through their range of uncertainty simultaneously in one analytical effort. A typical Monte Carlo analysis would calculate summary risk management data hundreds to thousands of times, each time with a different set of assumptions. On each iteration, every selected assumption is assigned a value within the range in which it is expected to lie. After each iteration, the resultant risk estimate or risk management estimator is recorded.[11]

Example 1 — A Typical Regulation

The probability of selecting any particular estimate for one of the assumptions is dependent on the assumption's probability distribution function. In simple terms, the probability distribution function is the shape of the uncertainty range around the assumption estimate. Figure 2 is such a plot. During the Monte Carlo analysis, the unit risk estimate could be any value of the risk on the x-axis. Most frequently, the estimate would be close to the maximum likelihood estimate. Less frequently, it would be on the tails of the curve. Typical distributions for risk, cost, exposure, and other critical assumptions for a typical EPA drinking water regulation are listed in Table 4. Figure 4 plots the outcome of the Monte Carlo run for the typical drinking water standard whose assumptions are those shown in Table 4.

The plot of the Monte Carlo outcomes indicates clearly the significance of the uncertainty within the analysis. For example, it indicates that the typical

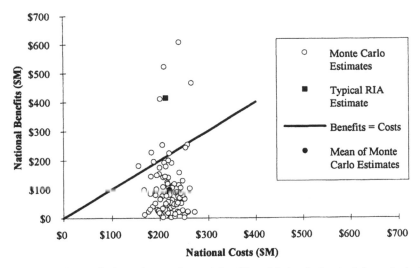

Figure 4 Monte Carlo estimates for a 1:1 million risk level standard for a typical carcinogenic drinking water contaminant.

Table 4 Typical Assumptions Used in Monte Carlo Analysis of a Drinking Water Standard

Assumption	Probability Function	Value Used in RIAs	Simulation Mean Value	Comments
Annual individual risk per μg	Lognormal $\mu = 1.43E^{-08}$ $\sigma = 1E^{-08}$	$50E^{-08}$	$1.52E^{-08}$	RIA uses upper 95% estimate
Ambient concentration	Normal $\mu = 100$ μg/l $\sigma = 40$ μg/l	100 μg/l	106 μg/l	Actual ambient levels are about 10 μg/l
Value of avoiding a case of cancer	Uniform min = $500,000 max = $10 Mil	$8 Mil	$5.4 Mil	
System population (e.g., size category one)	Uniform min = 25 max = 100	57	59.6	A similar assumption was used for each of 11 size categories
Household cost	Triangular min = $50 likeliest = $350 max = $500	$350	$310	An engineer-based assumption was used for each of 11 size categories
Systems above a standard of 5 μg/l (size category one)	Normal $\mu = 2,000$ $\sigma = 500$	2,000	1,912	A typical occurrence assumption was used for each of 11 size categories

EPA estimate is well above the vast majority of reasonable estimates. In fact, the point estimate of benefits, like those typically reported in regulatory impact analyses (RIAs), is greater than 6% of the estimates that might otherwise have been developed. In only 4 of 100 iterations would estimates of benefits and costs lead to a larger benefit–cost ratio than typically given in RIAs.

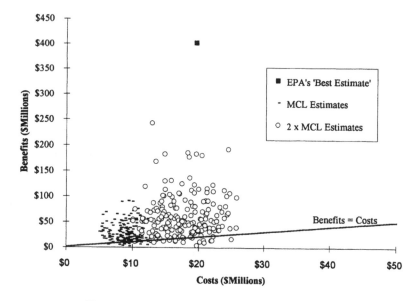

Figure 5   Benefits and costs of the nematicide MCL.

Another finding that can only come from full use of uncertainty data is that 91% of the time, benefits are expected to be less than costs. Finally, the most likely estimate of national benefits and costs for this regulatory alternative is in the midst of the highly dense grouping of estimates — a point where benefits are less than half the costs.

This first example is a very simple case. Using a 386 DOS 5.1 (IBM-like) personal computer and commercial software, it took less than 3 hours to create the spreadsheet, based on data available in regulatory impact analyses. The Monte Carlo analysis consisted of only 100 iterations and required a mere 101 seconds to perform. Preparation of the graphic required less than 15 minutes. The power of this presentation comes from the use of information on uncertainty, a type of data some forget is available for direct application. In the absence of a simple analysis and presentation like this one, the decision maker has no means to understand the quality of the risk management data provided. As the three examples below demonstrate, in some cases, uncertainties are small enough that the decision maker can be very comfortable using risk assessment and risk management analyses as the basis for a decision. In other cases, however, uncertainty is too great. Only through analysis like that presented here, however, will the decision maker know under which condition he or she is operating.

Example 2 — Nematicide Pollution Control

Figure 5 displays the analysis of national benefits and costs for two regulatory alternatives intended to control nematode pesticides found in drinking water. The actual spreadsheet used by EPA to estimated benefits and costs was used to develop the figure. Assumptions that were varied included: unit risk of cancer for the pesticides, the co-occurrence of the three pesticides under two different use patterns (six separate assumptions), the expected concentration of the pesticides in raw water, the value of avoiding a case of cancer, and a cost uncertainty multiplier. Each iteration of the spreadsheet required 4812 calculations. A Monte Carlo analysis for each of two regulatory alternatives produced the data displayed in Figure 5 in a total of 1 hour and 40 minutes. Each analysis consisted of 500 iterations. A subset of 200 randomly selected iterations for each alternative is plotted. The pattern of the analysis is adequately made by this plot.

The nematicide example provides two important insights. First, individuals lobbying for maximum control from these contaminants believed that the occurrence data available were inadequate to support a national risk assessment and subsequent national benefit and costs comparisons. Taking into account the reasonably well-understood uncertainty in occurrence produced an analysis that clearly demonstrates that benefits are likely to exceed costs under all but about 10% of the likely conditions. Hence, the concern that benefit–cost analysis would inevitably suggest less regulation does not hold up. Rather, a convincing case can be made that for some contaminants, a decision maker should regulate, even in the face of large uncertainties.

The second insight is that despite large uncertainties, these two regulatory alternatives can be distinguished from one another. The estimates of benefits and costs for the actual MCL are plotted with the estimates for an alternative that is half as stringent. While there is some overlap, the patterns are clearly separate, suggesting that the decision maker could tell these two alternatives apart. The fourth example, below, demonstrates a situation where this outcome does not arise.

A final note to be made about this example is that the so-called best estimate of national benefits found in the EPA regulatory impact analysis (and plotted on the figure) is nearly 10 times greater than the most likely estimate resulting from the Monte Carlo analysis. If EPA's so-called best estimates are routinely of this nature, and they appear to be in the drinking water regulatory program, then in the absence of this kind of analysis, the public and EPA management are routinely misinformed about the relationship of benefits and costs. A similar pattern is found in the next example.

Example 3 — Arsenic Removal

EPA has long been engaged in the development of regulations to control arsenic in drinking water. Figure 6 shows the outcome of a Monte Carlo analysis of three regulatory alternatives. Again, these data are derived from the EPA's own spreadsheet used to calculate risk management data for its regulatory impact analysis. That spreadsheet is very large, requiring 9313 calculations for each iteration. The Monte Carlo analysis for the three alternatives plotted required

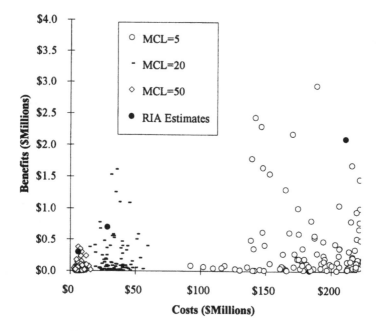

Figure 6   Benefits and costs of an arsenic MCL.

16 hours and nine minutes, total. Assumptions varied during the analysis included individual lifetime risk, the value of avoiding a case of cancer, the value of treating noncancer disease, the probability of fatality, the ambient concentration of arsenic in drinking water (and the variance of those concentration estimates) for each of four water systems size categories and sources, and a simple cost uncertainty multiplier. For two of the three alternatives, 100 iterations were done, while 200 were done for the most stringent alternative.

This example is important for one reason. It indicates that even when the most extreme assumptions are used, the benefits of regulation are not even 2/100 the size of the costs. While this could have been demonstrated by the so-called best estimates found in the draft regulatory impact analysis, such data remain open to the criticism that uncertainties are not adequately considered in such estimates, and that it might be possible to have much larger benefits. The Monte Carlo analysis provides compelling evidence that uncertainties for arsenic in drinking water are not so large as to permit an assumption that would result in a reasonable balance of benefits and costs. Clearly, regulation of arsenic in drinking water that would require use of the treatment technologies assumed in 1993 would be wasteful and unreasonable.

Example 4 — Alachlor

Those who seek a zero-risk environment will often argue there is no legitimate way to determine the value of a human life. Thus, any monetized

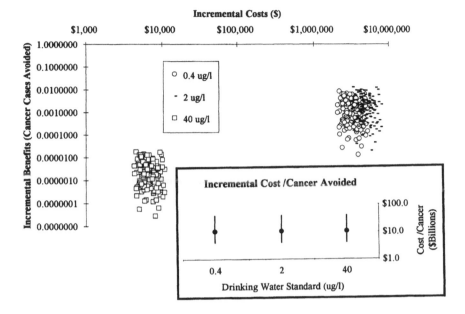

Figure 7   Incremental costs and benefits of three Alachlor drinking water standard alternatives.

comparison of benefits and costs is fatally flawed since it would have to assume some value per life. A simple way to avoid this complaint is to avoid use of a cost per life (or more correctly a cost per avoided case of cancer) assumption. Figure 7 plots an analysis of three alternative drinking water standards for Alachlor. The benefits are shown as cases of cancer avoided, rather than in dollar terms. Overlaid on the figure is a table that provides the imputed cost per cancer that would have to be assumed if costs were to equal benefits.

It is this imputed value analysis that exposes the underlying assumption that would be necessary to accept if cost per case were to be a part of the decision maker's mix of concerns. Note that regardless of the standard chosen, for benefits to equal costs it would be necessary to assume the society is willing to pay about $10 billion for each case of cancer avoided. Even taking uncertainty into account, 95% of the time society would have to be willing to pay over $3 billion per case avoided for benefits to equal costs. The overlaid table in Figure 7 thus allows the decision maker to understand the implications of the decision. Only then can he or she weigh the meaning of the social values implicit in the decision against his or her own or those of the executive branch.

Another important element of this final example is the use of incremental analysis. Since the decision maker is selecting from among alternatives, he seeks the answer to two questions. First, are the benefits in line with the costs? Second, which alternative is the most efficient? In consumer terms these two points may be stated: (1) is this worth buying; and (2) is it the best buy? In this case, because the incremental costs per case are so similar for the three alternatives, no alternative is a better buy than the other. Probably, none are worth buying.

Finally, Alachlor provides an example where the uncertainty is too large to permit selection between some alternatives. In Figures 5 and 6 above, plots of benefits and costs demonstrated that the alternatives were different. In this case, they are not. Two alternatives provide essentially the same incremental benefit at the same incremental cost. On the figure, it is difficult to distinguish between estimates for a 0.4 µg/l standard and those for 2 µg/l. The plotted estimates cover the same space, and on a statistical basis, they cannot be distinguished from each other. Thus, taking uncertainty into account, it is not always possible to distinguish between some alternatives, even those varying by a risk factor of 500%.

The decision maker needs to know when analysis will help and when it will not. Use of uncertainty analysis, as presented in the fourth example, indicates that if a selection between the two most stringent alternatives was required, it would have to be based on something other than incremental costs and benefits.

CONCLUSION

Responsible investment in environmental protection requires the best use of existing information. To abandon analysis of risks and economic impacts on the basis that there are uncertainties in the underlying assumptions is unnecessary. Analysis that incorporates knowledge about the uncertainties in the underlying data can provide far more powerful understandings about alternative actions than so-called best estimate analysis. It can expose the implied values of advocates. It can also define the actual cases where analysis is too uncertain to allow comparisons of alternatives.

REFERENCES AND NOTES

1. Dr. Schnare is former Chief of Economic, Legislative and Policy Analysis for the Office of Ground Water and Drinking Water of the U.S. Environmental Protection Agency. This paper was prepared while on temporary assignment to EPA's Office of Policy Planning and Evaluation. The views and opinions expressed herein are those of the author and do not necessarily reflect those of the Agency.

   This paper was peer-reviewed by Peter Cook, Deputy Director of EPA's Office of Ground Water and Drinking Water; Donn Viviani, Acting Director of the Waste and Chemical Policy Division of EPA's Office of Policy Analysis; and G. Wade Miller, President of Wade Miller Associates, a firm offering business and regulatory analysis services to government and the private sector for over a decade. Many of their comments and suggestions have been incorporated and the author extends his appreciation for their insight and assistance.

2. U.S. Environmental Protection Agency, Center for Resource Economics, as cited in the *Wall Street Journal*, p. B1, July 3, 1993.

3. U.S. House of Representatives Report 100–189, June 25, 1987.

4. A discussion of the risk assessment process as it reflects not only human welfare, but nonhuman ecologies as well, is presented in Schnare, D.W., Evaluating Engineering to Ensure a Sustainable Environment, in *Economic and Cleaner Production for Performance*, Misra K.B., Ed. (in press), chap. 4.

5. Cothern, R. and Schnare, D.W., The Limitations of Summary Risk Management Data, *Drug Metab. Rev.*, 17 (1 and 2), 145, 1986.

6. Eric Olsen, representing the National Wildlife Federation at an EPA-sponsored meeting on granting relief from national drinking water regulations, September, 1991; and in personal communications to the author in 1992.

7. Safety factors for some noncarcinogens include a factor of 10 for lack of threshold of effect data, another factor of 10 for nonhuman data, another factor of 10 for the absence of long-term studies, a factor of 5 for the lack of data on other source contributions, and a factor of 2 due to overestimation of water consumption. All together, this is a safety factor of 10,000. Some wags within the toxicology call this homeopathic toxicology, because these are dilutions so large as to be similar to those levels used for medicinal purposes in homeopathic medicine. To consider these safety factors anything less than overly conservative would require full confidence in homeopathic medicine, something even the liberal environmental activists have not openly endorsed.

8. U.S. EPA, Decision of the Administrator denying the petition of Lead Industries Association, Inc. for reconsideration of maximum contaminant level goal for lead under the Safe Drinking Water Act, October 12, 1992, p. 56.

9. Using the reported variation of the nonthreshold (logit) dose–response curve and the national distribution of blood-lead levels, a Monte Carlo simulation of dose–response data was generated. Regression of the estimated values against the nonthreshold model resulted in an R^2 value of 0.455. Fitting a threshold-based logistic curve resulted in an R^2 value of 0.451, indistinguishable from the curve used to generate the data.

10. Rabinowitz, M.B., Wang, J., and Soong, W., Apparent Threshold of Lead's Effect on Child Intelligence, *Bull. Environ. Contam. Toxicol.*, 48, 688, 1992.

11. For further information on use of this technique, see Morgan, G. and M. Henrion, *Uncertainty: A Guide for Dealing with Uncertainty in Quantitative Risk and Policy Analysis*, Cambridge University Press, 1990.

17

COMPARING APPLES AND ORANGES: COMBINING DATA ON VALUE JUDGMENTS

Resha M. Putzrath

CONTENTS

Introduction 245
Measuring Value Judgments 246
Combining Data 247
    Considerations on the Use of a Common Metric 247
    Effects of Nonlinearity 248
Comparing Issues on Their Own Metric 250
Conclusions 252
References 253

INTRODUCTION

The current model for risk assessment, based primarily on the paradigm in the National Academy of Sciences report *Risk Assessment in the Federal Government: Managing the Process*,[1] assumes value judgments will primarily be part of risk management decisions that are to be incorporated after the scientific risk assessment is performed. As a practical matter, however, some value judgments are an integral part of risk assessment. Selection of the type of analysis to be performed (worst-case, average, or typical) is one of the more obvious choices based on a value judgment that affects subsequent decisions as well as the outcome of the risk assessment process. Establishing severity scales that rank different toxic endpoints is another. Whether explicit on implicit, proper inclusion of value judgments in risk assessments depends on an understanding of the inherent attributes of such judgments, including how they change under a variety of circumstances. One characteristic of a value

1-56670-131-7/96/$0.00+$.50
© 1996 by CRC Press, Inc.

judgment that can affect its combination with other components of the risk assessment is the method by which these judgments are measured.

The explicit introduction of values and ethics into risk assessments rapidly expands the amount of information that could be included in the evaluation. Such an abundance of data raises additional issues. First, without some limitation on the amount of information to be included in the analysis, risk assessment will be cumbersome and perhaps unmanageable, which may in turn lead to unnecessary delays in decisions. Yet the choice of which data to include in any particular analysis is also based in part on value judgments. Second, once the information has been chosen, the method for incorporating this information must be selected.

The following discussion addresses some issues concerning the comparison and combination of information on value judgments that may become part of the risk assessment process. As the method by which value judgments are measured is critical to their accurate incorporation into risk assessment, this chapter focuses on the selection of the metric and methods of combining information.

MEASURING VALUE JUDGMENTS

Risk assessors are accustomed to numeric evaluation of each of the parameters used in their assessments. People hesitate, however, to provide a similar quantitative measure of their values. Some values may be inviolate for an individual in particular circumstances. In many, if not most, cases, however, people are ultimately able to make choices based on comparisons and therefore to set relative priorities.

While people are frequently reluctant to establish an absolute measure for each value, they are often willing to set relative or approximate measures: a mosquito is not generally valued as highly as a whale; neighborhood parks are usually preferred over municipal dump sites. Without at least some scale by which alternatives can be judged to be of more or less value, it is not possible to ask the question: which is the better or at least the less unacceptable alternative? Furthermore, if all choices remain of equal value for the decision maker, the ability to resolve issues diminishes. Thus, while few value judgments can be quantified precisely, at a minimum some relative indication of the level of concern can usually be assigned for each issue within a set of circumstances. Without the ability and willingness to make such decisions, an objective method for combining information may not be possible. With even an approximation of the level of concern and how it may change as the situation varies, however, a logical framework for making decisions on value judgments may be constructed. This analysis develops one such procedure.

Another consideration is that, while isolated estimates of risk have some utility, many risk assessments are used primarily for comparative purposes. Choice between two or more options is the crux of many societal risk analyses.

For example, comparative risk is often the basis for decisions regarding pollution prevention, life cycle analysis, selection of hazardous waste site remediation options, and change or substitution among alternative manufacturing processes. Such comparisons are difficult enough when the difference between the alternatives involves a choice between two issues within the same general topic, e.g., two environmental habitats or two adverse health effects. They become even more difficult when comparing across such subjects.

One advantage of comparative risk, however, is that it may not be necessary to provide an exact numeric value for all the information that might be included in the analysis. As discussed below, information on when the values become critical or below a level of concern may be sufficient. Three issues will be addressed in turn:

- What issues should be considered if the data are to be combined into one (numeric) comparison among options?
- Given that the values for each issue are not constant but vary with circumstance, how can these be measured and compared?
- If the data cannot or should not be combined, are there other methods that allow a comparison?

COMBINING DATA

As indicated by the title of the chapter, the first critical issue to be addressed is whether the data should be combined. Should high quality data be combined with low quality data? Can the probability that a chemical is a carcinogen be combined with its carcinogenic potency if it is a carcinogen? If so, how? Many methods are available for combining data. Use of analyses that weight data according to some objective or value scale can, to some extent, accommodate variation in factors such as uncertainty or quality. The primary question, however, of whether two or more specific sets of information should be combined by any of these mathematical procedures remains.

Considerations on the Use of a Common Metric

To ease combination and comparisons among different situations, all of the measurements may be converted into one set of units, i.e., a common metric. Some units of measure such as dollars can involve additional value issues, such as placing a monetary value on human life. To avoid such controversies, more positive metrics, such as number of lives saved, may be selected. Even if a value-neutral common metric could be found, however, the more fundamental issue still remains: under what conditions can or should diverse data be converted to one unit of measure?

To examine this issue more closely, consider some of the consequences that may result from use of any (even a value-neutral) common metric. Once all the issues to be compared are expressed in the selected metric, their associated numbers can be mathematically manipulated, e.g., summed, averaged, or set as ratios. Loss of the original units in which the element was measured eliminates the advantage of a check of the units at the conclusion of mathematical procedures, a useful scientific tool for determining whether the data have been combined in the proper manner.

The ability of numbers to be mathematically manipulated, moreover, does not guarantee that they should be combined. Consider, for example, that probabilities are expressed without units as a number between 0 and 1. All probabilities can be mathematically combined. The ability to combine numbers in itself is no guarantee that the numbers should be combined. As an obvious illustration of the potential dangers, it is possible to add the probability of dying of cancer with the probability of an audit by the Internal Revenue Service. Interpreting the result, even considering the inevitability of death and taxes, is more difficult. While the error in this example is trivially obvious, unitless exposure ratios for various toxic endpoints are often summed to create a hazard index for all potential noncarcinogenic effects at a hazardous waste site, even though the guidance document[2] for this procedure recommends evaluating each endpoint separately. Some of the potential inaccuracies resulting from "combining numbers" are discussed below.

Effects of Nonlinearity

Use of a common metric can conceal important information. Few items have the same incremental value in all situations. The value of an individual member of a species, for example, can depend on many factors. The last of a species may have great psychological value, but if it is not able to reproduce and propagate the species, it has little inherent value for the species or the ecosystem. One additional individual of a species near its critical mass for a sustainable population is likely to be considered of greater importance than an individual member of a thriving population. Similarly, people are less likely to be concerned about the thousand-fold change between a one-in-one-billion chance of becoming blind and a one-in-one-million chance than they are about the ten-fold change between a 9% and a 90% possibility.

The effect of a nonlinearity between changes in circumstance and differences in the level of concern on the use of a common metric can be substantial. Even if the metric were value neutral, use of a common metric inherently assumes that, once the conversion has been made, the units of measure are fungible for all of the items being measured. This assumption will not hold unless the value functions for the items being converted into the common metric are the same linear proportion of each other under all of the circumstances in the range of interest. As seen in the examples in the previous paragraph, this necessary condition does not hold for a number of elements that might be included in the risk assessment process.

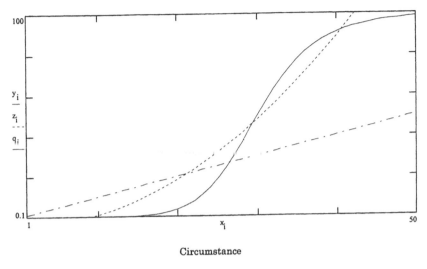

Figure 1   Plots of circumstances vs. level of concern for different value functions.

Consider the three hypothetical value function curves in Figure 1. The solid line represents a value, q, that is of little concern under the initial circumstances, rapidly increases in concern over some range of conditions, and becomes critical (or at least of maximal concern) at some point after which the concern remains about the same. Many values may take this form, e.g., when plotting level of concern vs. proximity to a person's residence for an undesirable land use. The second function, plotted as a dot-dash line, represents a value, y, that increases incrementally by the same linear rate over the range of circumstances of interest. While the value function is linear over that range, it may be nonlinear in ranges that are not of concern for the particular situation. For example, when planning a neighborhood park in a new development, the value function might be linear for parks in the range of size of a few city blocks, but might become nonlinear if the park is perceived as too small to provide basic amenities (swings, picnic tables) or large enough to include additional amenities (golf course, pool). The third value, z, that is plotted with a dashed line is a quadratic after an initial threshold below which there is no concern. The level of concern then rises more rapidly than value y and continues to rise over the range of interest.

While mathematical functions have been used for illustrative purposes and can be determined for some interactions of interest,[3] the actual curves may not be simple mathematical functions. Nonetheless, this example can be used to demonstrate difficulties that can arise when comparing level of concern when the value function curves differ in form. Note particularly that the respective rank order on the level of concern for each of the functions changes with the circumstances plotted on the x axis. At the second mark on the x-axis, y and z are of approximately the same concern (y slightly higher) while q is much lower. By the fourth mark on the x-axis, z is almost off the scale, q is very high, and y is substantially lower. Thus, the relative level of concern for these three value functions differs substantially in magnitude and relative priority depending on which set of circumstances is being evaluated. The magnitude of this disparity for nonparallel functions as circumstances change can be substantial, as has been calculated previously for the potential effect of toxicity dose–response curves on EPA's hazard index for mixtures.[4] While the effects of misestimating the numeric values is important, the observation that the rank order, i.e., the relative importance of the values measured, may also be incorrect may have even greater consequences, e.g., when establishing priorities.

This observed change in relative concern with circumstance illustrates another caution with regard to converting to a common metric. Under what circumstances should the conversion take place? Consider a situation where the y axis might be a measure of lives saved and the functions might represent different methods for disease prevention. Determining which method will save more lives depends on circumstances of implementation of each method, e.g., percent of population vaccinated or receiving nutritional supplements. If the conversion were to occur with 10% of the population and that relative value were to be used for all circumstances, one of the disease prevention methods is likely to be improperly valued at 90% of the population.

Additional cautions apply when the quality of information, uncertainty, and degree of risk aversion are included in the analysis. Three situations with the same upper-bound number could have quite disparate estimates of the most likely occurrence.[5] Similarly, three situations with the same estimate for the most likely occurrence could have quite different associated uncertainties. Should the estimates of the most likely occurrence be chosen for the conversion to a common metric? The worst case? If this choice substantially alters the decision, how should the elements be converted to a common metric? Such difficulties lead to consideration of alternate approaches.

COMPARING ISSUES ON THEIR OWN METRIC

As described above, converting all issues into a common metric inherently assumes a linear ratio between the issues over the entire region of potential concern. To the extent that the issues or values being measured do not fulfill this condition, use of a common metric may produce a less than optimal solution. If each issue is allowed to be measured by its own metric, however, by what mechanism can the apples and oranges be compared?

One pragmatic approach is to determine the circumstances under which each issue is critical, important, or not of concern for each individual or group involved with the particular situation being addressed. If such demarcations can be established, the amount of data to be included in the analysis can be limited to those elements most important for that choice. Moreover, the relative importance of each element of the decision process and the effect of large or small changes on that element, may be illuminated. Finally, while such a method will not result in one number for each option which can then be compared, the procedure will allow an analysis and comparison of the relative level of concern among several options.

To determine the critical demarcations for each issue, the following questions must be addressed by each individual or group involved in the process:

1. Is there a circumstance below which no (practical) concern exists?
2. Is there a circumstance above which action is crucial?
3. Can changes in the level of concern as the circumstances differ be assessed between these two levels?

Using the first two questions, a rank–order scheme for priority setting can be developed. Colors can be used for improving risk communication for the decision maker and other interested parties. All values below the level of (practical) concern can be designated as in the green zone. Those above a critical level would be in the red zone. The area in between would be designated the yellow zone.

This simple scheme has several advantages. All risk/value issues with estimates in the green zone, i.e., below that of (practical) concern, need not be considered further in the analysis. In particular, if there are two options within the green zone, they should be considered interchangeable. This method, therefore, allows maximum flexibility within an area of no practical or regulatory concern. Thus, this process also allows certain issues, no matter how important in other situations, to be considered of no consequence for a particular analysis.

All issues with estimates in their red zone must be included in the analysis, as the red zone defines those issues of critical concern for some individual or group involved with the particular situation being evaluated. Indeed, if the red zones are appropriately defined, a comparison of the number of issues in the red zone for each option is one crude measure of each alternative's relative level of concern. If all the options have one or more elements in the red zone, a decision process that focuses (almost) exclusively on these items may be sufficient to resolve the question. Note that the discussion at this juncture need not be strictly quantitative. By using this method, the most dire consequences, i.e., those that people are least likely to want to measure, may be simply designated as critical beyond a certain point. Indeed, locating the transition from yellow to red is likely to be more critical than defining the function in the red zone.

Those issues in the yellow zone could affect the analysis, in particular by modifying decisions based on those in the red zone. In fact, as indicated by the third question, the yellow zone itself should be subdivided if this procedure is to be practical. An estimate of the conditions for an issue that people are willing to live with, even though they are equally unwilling to consider it inconsequential, could be in a chartreuse zone. On the other hand, estimates of the level of concern for issues that are not quite crucial, but which cause great anxiety, could be in an orange zone. The chartreuse and orange zones could also be used as measures of uncertainty or to delineate the range between the worst case and the typical case. The width of these zones would serve as one measure of the extent to which small or large changes in circumstances would affect the conclusions.

Such a system allows clear identification of the most critical issues for each choice to be made. Selection of issues for comparison or priority setting is made specific to each situation. Furthermore, the procedure allows elimination from additional consideration of those issues that are not critical for that particular choice. The number of issues that must be considered in any individual situation is thus limited. Finally, allowing each issue to be measured in its own metric permits the risk/value functions to cross, i.e., that which is the most critical issue in one set of circumstances for a particular situation can become the least important in another.

CONCLUSIONS

No one method is likely to resolve all issues that can arise when assessing absolute or relative risk incorporating such desperate information as exposure, toxicity, and personal or societal values and ethics. Decisions based on such information must be made, however, and if methods cannot be found for comparing or combining such information, the amount of information that must be evaluated simultaneously may be overwhelming. Thus, approaches that allow a winnowing of less significant information and highlighting of crucial information should improve the appropriate relative ranking of the alternative choices.

While different types of information can be combined quantitatively, the procedures must be carefully examined. Analysts who use these procedures should understand their limitations, including the fact that information may be lost during the process. This understanding may not be fully conveyed to risk managers and stakeholders, however, leading to an improper use of the combined data. As illustrated, a conversion of several components of the analysis to a common metric that may be valid for one set of circumstances may not be valid for another situation. Even the priority rank of elements of the analysis can change as the circumstances in each option vary. The resulting effect on the decision-making process of these effects on the combined information could be substantial.

The alternative approach described in this analysis compares but does not combine the elements of the analysis. If each element is left on its own metric and those conditions that indicate a major change in response or attitude could be identified, the number of elements to be considered in a particular analysis is reduced. By identifying those elements that are most critical for the particular situation, efforts could be focused on refining the information needed to compare the critical issues more appropriately. Moreover, by keeping the individual elements separate, it is easier to identify and modify those elements that change with a change in the situation, e.g., the value of a different set of stakeholders in the decision-making process.

REFERENCES

1. National Research Council (NRC). *Risk Assessment in the Federal Government: Managing the Process,* 1983. National Academy Press, Washington, D.C.
2. U. S. Environmental Protection Agency (EPA). Guidelines for the Health Risk Assessment of Chemical Mixtures. *Fed. Reg.,* 51:34014–34025, September 24, 1986.
3. Clevenger, M. A., R. M. Putzrath, M. E. Ginevan, S. L. Brown, C. M. DeRosa, and M. M. Muntaz. "Risk Assessment of Mixtures: A Model Based on Mechanisms of Action and Interaction," in *Risk Analysis: Prospects and Opportunities,* C. Zervos, K. Knox, L. Abramson, and R. Coppock, Eds., (New York: Plenum Press, 1991), pp. 293–303.
4. Putzrath, R. M. and M. E. Ginevan. How the Concept of Benchmark Doses Demonstrates Some Failings of EPA's Hazard Index for Mixtures. Step 5 Working Paper 93-1 (Washington, D.C.: Step 5 Corporation, 1993).
5. Rodricks, J. V., S. L. Brown, R. Putzrath, and D. Turnbull. "Use of Risk Information in Regulation of Carcinogens," in *Determination of No Significant Risk under Proposition 65,*" H. E. Griffin and D. W. North, Eds., (California: California Department of Health and Welfare, 1987), pp. 18–44.

18

THE ETHICAL BASIS OF ENVIRONMENTAL RISK ANALYSIS

Douglas J. Crawford-Brown

CONTENTS

The Relationship of Ethics and Analysis 255
Ethics and the Definition of Risk 257
Risk Analysis as an Activity 262
The End 264
References 265

THE RELATIONSHIP OF ETHICS AND ANALYSIS

The chapters in this section,[1-4] as well as the earlier chapter by Norton,[5] are rooted in the claim that philosophical analysis, including ethical analysis, is central to ensuring the rationality of environmental risk analysis. This stands in marked contrast to the technocratic claim that risk analysis is a matter of method applied to well-established scientific data. Within this technocratic view, the rationality of a risk analysis is guaranteed by the use of the best scientific information interpreted through accepted modes of scientific reasoning. This is tantamount to arguing that existing scientific methods already embody any valid and necessary philosophical principles required by theories of rationality, even if individual analysts are not aware of those principles or the basis of their justification.

The technocratic view also usually separates questions of truth from questions of good, value, virtue, and morality. The role of the analyst becomes one of defining, to the degree possible given existing evidence, the objective characteristics of a situation that place it into the category of risky situations. This definition is claimed further to be an issue primarily of the ontology* of

---

* Ontology refers to the theory of the categories of existence.

risk, requiring only the tools of analysis needed to produce scientific explanation and prediction. The analysis is claimed to be more rational as the analyst provides a more truthful picture of such ontological properties as probability of effect, severity of effect, the etiology of an effect, and the antecedent conditions within which these causes may act.

If any further philosophical questions are raised in the technocratic view, they are those of epistemology*, arising in the course of analyzing uncertainty. For example, the technocratic risk analyst might ask questions about the degree to which an estimate of probability or severity is well established by existing evidence, and how this evidence leaves uncertainty that might be summarized as a confidence interval placed on point estimates of probabilities and severities. These questions arise from a concern for the quality of science underlying decisions and usually are treated within a framework of decision theory.

Questions of ethics typically are viewed by the technocrat as being outside the realm of risk analysis or, as in the view adopted by Schnare,[4] reducible to issues of human desires and needs. If ethical issues exist at all, they are claimed to be a part of risk management, where the manager must decide whether a risk is acceptable and whether anything should be done about it. Ethical analysis then is said to be separate from the ontological and epistemological analyses needed to characterize risk. Ethical analysis is seen to address the implications of a risk analysis for environmental decisions, but is not seen to be part of risk analysis per se.

The broad field of environmental ethics calls into question this distinction and separation. In particular, it asserts that issues of good, value, virtue, and morality (each of which is related to ethics) not only are a part of risk management, they necessarily structure the product and practice of risk analysis. This is the case even if individual risk analysts do not recognize that structure. Further, it is the lack of recognition of the role of ethics within risk analysis, the lack of critical discussion by the community of analysts, that ultimately weakens any claims to rationality.

If Bertrand Russell is correct and rationality is something like apt means for the highest ends, risk analysis cannot be claimed to be rational unless the full range of ends is defined clearly and completely and unless both the product and the conduct of a risk analysis are demonstrated to be appropriate in meeting those ends. Where those ends are defined by ethical analysis (and most ends are to some degree), the rationality of a risk analysis will hinge on ethical reflection. The present chapter provides a brief introduction to the link between ethical analysis and the treatment of ends within a risk analysis. The individual papers in this section,[1-4] as well as the paper by Norton,[5] contain more detailed explications of specific ethical positions.

---

* Epistemology is the theory of knowledge.

ETHICS AND THE DEFINITION OF RISK

It is one thing to suggest that risk analysis requires some basis in ethics, and quite another to show where that basis appears in specific acts of analysis. Where does risk analysis require any steps that draw on ethical considerations? Why can't the philosophical questions be limited to those related to the scientifically defined characteristics of risky situations (questions of ontology) and those related to the quality of scientific evidence and reasoning (questions of epistemology)? Cannot the ethical questions be dealt with after the risk analysis has been conducted?

For an introduction to these questions, the reader is referred to the important work by Shrader-Frechette.[5,6] She begins with the idea that risk analysts and managers should be rational in their decisions (without that claim, nothing that follows in the present chapter will have meaning). By this, she means that we should understand (1) how we reach our decisions; (2) how we define the goals of decision making and of actions based on those decisions; (3) the assumptions on which those decisions are based if the conclusions are to be said to be analytic rather than statements of opinion; (4) the quality of the evidence for those assumptions; and (5) the quality of our reasoning. It is through philosophical analysis of these components of a decision that rationality is created. This analysis lays bare the ethical assumptions that must be introduced, even if unrecognized by the analyst, if available information is to lead deductively to the conclusions of an analysis.

An obvious starting point in this philosophical analysis is the definition of environmental risk, which might look something like:

The total **possible** loss of **value** resulting from an environmental state or condition

We then ask: *where are the ethical dimensions in an analysis conducted under such a definition?*

By way of an answer, there are two senses in which this question can be approached. In the first sense, we can ask whether and how ethical reflection would influence the decisions made in conducting a risk analysis under this definition. This is an issue of the *intentions* of an analyst in providing estimates of risk. In the second sense, we can ask whether and how well the methodologies and/or product of that analysis reflect specific ethical positions which might be preferred by a risk manager, even if those positions are not understood by the analyst to be implicit in the rational justification of the methods and product. This is an issue of the *consequences* of an act of risk analysis. For example, a risk analysis which fails to address variability in a population is an adequate representation of the total risk required for rational risk management decisions only under an ethical position that equity is unimportant in such decisions. Ethical reflection on the need to consider equity might cause the analyst to adjust the analysis to include a calculation of variability (i.e., to *intend* that variability be considered). An analysis which does not include this calculation fails to provide an appropriately rational basis for policy decisions because it fails to address an important ethical end of that decision (i.e., it has the *consequence* of omitting variability from the analysis and leading to the creation of an inequitable state of risk).

The terms highlighted in the definition of risk given above might each indicate areas where ethical reflection could enter into discussions of the rationality of an analysis. The word "possibility" usually is taken to mean probability, likelihood, or frequency. Each of these terms seemingly is related more to epistemology and ontology than to ethics. The ontological position in risk analysis usually is that possibility refers to the actual frequency with which some event occurs, either across time in a given individual, across differences in the antecedent environmental conditions to which that individual might be exposed, or as a fraction within some population of individuals. This usually is referred to as the *objective* conception of risk. The epistemological position is that possibility refers to the state of knowledge of the analyst, with an adverse effect being "possible" if a reasonable person would have some nonzero rational confidence in its occurrence on the basis of existing evidence and norms of reasoning. This is the *psychologistic* conception of risk. A third position holds that "possibility" refers to the psychological state of dread in an individual and is termed appropriately *psychological* or *perceived* risk.

With respect to the word "possibility", then, ethics will enter the analysis only if there is an ethical basis for (1) selecting a specific meaning to the word, choosing an ontological, epistemological, or psychological slant and/or (2) selecting normative rules for forming confidence (in the psychologistic case) or appropriate psychological states. The chapter by Crawford-Brown and Arnold[1] and the work by Shrader-Frechette[3,6] argue for the position that there *are* normative constraints on the formation of rational confidence, and these constraints make up some sort of "epistemological ethics" rooted in the idea of intellectual obligation or intellectual virtue. These ideas bear some relation to earlier work by Longino,[7] Jasanoff,[8,9] and Giere.[10]

Individual analysts who violate these norms, perhaps basing their estimates of frequencies on insufficient evidence or on unreasonably conservative assumptions or outright lies, could be seen as not meeting their responsibility to depict correctly what is "possible". This would be simultaneously a professional, logical, epistemological, and ethical violation linked to intentions. Similarly, the analysis produced by this analyst then would "embody" this violation and would carry it forward into subsequent consequences such as policy decisions. Shrader-Frechette[3] provides an excellent example in the case of risk analyses for high-level radioactive waste disposal. The cited risk analyses are flawed ethically in that they necessarily require the assumption that a site is proven safe if there is no evidence to the contrary. Use of this flawed assumption (even if unrecognized by the analyst as necessary to establish the rationality of conclusions in the analysis) calls into question the ethical foundations of the analysis since it violates the responsibility of the analyst to employ only rules of reason that are likely to translate intentions (to protect public health) into the realization of desired consequences (a relatively safe environment).

The word "loss" can be interpreted through two ethical considerations. One is the ethical principle of *nonmalevolence*. In this interpretation, the analyst includes in the analysis only those adverse effects that occur as a result of some state of the environment. For example, the analyst might consider the appearance of cancers in the exposed population as a loss. Any possibility of these effects then would indicate the presence of a risk which the risk manager would feel ethically compelled to remove. An alternative (or supplementary) ethical principle is one of *beneficence*. In this interpretation, the population exposed to the environmental situation is conceived as having some potential for growth into a new and improved state of health. The environmental situation characterized by risk raises the possibility of thwarting this growth. Notice that in this latter case no adverse effect is produced, so there is no risk from the perspective of nonmalevolence. A more positive state of health is, however, thwarted. In a sense, there has been a lost possibility of an *increase* in health. To the degree the analyst works explicitly or implicitly under an ethical theory of nonmalevolence, only adverse effects will appear in the analysis. Only these will be present in the product of that analysis on which the risk manager will reflect. To the degree beneficence is considered, a loss of positive effects also will appear in the estimation of risk and be presented to the manager. Schnare[4] bases his conception of stewardship on a recognition of this distinction, calling for an ethic in which environmental decisions embrace the needs and desires of all individuals for an environment which fosters the development of human potential.

The most obvious place where an ethical theory is needed to analyze risk is in the idea of loss of "value", which is a general form of the severity of an effect. As the severity increases, more value is lost (the same may be said about increases in the frequency of an effect). What is valuable in the environment? We might assert that some part of the environment such as a human or a badger or a tree may be lost, but that is not yet a statement that there is a possible loss of *value*. The loss must be of something which is valued if a risk is to be present.

How is this value to be determined? Some possibilities spring immediately to mind. We might speak of the environment as having life support value; economic value; recreational value; scientific value; aesthetic value; genetic diversity value; historical value; character-building value; cultural symbolization value; diversity value; unity value; stability value; spontaneity value; religious value; or as a value stemming from life (or existence) itself.[11] These various values must be considered in an analysis so the analyst can determine whether a loss also is a loss of value, the possibility of which constitutes a risk. If any of these values are not reflected in an analysis, and if this omission can be linked to a failure in ethical reflection, then the intentions of the analyst are flawed ethically. And if these values are not reflected in the product of the analysis provided to a decision maker, the product is flawed and will weaken any consequent decisions.

The chapter by Norton[5] lays out quite neatly the argument that ethical concerns arise in the case of choosing the scale at which value is to be analyzed. Responsibilities of the risk manager to control adverse effects at different spatial and temporal scales require consideration of the appropriate scales at which frequencies of effects must be addressed and at which effects are to count as adverse. An analysis reflecting the ethical position that generations far into the future have ethical standing will differ from an analysis based on the position that only current individuals have standing. The latter position will cause the analyst to focus on possibilities of effects in current populations to the exclusion of future populations; and an analysis based on the ethical position that only individuals have standing will fail to address the possibility of adverse effects in ecosystems or landscapes (unless these latter effects lead to damage in individuals).

Perhaps we can avoid making this issue of value into an issue of ethics. We might, for instance, argue that values are not moral precepts; they simply state what is valued and what is not valued. Value is a statement about what the world is like (e.g., life-support and genetic value) or what humans hold dear (e.g., economic, recreational and cultural symbolization value). Ethics is concerned more with what we should do *given* that condition of the world and our desires (an idea as old as St. Augustine). This position is implicit in the chapter by Schnare,[4] who views all values as arising from humans and their needs or desires. His conception of stewardship is almost antiphilosophical, stemming from a claim that value is either biological (needs) or psychological (desires), and that responsibility to protect values is a product of a social contract between people. His entire chapter could be formulated with no reference to the need for ethical reflection; ethics would be replaced by political discourse and economic trading.

It does not necessarily follow that an analyst who states that value possibly will be lost is taking a position that such a loss ought to be prevented or that the basis for protection should be rooted in ethics. As Norton[5] makes clear, the value of an ecosystem may lie partially in its ability to sustain life over periods of time longer than the life span of an individual. Simple prudence would argue for including adverse effects to the ecosystem within a risk analysis focused on individuals. Ethics need not be invoked to include ecosystems into the ontology of an environmental risk analysis.

Still, the distinction is not so clear (and Norton is not insisting on such a distinction). Surely we "should" live our lives in a way which brings about the greatest good (an ethical concern), and bringing about the greatest good must include something akin to ensuring the preservation and creation of value in whatever form it is found. The analyst potentially contributes to this good in providing the analysis to a risk manager. If the analyst fails to address some key areas of value that are in danger of being lost, the product of that analysis will not provide an apt means for reaching the end of the risk manager, which is to preserve or create value, despite the otherwise noble intentions of the analyst. The consequent world "created" by the manager will be filled with less value than otherwise would be the case if the risk analysis had been more complete. Ethical reflection should stimulate the analyst to include consideration of all values which ought to be preserved or created by the manager.

Sharpe[2] continues this train of thought by looking at ethical theories for hints as to how the environment might become valued and included in our calculus of risk. She begins with the classical distinction between deontological (or rights-based) theories and utilitarian (or instrumental) theories. In the deontological theories, formalized by Kant in the 18th century,[12] parts of the environment take on value and are worthy of ethical respect because they have rights which we must protect if we are to claim to lead the good life.

A long-standing problem in these theories lies in deciding how a part of the environment comes to have ethical standing, to be worthy of respect either intrinsically or inherently.[11] Schnare rejects this position,[4] asserting the classical technocratic view that ethical standing is a result of a social contract in which nonhumans are excluded through their inability to participate in the negotiations. In any event, to the degree a risk analyst extends respect to parts of the environment, or is made to do so by professional or contractural or normative constraints, the possibility of damage to those parts will count towards an increased risk. Where no ethical standing is allowed, even the most extreme damage to a part of the environment, calculated to have a very high probability, will not count in the calculus of risk unless that damage affects some other part of the environment (such as humanity) which does enjoy ethical standing.

In the utilitarian theories on which Schnare bases his ethic of stewardship, the environment is viewed as the instrumental means by which happiness or good can be produced. Much of this theory stems from the work of Mill.[13] Parts of the environment count in the calculation of risk because the possibility of their damage leads to the possibility of a loss of happiness or good in humans. The major problem with such theories is deciding what is to be meant by happiness and good; whether these should mean only economic good, a view attacked by Sagoff,[14] pleasure (leading to a calculus of desire), fulfillment of biological needs, or some notion of higher noble interests (the view of Mill). Schnare bases his utilitarian conception of stewardship on the fulfillment of desires and needs, thereby avoiding what are often the more ethically interesting problems posed by Mill concerning the definition of "noble" interests.

Both Sharpe[2] and Norton[5] are concerned with the implications of these theories. First, Norton points out that even if utility is considered, there is still the difficult problem of deciding on the scale at which it should be measured. A large change in utility at a local geographic or temporal scale may have little impact in utility on a global or intergenerational scale. He argues for a form of ethical pluralism, in which what is "right" by either a utilitarian or deontological position changes when different scales of existence are considered. His point is that there may be no single "metric" for risk which captures the many scales at which ethical considerations must be addressed.

Sharpe is concerned that both ethical theories may fail to provide a morally compelling reason to strive for sustainability. She argues that the hope for a new ethical approach is a recognition that the "object" of ethical reflection must be entire ecosystems and landscapes in addition to individual organisms (particularly humans) living within those ecosystems. Only in this way can we be sure that environmental risk analyses will be expanded to include the possibility of damage to the ideal of sustainability. Her concerns are mirrored in Schnare's utilitarian ethic of stewardship.[4]

RISK ANALYSIS AS AN ACTIVITY

The discussion so far has focused on the definition of risk and how ethics might play a role in defining "possibility", "loss", and "value". We might step back and close with a discussion of the various senses in which ethical reflection is part of risk analysis as an *activity*. This focus on risk analysis as something in which people *engage* is the basis for the paper by Crawford-Brown and Arnold[1] which examines a separate strain of ethical theory. In contrast to deontological and utilitarian theories, placing environmental ethics within virtue theory calls our attention to the quality of humans and their attempt to lead the good life. It is in our *interactions* with the environment, with each other, and with our state of knowledge, that we express our moral nature. In an existential sense, we "create" our moral nature through our actions, including the actions of analysis.

As before, we might ask whether environmental risk analysis in particular, and environmental interactions in general, really are so laden with moral importance. Do we really express and create our virtues through such activities? Here, there is a need to identify how this might come about. We need to find aspects of those activities that say anything about our virtues and our moral nature.

We might look at our physical interactions with or through the environment, and how these are considered within an analysis of risk. Here we find three main areas in which ethical reflection within risk analysis might be important and might inform us of how we should act:

1. We interact with other humans *through* the environment. A polluter raises ethical problems because that pollution affects the health of other humans, and these other humans presumably have rights to be respected and protected. Ethically based risk analysis informs the policy maker of the sense in which an environmental action says something about the quality of interpersonal interactions.
2. We interact with the environment itself. To the degree that this environment is seen to have rights, the quality of this interaction must be judged from an ethical perspective if our interaction is claimed to be ethically justified. This

is not to say that ethics provides the only perspective (there also are issues of prudence and biological need), but it is one perspective. Ethically based risk analysis informs the policy maker of the sense in which an environmental action says something about the quality of human/nonhuman interactions.

3. We interact with ideals of human behavior while acting in the environment. For example, the polluter does not simply cause effects on parts of the environment. He or she also engages in activities that fail to satisfy ideals of human conduct. It is not, of course, only the polluter who runs the risk of these moral failings. The environmentalist or policy maker can have similar failings if protection of the environment is given undue weight in a calculus of risk that does not account for perfectly justified human needs, a point raised by Schmare.[4] Ethically based risk analysis informs the policy maker of the sense in which an environmental action says something about the quality of our striving towards goals of human perfection.

Each of these three areas will or will not be considered in a risk analysis, depending on the ethical position taken (explicitly or implicitly). The first area already forms the basis for most environmental risk analyses, where "loss of value" is taken to mean loss of human health or of economic utility. The environment becomes an instrumental means to maximize human welfare. The residual ethical questions are *which humans should be considered in estimating the potential loss of welfare* and *should equity of loss be considered*?

The second area (interactions with the environment) is on the horizon for environmental risk analysis and presents enormous difficulties. In this area, potential "loss of value" includes damage to parts of the environment other than humans. This raises issues of conceptual clarity *(what is meant by a measure of ecosystem health such as sustainability?)*; methodological rigor *(how do we measure such effects?)*; epistemological rigor *(how well can we understand and characterize such effects?)*; the availability of information *(how can the necessary information be made available to those who need to make decisions?)*; adequacy of scope *(how do we include all important measures of ecosystem health?)* and balance *(how do we weight the different effects that will occur at vastly different scales of organization, space, and time?)*. Each of these is a question of the methodology of risk analysis, but it also is a question of ethical thinking.

The third area, the area of virtue, lies most distant from current discussions of risk analysis. It is not a question of methodologies in risk analysis, but of how we express ourselves in choosing and employing those methodologies for specific ends. It says that what is "at risk" when we engage in environmental interactions is our own souls, however a soul is defined. This need not be a mystical idea or even one based in religion (it is not for this author). It can be an idea as simple as the persistent nagging feeling that our actions do not express who we are; the feeling that we are not living up to ideals of human conduct in the way we carry out our lives. A risk analysis might capture this by including "loss of virtue" as one the possibilities to be included in the calculus of risk.

The act of risk analysis might also raise issues of virtue. This might be because we have failed to conduct the analysis with the necessary rigor; have participated in a discourse about risk in a particularly nasty and demeaning fashion; have obtained scientific evidence through mistreatment of other parts of the environment (as in animal experiments, not that these always are ethically unjustified); have failed to provide in the analysis information needed to not only identify risks but to provide insights into courses of action; or have failed to consider the highest ends of the risk manager and of the society we serve in assigning value to parts of the environment, leading to an incomplete conception of stewardship.[4]

THE END

So what is the "end" or goal of risk analysis? I have suggested that it first must be to provide a truthful picture of the world and of the risk it contains (from whatever perspective that risk is defined). As St. Augustine said in the 5th century, "Hell is full of good intentions and desires." No amount of ethical reflection will improve the rationality of decisions based on risk analysis if the statements in that analysis are just plain wrong. Ethically sound intentions cannot produce ethically sound consequences in the absence of truth. Risk analysts cannot be faulted for spending so much time worrying about epistemology and ontology, since these will provide essential standards needed to pursue the truth. They should be applauded for this rigor just as Shrader-Frechette[3,6] should be applauded for insisting upon that rigor. It is a first step towards ethically sound analyses, and analysts are being virtuous in taking that step.

However, I also have suggested, based on the work of others in ethics, that the end of a risk analysis is to provide a picture of the total possible loss of value. Performing such an analysis requires deep reflection on what might be valued, why we should value it, and how it should be factored or weighted into the analysis and characterization of risk. Try as we might, we cannot separate the act of risk analysis from the act of valuing. And we cannot separate valuing from ethical reflection without doing damage to the ideal of leading "a good life" rather than simply "a prudent life", or "a life satisfying needs and desires", or "a life".

This is not to say that all of the ethical baggage needs to be loaded onto the shoulders of risk analysts who might be poorly equipped to analyze and bear that load. Most of the responsibility must be borne by risk managers and by others in society who use the product of a risk analysis. This society must engage in an ethical debate of the normative constraints under which the risk analyst acts. We can expect the analysts to calculate probabilities and severities with rigor and with the dignity attendant upon all human actions, but we cannot expect them to resolve or even recognize the ethical dilemmas posed by the words "possibility", "loss", and "value". Society as a whole must begin the task of defining the ethical perspective under which environmental risk will be judged as adequate or inadequate to the needs of risk managers, and to determine how this perspective will influence the conduct of a risk analysis.

REFERENCES

1. Crawford-Brown, D. and Arnold, J., "The Cardinal Virtues of Risk Analysis: Science at the Intersection of Ethics, Rationality and Culture", in *Handbook for Environmental Risk Decision Making: Values, Perceptions, and Ethics,* C.R. Cothern, Ed., Lewis Publishers, Boca Raton, FL, 1995, p. 279–289.
2. Sharpe, V., "Ethical Theory and the Demands of Sustainability", in *Handbook for Environmental Risk Decision Making: Values, Perceptions, and Ethics,* C.R. Cothern, Ed., Lewis Publishers, Boca Raton, FL, 1995, p. 267–277.
3. Shrader-Frechette, K., "Value Judgments Involved in Verifying and Validating Risk Assessment Models", in *Handbook for Environmental Risk Decision Making: Values, Perceptions, and Ethics,* C.R. Cothern, Ed., Lewis Publishers, Boca Raton, FL, 1995, p. 291–309.
4. Schnare, D., "The Stewardship Ethic — Resolving the Environmental Dilemma", in *Handbook for Environmental Risk Decision Making: Values, Perceptions, and Ethics,* C.R. Cothern, Ed., Lewis Publishers, Boca Raton, FL, 1995, p. 311–332.
5. Norton, B., "Ecological Risk Assessment: Toward a Broader Analytic Framework", in *Handbook for Environmental Risk Decision Making: Values, Perceptions, and Ethics,* C.R. Cothern, Ed., Lewis Publishers, Boca Raton, FL, 1995, p. 155–175.
6. Shrader-Frechette, K., *Risk and Rationality,* University of California Press, Berkeley, CA, 1991.
7. Longino, H., *Science as Social Knowledge,* Princeton University Press, Princeton, NJ, 1990.
8. Jasanoff, S., *The Fifth Branch: Science Advisors as Policymakers,* Harvard University Press, Cambridge, 1990.
9. Jasanoff, S., "Acceptable Evidence in a Pluralistic Society", in *Acceptable Evidence: Science and Values in Risk Management,* D. Mayo and R. Hollander, Eds., Oxford University Press, New York, 1991, pp. 29–47.
10. Giere, R., "Knowledge, Values and Technological Decision: A Decision Theoretic Approach", in *Acceptable Evidence: Science and Values in Risk Management,* D. Mayo and R. Hollander, Eds., Oxford University Press, New York, 1991, pp. 183–203.
11. Rolston, H., "Humans Valuing the Natural Environment", in *Environmental Ethics: Duties to and Values in the Natural World,* Temple University Press, Philadelphia, 1988, pp. 1–20.
12. Kant, I., *Foundations of the Metaphysics of Morals,* L. White Beck, trans., Macmillan, New York, 1971.
13. Mill, J., *Utilitarianism,* Harmondsworth: Meridian, London, 1967.
14. Sagoff, M., *The Economy of the Earth,* Cambridge University Press, New York, 1988.

19 ETHICAL THEORY AND THE DEMANDS OF SUSTAINABILITY

Virginia A. Sharpe

CONTENTS

Introduction 267
Values and Risk Assessment 268
Ethical Theory: Obstacles to Sustainability 270
Rights-Based Moral Theory 270
Utilitarianism 272
Conclusion: Ethical Theory — Meeting the Demands of
    Sustainability 275
References 276

INTRODUCTION

In 1990 the Relative Risk Reductions Strategy Committee (RRRSC) of the EPA's Science Advisory Board issued its report *Reducing Risk: Setting Priorities and Strategies for Environmental Protection*.[1] In that report, the RRRSC called on policy makers to take a more comprehensive approach to environmental risk assessment and reduction. Notably, the Committee recommended that assessment and remediation should be broadened beyond the traditional focus on risks to human health and welfare to include attention to ecosystemic health both as a good in itself and a necessary condition for human well-being. The RRRSC supported this recommendation with an empirically based ranking of relative risk to the natural ecology and human welfare. The findings expressed in this ranking are as revealing as they are daunting. The hazards that tend to capture public attention (and therefore public funds): oil spills, groundwater pollution, herbicides, and pesticides, are those that pose only relatively low to medium risk. The highest risk, however, is associated with the seemingly remote problems of habitat destruction, species extinction, stratospheric

1-56670-131-7/96/$0.00+$.50

ozone depletion, and global climate change. In short, the Committee's findings emphasize that the most serious risk we face is the ultimate risk: the depletion of the biosphere and the earth's ecosystems.

This paper is a response to the Committee's call for attention to ecosystemic and biospheric sustainability. In particular it looks at some of the philosophical work that will need to be done if cultural priorities are to shift in this direction.

Jim MacNeill is the former Secretary-General of the World Commission on Environment and Development and principal author of the report *Our Common Future*. His work has been vital in advancing the concept of sustainability. "The obstacles to sustainability," he says, "are not mainly technical; they are social, institutional and political."[2] In other words, many powerful forces and habits of mind resist the notion that human activities must sustain the integrity of the environment on which they depend. Whereas nature used to be the sole determinate of earth's future, a — or perhaps *the* — principle determinate is now human culture.[3]

In this paper, I will look at ethical theory as one very significant aspect of culture. I will discuss how the structure of ethical theory may actually act as an obstacle to sustainability and, alternatively, how ethical theory might be structured to be conducive to the demands of sustainability. I will begin with some prefatory remarks about the evaluative nature of human judgment.

VALUES AND RISK ASSESSMENT

No structured assessment, whether political, economic, or scientific is normatively neutral. Political and economic theories express assumptions about the proper location and distribution of wealth, privilege, and power. Scientific theories are based on assumptions about what is worth knowing. Risk assessment, as a scientific methodology, expresses either implicitly or explicitly its normative commitments in the following ways: most fundamentally, the concept of harm presupposed by risk assessment is itself evaluative. It requires us to identify the things that we value as good and disvalue as bad — those things that we care about securing and avoiding. Not incidentally, our sensory and cognitive perceptions play a key role in our discernment of what matters. In other words, we care about what — often immediately — affects us, and much of this is brought to our attention by sight, smell, and the other senses, and by the experience of pleasure and pain. Elaborating on the tendency to care about what immediately affects us, the Scottish philosopher David Hume states that in everyday life humans are

> ...principally concerned about those objects which are not much remov'd either in space or time, enjoying the present and leaving what is afar to the care of chance and fortune. Talk to a man of his condition thirty years hence, and he will not regard you. Speak of what is to happen to-morrow, and he will lend you attention. The breaking of a mirror gives us more concern when at home, than the burning of a house when abroad, and some hundred leagues distant.[4]

Value commitments inform every level of risk assessment from assumptions about who is harmed and what counts as a harm to assumptions about the relevant temporal and spatial extent of harm's probability and magnitude.

In risk assessment, normative assumptions about what matters as a harm to be avoided find expression in "outcome measures" or "endpoints." Normative assumptions about who or what should be protected from harm are manifested in the identification of target populations for risk estimation. Traditionally, the target population for environmental risk assessment has been what is nearest to us: the human population, with cancer and other harms to human health and welfare supplying the majority of outcomes measures. Although human health and welfare are values that few would challenge, they are values nonetheless. In Western culture, they have gained prominence over the value of environmental health and harmony that we find, for example, in traditional Native American approaches to the environment. Further, these values have tended, until recently, to be extended only to immediate posterity. This poses another contrast to Native American worldviews where every decision is weighed in view of its impact on the next seven generations. This is the scope of what they value. In today's *comparative* risk assessment, the values of human health and welfare are often balanced against other values, such as economic development, that are promoted by diverse laws and regulatory agencies.

In debates about whether risk analysis can be purged of normative or what are often called "subjective" elements, I would place myself with Dreyfus, MacLean, and Shrader-Frechette,[5] all of whom argue that values are inherent to risk assessment and that they should be the subject of public deliberation, not methodological eradication.

In its report, *Reducing Risk*, the RRRSC grants that "subjective" values do and should influence comparative risk assessment. However, the RRRSC also seems to suggest that these values are potentially corrupting influences which must be constrained by the albeit imperfect tools of the objective science of risk analysis.[6] This conviction has latent within it an aspiration that Shrader-Frechette has elsewhere identified as "physics envy;"[7] namely, that scientific methodology should ideally be not only wholly verifiable but also normatively neutral.

I would argue that acknowledging the values that inform views about what constitutes harm, about who or what is harmed, about comparative harms and countervailing benefits is not a defect, but rather, is a representation of the reality of human concerns. As Amelie Rorty has pointed out, the observation that theories and methodologies are normatively laden is not a charge against them. Nor does this observation unmask emotive commitments that otherwise preclude critical rationality. Rather, this observation "constitutes a plea for theorists to examine the ways that their views presuppose, express, and reinforce their [normative commitments]."[8] Explicating these value commitments means not only recognizing our motivations but making them available for public discourse. If the legitimacy of human concern is suspect, then let it be exposed to critical scrutiny, not dismissed as a distorting feature of method and therefore an obstacle to be overcome.

With this in mind, I would like to point out that my focus on sustainability — the idea that human activities whether large- or small-scale must sustain the integrity of the environment on which they depend — is itself an expression of a complex of normative assumptions. Among them: that value extends beyond the immediate present, that balanced and thriving ecosystems are valuable both intrinsically and as conditions for continued human and nonhuman existence, that environmental depletion (through unsustainable practices of habitat destruction, species extinction, stratospheric ozone depletion, and air and water pollution) should have a primary place in environmental risk assessment and reduction, and finally, that unsustainable practices are remediable by the choices of those living today.

ETHICAL THEORY: OBSTACLES TO SUSTAINABILITY

In what follows, I will look at the ways in which the structure of Western ethical theory stands as an obstacle to sustainability. In particular, I will focus on the construal of the individual and the community in rights-based theory and two versions of utilitarianism. Again, sustainability is the notion that human activities, whether large- or small-scale, must sustain the integrity of the environment on which they depend.

RIGHTS-BASED MORAL THEORY

Deontological moral theory, which originated with Immanuel Kant in the 18th century, is based on a metaphysical conviction that the autonomous individual is the locus of morality. The fact that individuals can be literally self-legislating by the use of rationality dignifies them with intrinsic worth. It is on the basis of this inherent moral value that the individual is inviolable and, as Kant famously says, must be treated "always as an end and never as a means only."[9] In other words, autonomous persons are not to be regarded solely in terms of their usefulness; as creators of the moral law, they are ends in themselves. This account of moral personality has been a cornerstone of liberal democratic political theory — both contractualism and libertarianism. In both, Kant's construal of human rationality has provided the basis for the attribution of equal rights and liberties.

The civil rights movement is Kant's legacy in the sense that it argues for moral equality, not on the basis of any arbitrary social, economic or personal attributes, but on the basis of the essential attribute of humanity alone.[10] The notion, that as rational agents we each have intrinsic worth, together with its application in the domain of civil rights (seen in the extension of rights in the Emancipation Proclamation, the 19th Amendment, and the Indian Citizenship Act) has been so compelling that it has also come to provide a model for animal and environmental rights theory as well. Following familiar civil rights rhetoric, these theorists claim that anthropocentric or human-centered morality is "speciesist"[11] and a manifestation of "human chauvinism."[12] The overriding claim is that moral theory cannot ignore the rights of nonhumans. They also have intrinsic worth and deserve our respect.

The novelty of arguments for animal or environmental rights is that they extend the liberal paradigm beyond its original target group of humans. They do so, however, only by embracing the basic assumption of liberalism: that certain characteristics possessed by individuals (though, in this case, not necessarily rationality) are the basis for moral equality.

Although the question of just what rights are possessed and how they are expressed strains the traditional rights model when it is applied to nonhumans, the use of this model nonetheless reinforces certain assumptions about self and community and the job that morality is supposed to do. It is these assumptions that act as obstacles to sustainability. I discuss them below.

Implicit in any moral theory is a view of the problem that morality is supposed to solve. To the extent that a rights-based theory gives the individual pride of place, it motivates the view that the essential problem for morality is the problem of "freedom and how to achieve it given that other freedom aspirants exist and conflict between them is likely."[13] Embedded within this view is the belief that the requirements of the individual are paramount and are *prima facie* legitimate unless they unjustly transgress the rights of other individuals. Rights-based theory is thus exclusionary. Its individualistic focus makes it indifferent to the requirements of collectivities, for example, ecosystems, habitats, and species. The demands of sustainability fundamentally challenge this assumption by placing individual requirements within the broader context of biospheric requirements.

A related matter is the construal of morally salient *relationships* that one finds in the rights model. Here, morally relevant associations are based on external mechanisms such as contracts and agreements — mechanisms that are designed to safeguard the freedom of *individual* parties while simultaneously advancing their interests. This is one reason why rights theory is problematic in its extension to nonhumans or those who cannot enter into contracts. Nonetheless, these attempts to extend the rights model leave unchallenged the view that individuals are only externally related atoms with interests that can be weighed independently of one another. The lesson of ecology, however, is quite the opposite. Individual interests are precisely not independent, nor are they connected through contractual agreement alone. Living systems never occur in isolation; they are part of an intricate web of unchosen interdependencies, more far reaching than any meeting of the minds.

Although rights theory has enabled breathtaking progress in the spread of social justice, its fundamentally individualistic orientation is at odds with an ontology of interdependence. Because rights-based moral theory makes no theoretical provisions for the fact of ecological interdependence, paradoxically, it cannot account for the continued existence of the individuals that it assumes. As Annette Baier has observed, "if the morality the theory endorses is to sustain itself, it must provide for its own continuers, not just take out a loan on ... the enthusiasm of a self-selected group of environmentalists, who make it their business or their hobby to be concerned with what we are doing to mother earth."[14] "A decent morality," she says, "will *not* depend for its stability on forces to which it gives no moral recognition."[15]

UTILITARIANISM

Risk evaluation is guided by the values of efficiency and equity in the reduction of harms. Utilitarianism is often the moral theory believed to be operational in meeting these goals. It can provide a framework by which "(1) to maximize the benefits of government expenditures for health and safety and (2) to promote equity and consistency in allocation and funds among safety programs."[16]

Utilitarian theory is based on a psychology of human choice and action. According to a utilitarian analysis, humans are identified with our interests. In individual decision making we attempt to maximize those things that we prefer and minimize those thing that we eschew. Given human psychology and the fact of collective social existence, the job of morality, then, according to utilitarianism, is to aggregate interests to produce the best overall outcome. In other words, to produce the "greatest happiness" or "greatest good for the greatest number." The judgment of the best overall outcome is made from an "impersonal standpoint which gives equal weight to the interests of everyone."[17]

From its inception, utilitarianism has understood the notion of interests broadly both in character and scope. First, J. S. Mill distinguished between higher and lower interests, establishing that noble interests — those under the impulse of human dignity — inevitably do more to advance the general welfare than lower interests, such as the pleasures of mere sensation. Thus, "it is better to be Socrates dissatisfied than a fool satisfied."[18] Second, unlike Kant, who regarded rationality as the dividing line for the attribution of moral equality, Bentham and Mill identified the *capacity to suffer* as the criterion of interest and thus as the criterion of inclusion into the moral calculus. The utilitarian standard of morality, says Mill, "that existence should be exempt as far as possible from pain, and as rich as possible in enjoyments [extends] ... to the whole sentient creation."[19]

There are a number of points to be made about how utilitarian theory stands relative to the demands of sustainability. Such a theory is not, strictly speaking, anthropocentric, because it extends the domain of interests to include all sentient beings. Appealing as this may be to those who advocate for animals, the utilitarian notion of the greatest good is a view informed primarily by the ideals of human civilization and has nothing to do with the functioning of natural ecosystems and the essential role of the predator–prey relationship in the balance of nature. It is in nature, after all, that the pain and insecurity of animals is most acute. A utilitarian calculus that intended to provide for the security and subsistence of animals would, as Sagoff has pointed out, require removing them from the wild and placing them in more humane, managed environments such as zoos and botanical gardens.[20] In short, by identifying as the goal of morality the maximization of pleasure and the minimization of pain for all sentient creatures, utilitarian theory diverts attention away from ecosystemic sustainability. It sacrifices the integrity and complexity of ecosystems to the protection and gratification of animals (human and nonhuman). This philosophical discontinuity is one reason why "the S.P.C.A. does not set the agenda for the Sierra Club."[21]

A second issue with regard to the utilitarian notion of interests is the way in which it functions in "preference utilitarianism," a reigning theory of welfare economics. In his book, *The Economy of the Earth*, Mark Sagoff has adroitly demonstrated that preference "utilitarianism" is in fact a misnomer, as it is fundamentally incompatible with the theoretical constraints of true utilitarianism.[22] Nevertheless, the influence that this doctrine continues to have in economics and environmental policy makes its investigation imperative with regard to sustainability.

Put simply, for the preference utilitarian, the notion of interest is reducible to an individual's wants, desires or preferences. Contrary to Mill, the preference utilitarian makes no *qualitative* distinctions between desires. Each preference is given equal weight within a strictly quantitative assessment. Utility, then, is defined, not in terms of the well-being or happiness that proceeds from the achievement of our desires (because this would introduce a qualitatively superior norm), but from the quantitative maximization of sheer preference satisfaction. The calculation of utility is accomplished through cost–benefit analysis and preference is standardly assessed in terms of "willingness to pay".

The concept of value on this economic model is implausibly narrow. By equating what is valuable with what is desired, such a theory ignores the fact that we do value some desires more highly than others. Manifestly, we identify some desires as base and some as noble. This being the case, there must, logically, be some standard of evaluation beyond the strictly desirable. Additionally, only those that either express preferences or are the object of preference are valued. Thus, if no one is willing to pay for the preservation of a habitat, the habitat has no value. As Aldo Leopold puts it, this view assumes that only the commercial parts of nature are valuable and that the "economic parts of the biotic clock will function without the uneconomic parts."[23]

Although preference utilitarianism is arguably an effective basis for free-market consumerism, public policy is intended, among other things, to protect and promote values that the market ignores.[24] Social regulation of this sort is at the basis of the existing mandate for agencies like EPA and OSHA. As Sagoff points out, these agencies are given authority "to achieve stated ethical, aesthetic, and cultural objectives such as a cleaner environment and a safer workplace ... [not, as a rule, to achieve] efficiency of markets."[25] If risk assessment is to serve public policy in this regard and to meet the demands of sustainability, it cannot be founded on a theory of preference utilitarianism.

It should be pointed out that these deficiencies of preference utilitarianism are not shared by bona fide utilitarian theories. Despite classical utilitarianism's humanistic bias toward the conditions of civilized life (rather than life in the wild), utilitarianism does show some promise vis à vis the issue of sustainability. On this model, interests must be educated by custom in order to take on the disinterested character of the moral point of view. In other words, preferences are qualitatively distinguishable according to the degree to which they achieve identified social goods. With the aid of reflection, which involves public deliberation, we correct our naive desires in accordance with what Mill identifies as our powerful natural sentiment to "be in unity with our fellow creatures."[26] Thus morality expands by a "progress of sentiments":[27]

> Not only does all strengthening of social ties ... give to each individual a stronger personal interest in practically consulting the welfare of others; it also leads him to identify his feelings more and more with their good and ... he comes as though instinctively, to be conscious of himself as a being who *of course* pays regard to others.[28]

In classical utilitarian theory, morality emerges out of our social and interdependent existence and our affective responses to others. It is based, therefore, on the primacy of community and, as such, is amenable to the idea that the notion of community may be expanded to include what Aldo Leopold called "the land". If we allow the lessons of ecology to inform our natural sentiment of unity, we will recognize an affinity with "the land". Consequently, social approbation will evolve around behaviors that exhibit cooperation with nature. In this way, the "progress of sentiments" informs and broadens the tendency mentioned at the beginning of this paper to care about what immediately effects us. It does so by bringing closer what previously seemed remote. In risk assessment, this would involve the evolution of a commitment to understand the broader spatial and temporal dimensions of environmental risk: those that are global and long term or irremediable. As Callicott articulates it:

> Once land is popularly perceived as a biotic community — as it is professionally in ecology — a correlative land ethic will emerge in the collective cultural consciousness.[29]

In sum, the structure of classical utilitarian theory allows that the moral quality of our desires and actions be assessed in terms of our membership in the community of nature. The difficulty of extending the principle of utility (the principle of the greatest good) throughout the biotic community should not, however, be underestimated. For this is no less than the question of how human beings may, under conditions of uncertainty, sustain the balance of nature. Aldo Leopold recognized the particular challenge that utility poses for the economic biologist. His observation might equally pertain to the risk analyst:

> Ecology is a new fusion point for all the sciences.... The emergence of ecology has placed the economic biologist in a peculiar dilemma: with one hand he points out the accumulated findings of his search for utility, in this or that species; with the other, he lifts the veil from a biota so complex, so conditioned by interwoven cooperations and competitions, that no man can say where utility begins or ends.[30]

CONCLUSION: ETHICAL THEORY — MEETING THE DEMANDS OF SUSTAINABILITY

It is axiomatic that the way we understand ourselves and our relationships with others will influence how we understand the problems that morality is intended to solve. If we understand ourselves fundamentally as autonomous individuals, we will understand it as morality's primary function to provide the conditions for the equitable expression of freedom. We will ask what is owed to me and to others as rights bearers. If we fundamentally identify ourselves with our personal desires, then morality will be seen to have the task of maximizing preference satisfaction. We will ask how I and others can best achieve our desires. However, if we understand ourselves fundamentally as "members of a community of interdependent parts"[31] (related to others by bonds of kinship, friendship, work, mutual security and dependence, common purpose) then we will understand it as morality's job to maintain, sustain, and cultivate the relationships that sustain us and form our identities. We will ask: what are my obligations to the community and to those with whom I share the community?

Ethical theory that is conducive to the demands of sustainability will first and foremost take seriously the lessons of ecological interdependence. It will be "communitarian" in emphasis and it will extend this notion broadly to the *biotic* community. Such a theory will also be respectful of the demands placed on us by the other communities of which we are a part. In particular, in order to address the demands of sustainability, such a theory must address the demands of social justice in the global human community.

Ecological interdependence also provides an apt metaphor for the relationship between risk assessment and ethics. In order to incorporate the goal of sustainability into risk assessment, we need an expanded ethic that reflects a shift in cultural values. Conversely, in order to expand our ethical orientation, we need to determine the probability and magnitude of harms that would jeopardize sustainability. Informed by this broadened set of values, risk analysis can help us respond appropriately to the needs of the widest community of which we are a part.[32]

REFERENCES

1. U.S. Environmental Protection Agency. Science Advisory Board. *Reducing Risk: Setting Priorities and Strategies for Environmental Protection*. SAB-EC-90-021 (September 1990).

2. MacNeill, J., P. Winsemius, and T. Yakushiji. *Beyond Interdependence: The Meshing of the World's Economy and the Earth's Ecology*. (New York: Oxford University Press, 1991), p. 19.

3. Rolston, H. Forward to *Environmental Ethics: Readings in Theory and Application*. L. Pojman, Ed. (Boston: Jones and Bartlett, 1994), p. xv.

4. Hume, D. *A Treatise of Human Nature*, L. A. Selby-Bigge, Ed., 2nd ed. (Oxford: Clarendon Press, 1978), p. 429.

5. Shrader-Frechette, K. "Values, Scientific Objectivity, and Risk Analysis: Five Dilemmas," in *Quantitative Risk Assessment*, J. F. Humber and R. F. Almeder, Eds. (Clifton, NJ: Humana Press, 1987), pp. 149–170.

6. U.S. Environmental Protection Agency. Science Advisory Board. *Reducing Risk: Setting Priorities and Strategies for Environmental Protection*, SAB-EC-90-021 (September 1990), pp. 8, 16.

7. Shrader-Frechette, K. "Values, Scientific Objectivity, and Risk Analysis," in *Quantitative Risk Assessment,* J. F. Humber and R. F. Almeder, Eds. (Clifton, NJ: Humana Press, 1987), p. 151.

8. Rorty, A. O. *Mind in Action: Essays in the Philosophy of Mind*. (Boston: Beacon Press, 1988), p. 21.

9. Kant, I. *Foundations of the Metaphysics of Morals*, L. White Beck, trans. (New York: Macmillan, 1971), p. 47.

10. Sharpe, V. A. "Justice and Care: The Implications of the Kohlberg-Gilligan Debate for Medical Ethics." *Theor. Med.,* 13:295–318, 1992.

11. Singer, P. "Animal Liberation" in *Environmental Philosophy: From Animal Rights to Radical Ecology*, M. E. Zimmerman, J. B. Callicott, G. Sessions, K. J. Warren, J. D. Clark, Eds. (Englewood Cliffs, NJ: Prentice Hall, 1993), pp. 22–32.

12. Routley, R. and V. Routley. "Against the Inevitability of Human Chauvinism," in *Ethics and the Problems of the 21st Century*, K. E. Goodpasture and K. M. Sayre, Eds. (Notre Dame: University of Notre Dame Press, 1979).

13. Baier, A. "Hume. The Women's Moral Theorist?," in *Women and Moral Theory*, E. F. Kittay and D. T. Meyers, Eds. (Savage, MD: Rowman and Littlefield, 1987), p. 45.

14. Baier, A. "The Need for More than Justice." *Can. J. Philos.,* 13(Supp.), p. 53–54, 1987.

15. Baier, A., quoted in O. Flanagan, and K. Jackson,"Justice, Care and Gender: The Kohlberg-Gilligan Debate Revisited."*Ethics,* 97:630, 1987.

16. Shrader-Frechette, K. "Values, Scientific Objectivity, and Risk Analysis," in *Quantitative Risk Assessment,* J. F. Humber and R. F. Almeder, Eds., (Clifton, NJ: Humana Press, 1987), p. 152.

17. Scheffler, S. *Consequentialism and Its Critics*. (Oxford: Oxford University Press, 1988), p. 1.

18. Mill, J. S. *Utilitarianism.* (Harmondsworth: Meridian, 1967), p. 260.

19. Mill, J. S. *Utilitarianism.* (Harmondsworth: Meridian, 1967), p. 263.

20. Sagoff, M. "Animal Liberation and Environmental Ethics: Bad Marriage, Quick Divorce" in *Environmental Philosophy: From Animal Rights to Radical Ecology*, M. E. Zimmerman, J. B. Callicott, G. Sessions, K. J. Warren, J. D. Clark, Eds. (Englewood Cliffs, NJ: Prentice Hall, 1993), pp. 84–94.
21. Sagoff, M. "Animal Liberation and Environmental Ethics: Bad Marriage, Quick Divorce" in *Environmental Philosophy: From Animal Rights to Radical Ecology*, M. E. Zimmerman, J. B. Callicott, G. Sessions, K. J. Warren, J. D. Clark, Eds. (Englewood Cliffs, NJ: Prentice Hall, 1993), p. 87.
22. Sagoff, M. *The Economy of the Earth.* (New York: Cambridge University Press, 1988), chap. 5.
23. Leopold, A. "The Land Ethic," in *Environmental Philosophy: From Animal Rights to Radical Ecology*, M. E. Zimmerman, J. B. Callicott, G. Sessions, K. J. Warren, J. D. Clark, Eds. (Englewood Cliffs, NJ: Prentice Hall, 1993), p. 102.
24. An interesting example is the move to bring "unowned public goods" such as clean air and water — previously known as "externalities" — into overall accountings of the cost of business.
25. Sagoff, M. *The Economy of the Earth*, (New York: Cambridge University Press, 1988), p. 14.
26. Mill, J. S. *Utilitarianism,* (Harmondsworth: Meridian, 1967), p. 284.
27. Baier, A. *A Progress of Sentiments: Reflections on Hume's Treatise,* (Cambridge, MA: Harvard University Press, 1991).
28. Mill, J. S. *Utilitarianism,* (Harmondsworth: Meridian, 1967), p. 284.
29. Callicott, J. B. "The Conceptual Foundations of the Land Ethic," in *Environmental Philosophy: From Animal Rights to Radical Ecology*, M. E. Zimmerman, J. B. Callicott, G. Sessions, K. J. Warren, J. D. Clark, Eds. (Englewood Cliffs, NJ: Prentice Hall, 1993), p. 116.
30. Leopold, A., quoted in Callicott, J. B., "The Wilderness Idea Revisited: The Sustainable Development Alternative," in *Reflecting on Nature: Readings in Environmental Philosophy*, L. Gruen and D. Jamieson, Eds. (New York: Oxford, 1994), p. 255.
31. Leopold, A. "The Land Ethic," in *Environmental Philosophy: From Animal Rights to Radical Ecology*, M. E. Zimmerman, J. B. Callicott, G. Sessions, K. J. Warren, J. D. Clark, Eds. (Englewood Cliffs, NJ: Prentice Hall, 1993), p. 96.
32. Thanks are due to Susan Stocker who provided invaluable insight along the way.

20

THE CARDINAL VIRTUES OF RISK ANALYSIS: SCIENCE AT THE INTERSECTION OF ETHICS, RATIONALITY, AND CULTURE

Douglas J. Crawford-Brown and Jeffrey Arnold

CONTENTS

Introduction 279
Why Consider Virtue in Risk Analysis? 281
Where Is Virtue Found in Risk Analysis? 282
Intellectual Virtues and Judgments of Epistemic Status 285
Conclusion 288
Acknowledgment 289
References 289

INTRODUCTION

Risk analysis has been proposed as a tool capable of bringing heightened rationality to public policy decisions intended to preserve or enhance value in the world. This value can derive from economic, recreational, life-supporting, scientific, aesthetic, character-building, or other considerations from a long list of characteristics cherished by humanity. These values need not be thought of as products of human preferences, however, but may be thought of as existing independent of human thought. Regardless of the source of those values, risk concerns the possible loss or reduction of them from some perspective (which usually is taken to be a human perspective). In this regard, risk analysis might be defined as *the rational determination of the total loss of value in the world that might be realized in some well-defined situation or event.*

Philosophical discussions of risk analysis generally have explored normative and methodological rules by which a risk analysis entails both a conceptually complete and methodologically rigorous picture of the probabilities (possibility) and effects (loss of value) associated with a "risky" situation. Specific individuals or groups then are seen to arrive at different positions on "the risk" due to differences in the body of evidence examined, the theoretical framework within which the evidence is interpreted, the treatment of uncertainty, the characteristics of risk that must be depicted in risk characterization, the location of value in the world, and the role of risk analysis within decision making.

The ethical dimension of these discussions has been focused either on the role of human values in assessing and ranking risks within a framework of decision theory, or on the conflict between utilitarian and Kantian ethical assumptions. Much less attention has been directed towards virtue theory in locating the ethical foundations of risk analysis.

This lack of attention to virtue and culture is not surprising, given the historical development of risk analysis within primarily technical communities where "risk" is conceived as an objective property of the world characterized primarily by probability and severity. In such communities, philosophical discussion generally is restricted to the rationality of beliefs and, to some degree, the rationality of means. This leads to a focus on epistemology in response to concerns over the rationality of belief and the calculation of probability.

Human values and ethics are assumed to be pertinent only in judging the rationality of ends as these appear in assigning a measure of severity (although other authors have explored links between ethics and epistemology[1-3]). These values are seen to arise from the larger society within which the risk analyst must work, with the analyst being concerned only with the calculation of probabilities and severities once the values have been given. The rationality adopted by such technical communities is one focused on epistemology since their goal is to produce the most truthful picture of any probabilities and severities underlying risk. These probabilities and severities as depicted in the risk analysis are measured from the perspective of an ethical stance taken as a given from the larger policy community concerned with risk management.

The present paper extends the discussion of ethics in risk analysis by focusing on the qualities of the individual risk analyst rather than solely on the value possibly lost in a risky situation. We will use the term "risk analysis" as a verb signifying the analyst's act of depicting risk. This does not mean we are uninterested in the correspondence between statements of risk and objective properties of the world, since the "act" conducted by the risk analyst is directed precisely towards producing such statements. However, we are more interested in the questions of whether there is any sense in which a risk analyst displays virtue in the conduct of a risk analysis, and whether anything useful can be said about the way in which virtue might be brought into risk analyses differently from different ethical perspectives.

WHY CONSIDER VIRTUE IN RISK ANALYSIS?

We begin with a conception of virtue as some form of human excellence, or as the quest to exemplify this excellence through concrete acts in life. The risk analyst might exemplify virtue both in choosing the ends of an analysis and in striving to reach those ends through the highest ideals of human contemplation and action. The second set of ideals links virtue theory to theories of aim-oriented rationality,[4] since the latter insists that rationality brings more than "truth" into the world: it brings a form of truth useful in reaching the highest ends of human existence (which we take to mean some realization of virtues such as *justice, fairness, equity*, etc.). In the same way that aim-oriented rationality provides standards to judge the results of a risk analysis directed towards stated aims, virtue theory provides standards to judge the individual analyst working towards those aims.

In both utilitarian and Kantian ethical theory, the risk analyst is drawn to ethical questions in considering where value may be lost in a risky situation. For the utilitarian, the preservation of value is achieved through the maximization of utility. For the Kantian, it is achieved through the maximal preservation of rights.

In virtue theory, the ethical focus shifts from the risky situation to the act of analysis and to the analyst carrying out that act. The risk analyst "ought" to conduct the analysis in a manner exemplifying the highest ideals of human conduct, however these are conceived. The analyst "ought" to strive towards a perfect conception of the proper mode to confront, analyze, and characterize "risk". To do otherwise is not wrong simply because it is an unreliable way to find truth and to preserve value in the world (although it is also wrong for these reasons). It is wrong because it violates a culture's deepest conception of the good person and the good life. To do (or be) otherwise is wrong because the analyst has failed to make full use of intellectual gifts, letting base desires (or vices) cloud rational justification of belief.

Is risk analysis really so fraught with issues of virtue? Why should we speak of prudence, hope, charity, temperance, prudence, fortitude, or any of the other multitude of virtues when thinking of risk analysis? Why should the analyst consider anything other than the truthfulness of the risk characterization, its correspondence with the "true" probabilities and severities?

Consider the debate over the risks of smoking. Scientists at numerous health organizations have argued that active smoking is addictive and causes lung cancer. Tobacco company representatives (including scientists) argue otherwise in the face of what seems overwhelming evidence. Walker Merryman, spokesman for the Tobacco Institute, states,[5] "smoking has been identified as a risk factor for various diseases but has yet to be proven to have a causal role in the development of those diseases." Epistemological analysis suggests he is basing his position on a misconception of proof, ignoring the utility or necessity of probabilisitic conceptions. Analysis of rationality suggests he is basing his position on an improper conception of "sufficient reason", a conception unlikely to act as an apt means to reach the highest end of protecting health with reasonable confidence.

Epistemology and rationality already provide tools for judging the reasonableness of the position taken by Merryman and the tobacco industry. Characterizations of risk might be said to be true or false, effective or ineffective, but not virtuous or unvirtuous. What is to be gained by focusing on virtues of the person behind that position? One gain is that "the person" might be the precise locus of considerations on virtue. An ethical analysis of the statements from the tobacco industry might suggest that the representatives are not simply employing faulty reasoning about risk, but deliberately violating epistemological and rational norms for ignoble ends. It is this violation of norms governing the act of risk analysis *by an individual providing testimony* that might be the focus of ethical judgments.

Ignoring virtue might also lead to dysfunctional behaviour by the analyst, weakening the rationality of means. One of the ends of the risk analysis is to produce a truthful depiction of "the risk". Epistemology and rationality can provide intellectual tools for judging whether an analysis is capable of producing truth reliably.[6] What these tools cannot do is compel an individual analyst to conform to the rules. The tools do not necessarily stimulate the will to do what is right rather than what is merely expedient. A focus on virtue provides a ethically compelling reason for the individual analyst to realize the epistemological and rational norms even when there is no chance of being "caught" doing otherwise through professional practice.

WHERE IS VIRTUE FOUND IN RISK ANALYSIS?

Five areas seem most likely as candidates for virtue in risk analysis, even if that analysis is conceived as being primarily an act of scientific analysis (which we take it to be here).

Virtue might be found in judging the epistemic status of any claims made in the analysis.

Epistemological reflection is one of the central *desiderata* of rationality[7] (others being *logicality, methodological rigor, practicality, valuation, ontological reflection* and *conceptual clarity*). Discussions of epistemic status arise in part from the presence of the term "possible" in the definition of risk (i.e., total *possible* loss of value). One conception of possibility is psychologistic possibility, in which loss of value is "possible" if there is evidence to support belief in that possible loss.[8] Philosophies of epistemology and rationality provide tools for analyzing epistemic status and for bringing this analysis to bear on characterizations of risk within an ontology of psychologistic risk.

An ethic of virtue elevates *truthfulness, rationality,* and *epistemological reflection* to central intellectual gifts expressing a culture's conception of the best of human thoughtfulness. This requires a conception of ideals for confronting and depicting the strengths and limitations of human understanding, and for reflecting these in the act of risk characterization. A virtuous analyst adopts a position of *humility,* admitting freely to epistemological concerns. In Popperian terms, this humility appears in the organized skepticism of science and other forms of rational activity.[9] The absence of the virtue of humility as realized in organized skepticism is why Leonard Cole found the attempt by the U.S. Environmental Protection Agency (U.S. EPA) to discourage discussions of uncertainty in the risks of radon both politically and morally offensive.[10] The agency's position seemed to counteract the virtues of epistemic reflection central to the quest for a good life based partially on a love of rationality and obedience to its standards.

Virtue might be found in the discourse within which risk analyses are set.

Individual risk analysts often work together in teams or appear on Science Advisory Boards. Debates develop between parties with a stake in a particular risky situation. Analysts having the discourse may be characterized by a willingness to listen to other views, to give *respect* to dissenting positions while remaining *cool* and *dispassionate.* Or they may be characterized by intolerance, a position of *vanity* or inappropriate *superiority.* Such respect is necessary both for reasons of rationality (ensuring that analyses from different perspectives are aired and assessed, thereby broadening the conception of *possible* losses of value) and for reasons of virtues such as *charity* related to the interactions between people.

Both virtue and rationality require *conceptual clarity,* compelling each analyst to make his or her terms clear to others engaged in a debate so the terms can form a common linguistic base through which analysts may search for intersubjective agreement in judgments. Virtue requires a stance of *openness,* a humble recognition that one's own position on risk might be improved by opening it clearly to examination by others.

Virtue might be found in the process of generating scientific information used in risk analysis.

This process might involve interactions between a principal investigator and her subordinates, as when credit is given properly (with *charity*) for contributions to research, or when subordinates are treated with *justice.* The interactions might be between a principal investigator and his funding source. Virtue, for example, is ignored when the same work is billed to several sources. It might involve virtues of interaction between researchers and human subjects, such as in the provision of informed consent based on *respect for autonomy*. It might also involve interactions between humans and other parts of the world, as when experiments are performed on other species or when use of resources is characterized by *prudence* or waste. Each of these issues of virtue arises even if the results of the scientific research meet all of the criteria of epistemology and rationality.

Virtue might be found in the quest to make the results of risk analysis useful, or to make them useful in a particular way.

Practicality is both one of the *desiderata* of rationality and a virtue if leading a good life also requires the realization of value in the physical world. Consider, for example, the case of exposure to radon in drinking water. A risk analysis might include estimates of the average probability of cancer in the exposed population, leaving the policy maker with the option only of lowering exposures to radon if this average probability is to be reduced. Alternatively, the analysis might divide the estimates of probability between smokers and nonsmokers, showing that smokers are much more sensitive to the effects of the radon. This gives the policy maker a second option, namely, controlling risks by controlling cigarette smoking, which in turn removes the conditions under which radon can cause lung cancer.

The practicality of an analysis is an issue of rationality when the focus is on the outcome of an action. Practicality is an issue of virtue when one judges why the analyst has deliberately chosen to emphasize a particular set of possible actions, or to hide some from public view. Virtue is evident when an analyst shows *temperance* in finding a balance between purely methodological inquiries (science for its own sake) and research guided by a direct quest to improve the condition of value in the world. Virtue is evident when an analyst shows *charity* by dedicating her work to the generation of insights into both the cause and solution of suffering.

Finally, virtue might be found in the selection of effects (possible losses of value) to be considered in a risk analysis.

For the purpose of rationality, these effects should correspond to the highest values of any community using the analysis in reaching decisions (valuation being one of the *desiderata* of rationality). The risk analyst might choose to focus society's attention onto values which will be damaged by one outcome of a risky situation (demonstrating concern for *nonmalevolence*) and/or onto values that will not be damaged but simply will not be fostered to appear (demonstrating concern for *beneficence*). The focus is not on whether the "risky" situation actually contains markers of beneficence and nonmalevolence, but on the degree to which the analyst has chosen to help society locate and foster these markers. Virtue enters because the virtuous person strives to bring embodiments of virtue into the world. Similar remarks can be made about risk analysts who choose to focus on measures of environmental *justice* or intergenerational *equity*.

INTELLECTUAL VIRTUES AND JUDGMENTS OF EPISTEMIC STATUS

We close with consideration of intellectual virtues in the daily practice of risk analysis. These virtues are related to the rationality of belief, an assessment of the degree to which an analysis provides a truthful picture of the various characteristics of risk. Even within this highly narrowed area of rationality, virtues may be found related to the manner in which an analyst assesses the epistemic status of claims about risk.

Virtue is linked to epistemological reflection through rationality. Rationality requires epistemological reflection to ensure beliefs are well grounded or justified. The analyst is called upon to show that methods or principles used to assess the quality of belief (such as the quality of an estimate of the probability of cancer) lead reliably to truth and/or meet intellectual obligations.[6] Virtues of scientific rationality are concerned most directly with the epistemological position of intellectual obligation, including the obligation to use methods proven reliable in the pursuit of truth. A virtuous analyst strives to meet intellectual obligations in the justification of beliefs.

Intellectual virtues are related to intellectual values held in esteem by a community or culture, including the scientific culture charged with estimating probabilities and severities. A virtuous analyst scans the culture to locate candidate values (in the form of intellectual obligations) and ensures these values are reflected in the act of analysis. Each of these intellectual acts (the act of searching for candidate values and the act of adhering to them) is part of the quest for virtue. For our interest here, the values are those associated with the pursuit of truth (classical rationality) and/or the pursuit of useful truth (aim-oriented rationality). The virtuous analyst feels a sense of *obedience* or *piety* towards these values, and strives to keep that sense unswayed by self-serving passions. This is not to say that passions are removed, only that they are brought into the service of the higher goal of obedience to a culture's values.

Longino[11] has separated intellectual values into three forms: *bias*, *contextual* and *constitutive* values. Since bias seems to be a vice rather than a virtue, we will speak here of the virtue of completeness rather than of bias. Completeness values concern the manner in which the analyst deliberately presents a balanced picture of the risk. They might be violated by considering only a subset of the effects or losses of value stemming from a risky situation. They might be violated by choosing to ignore uncertainty, as when only a "worst-case" or "maximum likelihood" assumption is examined, rather than depicting conclusions drawn from all possible sets of assumptions. They might be violated by deliberately falsifying data, or ignoring data and interpretive theories that lead to positions other than those held by the analyst. The analogues in virtue are *justice* (a weighing of all perspectives on a risky situation) and *honesty* (a willingness to truthfully reveal one's inner state of knowledge).

Contextual values arise from the context within which a risk analysis is conducted. They are reflected in the struggle to ensure that judgments made during the course of an analysis respect the *integrity* of the whole problem under analysis. They are needed because the quest for perfection in any single part of the analysis may be at the expense of other parts. This is a consideration of *situational rationality*.

Consider, for example, the U.S. EPA attempt to regulate radon in drinking water. To perform the risk analysis, there is a need for calculations of exposure conditions, intake rates, lung depostion fractions, effects of smoking, etc. If the analyst were to devote excessive time to any one aspect (such as the deposition fractions), this would leave less time to consider the other factors needed to calculate risk. In other words, the situation provides limited time within which the complete act of analysis must take place, and the analyst's highest goal must be to ensure that all aspects of the analysis are given reasonable consideration.

The issue of contextual values can be seen in the Science Advisory Board review of the radon risk analysis. It is important to bear in mind here that these reviews are part of the context of regulation, and that this context specifies some maximal length of time to be devoted to the risk analysis. The SAB contains individuals with highly specialized areas of expertise, each of whom will tend to focus on a small area of the analysis and insist on high standards of rigor in that area. This insistence, however, eventually must give way to the need for review of the entire risk analysis within the context of this limited time.

Constitutive values are those held by the individual analyst as philosophical positions on the nature of evidence and reason. While these may be affected by bias and context, they need not be. They form the "internal compass" by which the analyst assembles evidence and reasons towards a conclusion. The most obvious virtues with which these values are connected are *veritas* and *temperance* (not being overly committed to a single epistemology when other valid positions are available), with both of these supported by *fortitude* (moral strength in the face of external pressure) and *discipline* (the adherence to approved methods so as to remain coolheaded and unswayed by improper social pressures).

Logicality and methodological rigor, therefore, become important virtues because they ensure the analyst follows professed beliefs (including beliefs in methods) to their proper conclusion. Otherwise, beliefs might be invoked for nothing more than strategic reasons in a particular context. In a recent analysis, the authors examined the epistemic status of the belief that formaldehyde should be classified as a carcinogen.[12] This was an attempt to explore the degree to which a coherent set of intellectual values (and, hence, virtues) could be used to structure a risk analysis.

The analysis was based on five separate potential lines of reasoning, each referred to as a *relevance strategy* because it is one strategy of reason by which an observation is made relevant to a particular belief. These relevance strategies were:

1. *Direct empirical* — an instance in which the incidence of cancer was measured following exposure to formaldehyde at the concentration of interest.
2. *Semiempirical extrapolation* — an instance in which the incidence of cancer was measured following exposure to formaldehyde at concentrations in excess of the concentration of interest.
3. *Empirical correlation* — an instance in which the incidence of cancer is not measured, but some property (such as *in vitro* mutation) correlated with cancer is measured.
4. *Theory-based inference* — an instance in which the incidence of cancer is not measured, but phenomena (such as cellular proliferation) existing as part of the etiologic chain leading to cancer are observed.
5. *Existential insight* — an instance in which the analyst makes a purely subjective judgment that experience supports the claim that formaldehyde is a carcinogen.

We make no attempt here to summarize the findings of this epistemic analysis. Suffice it to say that data were assembled in six contexts (mice exposed to formaldehyde in air; rats exposed to formaldehyde in air; human cells exposed *in vitro*; mammalian cells exposed *in vitro*; and human populations exposed to formaldehyde in air at high and at low concentrations). For each context, summary judgments were drawn as to whether the evidence assembled for that context supported the claim that formaldehyde is a carcinogen *in that context*. These judgments were based on a wide variety of data falling into any and all of the five relevance strategies listed above. Judgments then were extrapolated between contexts, with a focus on the context of human exposures at low concentrations (below 2 ppm).

The conclusion of the analysis is not important here, particularly since the study was intended as a test of methodology rather than a full epistemic analysis. What is important is that even a rather formal methodology for assessing the epistemic status of claims in risk analysis called repeatedly for judgments in which different kinds of information and different modes of reasoning (relevance strategies) were "weighted" into a final judgment of carcinogenicity. It is in these judgments that issues of constitutive values were most evident.

It was necessary to assign some measure of intellectual obligation to the different relevance strategies. Otherwise, how could a conclusion of "not carcinogenic" by one strategy be balanced against a finding of "carcinogenic" from another strategy? It was necessary to balance judgments formed from an epistemological perspective of correspondence and foundationalism against those formed from a perspective of coherence. It was necessary to consider how scientific theories are to be judged; how to value principles of verification, falsification, research velocity, conceptual success, scope, proceduralism, and so on. Further, it was necessary to consider how understanding was to be balanced against observational statements; whether understanding of causal mechanisms was prerequisite to the rational formation of belief when correlations and associations were found.

Each of these considerations clearly is an issue of epistemology requiring assignments of intellectual obligation. They are issues of rationality to the degree epistemological reflection is necessary for the rational formation of belief. They are issues of human values to the degree constitutive values are required to determine the intellectual obligations imposed on an analyst. And they are issues of virtue to the degree an individual analyst has conformed to ideals of *veritas* in choosing the constituitive values to be honored in the analysis.

CONCLUSION

Philosophy was highly unified in the 12th century. The virtuous person led a good life. The good life was a life spent in accordance with the laws of God. These laws were found through the pursuit of truth, and much of this truth was to be found in an examination of nature. Nature carried a message and embodied virtue. This truth was beautiful and part of the foundation of ethics. To search for truth, beauty, good, value, morality, and virtue was to engage in a single act.

The situation is much different today. Central philosophical questions have been separated into distinct areas of study. Issues of truth are relegated to epistemology and science. Issues of beauty are isolated in aesthetics. Issues of morality are assigned to ethical theory or theology or sociology. The result of separating truth from the world and placing it entirely into the human mind has made the concept of virtue seem quaint in the face of concerns over scientific methodology and procedure.

This paper has explored the issue of whether virtues really are so remote from the practice of scientific risk analysis. The conclusion is that they are not. They lie just below the surface of judgments and the practice of risk analysis, informing the analyst of how deeply held values are reflected in the act of conducting and reporting an analysis of risk. They appear whenever society turns towards a risk analysis and asks in addition to whether the analysis is truthful or useful, whether the analyst has embodied the highest ideals of human inquiry in producing that analysis. They appear when society insists that an analysis not simply reflect probabilities and severities, but also reflect a vision of who we are as people. Virtue theory provides powerful ethical insights into the relationship between an individual, the beliefs he holds, the decisions he makes, the actions he takes, the testimony he gives, and the conception of a good life nurtured by the surrounding culture.

ACKNOWLEDGMENT

The authors wish to thank the U.S. EPA for generous financial support in developing the ideas on risk and epistemic analysis.

REFERENCES

1. K. Shrader-Frechette, *Risk and Rationality*, University of California Press, Berkeley, 1991.
2. R. Giere, "Knowledge, Values, and Technological Decisions: A Decision Theoretic Approach," in *Acceptable Evidence*, D. Mayo and R. Hollander, Eds., p. 183, Oxford University Press, Oxford, 1991.
3. D. Mayo, "Sociological versus Metascientific Views of Risk Assessment," in *Acceptable Evidence*, D. Mayo and R. Hollander, Eds., p. 249, Oxford University Press, Oxford, 1991.
4. N. Maxwell, *From Knowledge to Wisdom*, Blackwell, Oxford, 1984.
5. Interview in the *New York Times,* June 17, 1994.
6. W. Alston, "Concepts of Epistemic Justification," in *Empirical Knowledge: Readings in Contemporary Epistemology*, p. 23, P. Moser, Ed., Rowman and Littlefield, New Jersey, 1986.
7. M. Bunge, "Seven Desiderata for Rationality," in *Rationality: The Critical View*, J. Agassi and I. Jarvie, Eds., Martinus Nijhoff, Dordrecht, 1987.
8. D. Crawford-Brown and J. Arnold, "Theory Testing, Evidential Reason and the Role of Data in the Formation of Rational Confidence Concerning Risk," in *Comparative Environmental Risk Assessment*, C. Cothern, Ed., Lewis Publishers, Boca Raton, FL, 1992.
9. K. Popper, "On the Sources of Knowledge and of Ignorance", *Proc. Brit. Acad.,* 46, 1960.
10. L. Cole, *Element of Risk: The Politics of Radon*, AAAS Press, Washington, 1993.
11. H. Longino, *Science as Social Knowledge*, Princeton University Press, Princeton, NJ, 1990.
12. D. Crawford-Brown, J. Arnold and K. Brown, "Hazard Identification in Carcinogen Risk Analysis: An Integrative Approach. Part III. An Application of the Methodology: Formaldehyde in Air," report to the U.S.E.P.A., 1994.

21 VALUE JUDGMENTS IN VERIFYING AND VALIDATING RISK ASSESSMENT MODELS

Kristin Shrader-Frechette

CONTENTS

Value Judgment AI: Failure to Prove Site Unsuitability Equals
Site Suitability 292
A Two-Value Decision Framework Underlies Value Judgment AI..... 298
Value Judgment CVV: Using Computer Verification and
Validation 302
Conclusion 305
References 306

The hallmark of science is objectivity. Unlike many other areas of knowledge, science is supposed to be the most testable and reliable form of learning that we humans have. What is not testable and reliable often loses its status as science and therefore its claim to hegemony in policy and politics, in courtrooms, classrooms, and laboratories.

If science is, ideally, testable and reliable, then what happens in problematic areas where data are inadequate to ensure testability and reliability? Either we attempt to obtain adequate data, or we relegate problematic areas of knowledge to the realm of metaphor and metaphysics, or we develop a new account of what it means for knowledge to be testable and reliable. This chapter investigates two recent and prominent attempts of mathematical modelers, especially hydrologists and geologists, to develop a new account of testability and reliability applicable to research where data are lacking. To evaluate these attempts, we investigate risk assessments of Yucca Mountain, Nevada, proposed as the site of the world's first permanent, geological repository for high-level nuclear waste and spent fuel. We argue that at least two of the value judgments central to the 1992 Department of Energy (DOE) conclusion (about site suitability) are highly questionable. These value judgments are (AI) that if the site cannot be proved unsuitable, scientists should assume that it is suitable, and (CVV) that if data are lacking, computer models are adequate to verify or validate the required 10,000-year safety of the site. After investigating AI and CVV, we argue that these two claims are questionable on grounds of reliability and testability and that they exhibit problematic value judgments in risk assessment.

VALUE JUDGMENT AI: FAILURE TO PROVE SITE UNSUITABILITY EQUALS SITE SUITABILITY

Scientists evaluating the suitability of Yucca Mountain, over the 10,000 years during which the high-level wastes and spent fuel will be most dangerous, obviously cannot test and prove that the site will be safe. Because they cannot, they have adopted the value judgment (AI) that, if they cannot prove the site is unsuitable, then they may conclude that it is suitable:

> If ... current information does not indicate that the site is unsuitable, then the consensus position was that at least a lower-level suitability finding could be supported (Younker et al., 1992a, p. E-11; Shrader-Frechette, 1993, pp. 105–114).

This value judgment, from the 1992 DOE *Early Site Suitability Evaluation* (ESSE) amounts to the claim that, because one does not know of a way for repository failure or unacceptable radionuclide migration to occur, therefore none will take place. As such, the DOE value judgment AI is an example of the appeal to ignorance. Inferences known as "appeals to ignorance" are problematic because from ignorance, nothing follows. One's ignorance about potential problems — and admittedly such ignorance is always only partial — is not a sufficient or valid condition for asserting that potential problems are not significant. Of course, many conclusions in science are not based on deductively valid inferences. They rely instead on good reasons and on inductive support. The problem with risk assessors' use of the appeal to ignorance, in Yucca Mountain analyses, is that there are inadequate inductive grounds for many of their conclusions. For example, they have insufficient data regarding predictions of future volcanism and seismicity. Hence, in the face of incomplete data, the assessors often make the value judgment that they can accept an invalid inference, the appeal to ignorance. Repeatedly in its Yucca Mountain studies, U.S. Department of Energy (DOE) risk assessors appeal to ignorance. They have argued, for example:

> ...no mechanisms have been identified whereby the expected tectonic processes or events could lead to unacceptable radionuclide releases. Therefore ...

the evidence does not support a finding that the site is not likely to meet the qualifying condition for postclosure tectonics (U.S. DOE, 1986b, Vol. 2, p. 6–280).

Similarly, the DOE has argued:

...the Yucca Mountain site has no known valuable natural resources...Therefore, on the basis of the above evaluation, the evidence does not support a finding that the site is not likely to meet the qualifying condition for postclosure human interference (U.S. DOE, 1986b, Vol. 2, p. 6–292).

Likewise, the DOE has concluded.

...no impediments to eventual complete ownership and control [of Yucca Mountain] by the DOE have been identified. Therefore, on the basis of the above evaluation, the evidence does not support a finding that the site is not likely to meet the qualifying condition for post-closure site ownership and control (U.S. DOE, 1986b, Vol. 2, pp. 6–12, 6–25).

Accepting the ESSE value judgment AI [If...current information does not indicate that the site is unsuitable, then the consensus position was that at least a lower-level suitability finding could be supported" (Younker et al., 1992a, p. E-11)] virtually guarantees that, despite serious uncertainties regarding the site, risk evaluators will judge it suitable. Indeed, only an invalid inference, like the appeal to ignorance, could allow one to conclude that a site is suitable, despite massive and widespread uncertainties about it. The ESSE peer reviewers for the DOE warned that there were substantial, nonquantifiable uncertainties regarding "future geologic activity, future value of mineral deposits and mineral occurrence models...rates of tectonic activity and volcanism...natural resource occurrence and value" (Younker et al., 1992b, p. B-2). Because (1) there is uncertainty regarding crucial site factors (see the previous quotation), and because (2) this uncertainty precludes proving the site unsuitable, the value judgment AI [that the absence of a disqualifying condition is sufficient to guarantee lower-level site suitability (Younker et al., 1992a, p. E-5)] virtually guarantees that the site will be found suitable.

By assuming AI, that the failure to prove site unsuitability is sufficient to support a finding of site suitability (Younker et al., 1992a, p. E-11), the ESSE Team not only appealed to ignorance but also placed the burden of proof on those arguing for site unsuitability. Placing the burden of proof on one side of a controversy is ethically questionable because it treats the two sides inequitably. On the contrary, it is arguable that risk assessment/evaluation — like that in civil cases, tort cases — ought to follow the decision rule of supporting the side having the greater weight of evidence on its side (Shrader-Frechette, 1991, pp. 133–145). If so, the burden of proof in the ESSE is both ethically inconsistent with standard civil-case procedures and inequitable in its placement of evidentiary burdens. Hence, to the degree that the DOE's ESSE conclusions about site suitability presuppose AI, they are problematic.

Of course, it is arguable that there are pragmatic reasons for searching exhaustively for contrary evidence (about site suitability) and, after finding none, concluding that the site is suitable. In the Yucca Mountain case, however, such conclusions based on AI are inductively problematic because they often rely on studies that are neither as fully representative nor as comprehensive as they reasonably could be. For example, several risk assessors admitted that "measurement of infiltration [of water] into Yucca Mountain has not been performed." They also admitted that they had not considered fracture flow, even though it could cause rapid migration of radwaste at the site. Despite these two significant areas of ignorance, the assessors decided that radioactive releases at the site would be "significantly less" than those prohibited by government standards. They also concluded that there would be less than one health effect every 1400 years caused by Yucca Mountain (Thompson et al., 1984, pp. 7, 47, i, v–vi).

Other DOE risk assessors also based their conclusions about site suitability on studies that are not representative. One group of researchers, for example, used a computer model to simulate radionuclide transport. The computer model was based on a number of conditions that were either unknown or counterfactual (unrealistic) — such as that the groundwater flow was one-dimensional, the transport was dispersionless, the geologic medium was homogeneous, and the sorption was in a constant velocity field. Despite these unknown or counterfactual conditions, the DOE assessors used the appeal to ignorance to conclude that their model "was found to be an effective tool for simulation of the performance of the repository systems at Yucca Mountain" (Lin, 1985, pp. i, 1). How could a model, especially a model based on highly unrealistic conditions not applicable to the specific site, be found effective, short of some actual empirical testing? The same question arises for other Yucca Mountain risk assessors, DOE consultants, who write: "for the rock mass, it was assumed that nonlinear effects, including pore water migration and evaporation, could be ignored. In practice, nonlinear effects and the specific configuration of the canister, canister hole, and backfilling material would strongly influence very near field conditions" (St. John, 1985, p. 2). Why did the DOE assessors assume counterfactual conditions known not to be applicable at Yucca Mountain in specifically doing a Yucca Mountain study?

Employing value judgments that involve problematic inferences, such as the appeal to ignorance (AI) are not limited to a few studies. Indeed, they are found throughout DOE assessments of many different repositories. At the proposed permanent facility at Hanford, Washington, for example, the DOE assessors made a number of problematic appeals to ignorance, including the following:

A final conclusion on the qualifying condition for preclosure radiological exposures cannot be made based on available data...it is concluded that the evidence does not support a finding that the reference repository location is disqualified (U.S. DOE, 1986a, Vol. 2, pp. 6–75).

Making a similar appeal to ignorance, other DOE assessors, for example, claimed that the data were insufficient to allow them to state that offsite migration of radwaste would never occur, but then they concluded that there was only a small chance of contaminating public water supplies (Borg et al., 1976). If they were ignorant about whether offsite migration would occur, how could they know that contamination of water was unlikely? Still other DOE assessors, after noting that changes in groundwater flow "are extremely sensitive to the fracture properties", concluded that they could simulate partially saturated, fractured, porous systems like Yucca Mountain "without taking fractures into account" (Wang and Narasimhan, 1984, 1986). Because this conclusion was based neither on empirical work nor on application to the Yucca Mountain site, it is another classical example of an appeal to ignorance. Indeed, any conclusions that one can simulate adequately a fractured medium may rely on an appeal to ignorance, especially because some DOE assessors warn that simulation models of transport in fractured porous tuff

> ...demonstrate that the validity of the effective continuum approximation method cannot be ascertained in general terms. The approximation will break down for rapid transients in flow systems with low matrix permeability and/or large fracture spacing, so that its applicability needs to be carefully evaluated for the specific processes and conditions under study (Pruess et al., 1988).

Yet many of the experiments proposed for sites like Yucca Mountain have never been done before in an unsaturated, fractured medium (Nevada NWPO, 1988b). Very little study of similar hydrogeologic systems has been done prior to proposing Yucca Mountain as a permanent repository for U.S. high-level waste (Nevada NWPO, 1988a). If the validity of the radwaste-migration methods cannot be ascertained, and if most understanding of radioactive leaching and transport is based merely on laboratory experiments and simulations (Jantzen et al., 1989), then it is impossible to guarantee that the repository will prevent dangerous offsite migration of radionuclides.

One area of risk assessment that is most susceptible to value judgments like AI is evaluation of human error and interference. In its 1992 *Early Site Suitability Evaluation* (ESSE), for example, the DOE admitted:

> ...the performance analyses did not quantitatively evaluate the potential for adverse effects on repository performance by disruptive processes or events such as faulting or human intrusion (Younker et al., 1992a, p. 2–150).

Instead, the authors said assessments that "address these processes 'uncovered no information that indicates that the Yucca Mountain site is ... likely to be disqualified'" (Younker et al., 1992a, p. 2–150). Such a response is problematic, however, because it is based on the value judgment AI. The judgment errs because the absence of information that the site is likely to be disqualified provides no justifiable inference that the site is suitable, particularly if no precise probabilistic studies have been done. This appeal to ignorance is all the more questionable because some of the greatest uncertainties regarding a repository have to do with future disruptive events and human interference. Indeed, in 1990 when Golder Associates studied repository performance, they found "that disruptive processes that cause direct releases to the accessible environment provide the only conditions under which the EPA standards might not be met" (Younker et al., 1992a, p. 2–157). This means that the one occurrence most likely to present a radiation hazard at Yucca Mountain is precisely the threat that the DOE team did not (and likely could not) evaluate quantitatively when the DOE drew its conclusions about site suitability.

Also appealing to the value judgment AI, the ESSE authors admitted that they did not take into account "the probability of occurrence of the scenarios" when they were "estimating the probability of exceeding the [radiation] release limits" set by the government (Younker et al., 1992a, p. 2–155). If the probability of various occurrences was not taken into account, then how could one determine whether radiation-release standards would be met? Likewise, because the ESSE team admitted that it did not know the precise materials and design for the waste containers (Younker et al., 1992a, p. 2–155), how could the team conclude, as it did, that the waste-package containment would meet regulatory criteria for postclosure system performance? The team must have appealed to ignorance.

Of course, because future human behavior is so difficult to predict, it is easy to see how assessors might rely on value judgments like AI. Often, however, assessors use AI because they rely on assumptions rather than evidence. For example, several DOE risk assessors listed 11 assumptions that they had made about the Yucca Mountain site, assumptions such as that the flow path from the repository to the accessible environment would be vertically downward. [Assessors also made this assumption at the Maxey Flats facility for low-level nuclear waste; the assumption was proved false when offsite radionuclide migration occurred through horizontal fractures and bedding planes at the site (Meyer, 1975, p. 9; Pacific Northwest Laboratory et al., 1980, pp. v, I-1, I-2, I-14, IV-6, IV-9, V-7ff.).] After making these 11 assumptions, some of which are quite questionable, the assessors concluded:

> Given the general assumptions and boundary conditions listed above, it is not necessary to use sophisticated groundwater flow models or complex contaminant-transport equations to estimate radionuclide transport times and amounts at a repository site.... [E]ven without engineered barriers, Yucca Mountain would comply with NRC requirements for slow release of wastes (Sinnock and Lin, 1984, pp. 7, 37).

In appealing to the value judgment AI, the DOE risk assessors have attempted to guarantee the impossible. It is impossible for any scientist to guarantee regulatory compliance for 10,000 years, as these assessors have done, because of possible climate change, volcanic activity, and so on. The only way assessors could have formulated their conclusions in a nonproblematic way would have been to make an "if ... then" claim, such as: "if our assumptions about Yucca Mountain are reliable for the centuries required, then the site would comply with NRC requirements."

The same DOE assessors also rely on the value judgment AI and the appeal to ignorance in estimating the failure rate of waste canisters (Sinnock and Lin, 1984, p. 47) — even though the final canister design/composition has not been approved, even though there has been widespread canister failure in the past at DOE radwaste facilities, and even though (in some experiments) all canisters of the required reference material have failed (within a year) because of stress-corrosion cracking (Pitman et al., 1986). From partial ignorance about the final design/composition of the canister, no reliable conclusion about failure rate can be drawn. In particular, no valid conclusions about failure rate can be drawn on the basis of mere simulations. One group of DOE assessors, for example, used Monte Carlo simulation models for waste-package reliability (Sastre et al., 1986, p. 22), and then they concluded that the general method of probabilistic reliability analysis is "an acceptable framework to identify, organize and convey the necessary information to satisfy the standard of reasonable assurance of waste-package performance according to the regulatory requirements during the containment and controlled release periods" (Sastre et al., 1986, p. 65). However, immediately after drawing this positive conclusion about the acceptability of their nonexperimental simulation — a conclusion that appeals to ignorance — the DOE assessors contradicted their own conclusion of acceptability. They wrote: "This document does not show how to address uncertainties in model applicability or degree of completeness of the analysis, which may require a survey of expert... opinions" (Sastre et al., 1986, p. 66). If the document does not address uncertainties in the model or the completeness of the analysis, how can the DOE authors conclude that the model used in the document is "an acceptable framework" for Yucca Mountain? Again, value judgment AI appears to be at work.

A similar inconsistency, combined with an appeal to ignorance, occurs in another important Yucca Mountain risk assessment. The DOE scientists admit that "in most cases, hydraulic data are insufficient for performing geostatistical analyses. Site-characterization studies should provide the hydrogeological data needed for modeling the groundwater travel time based on site statistics" (Sinnock et al., 1986, p. 58). They also admit that they may "have underestimated cumulative releases of all nuclides during 100,000 years, by an amount that is unknown, but probably insignificant" (Sinnock et al., 1986, p. 77). If the hydraulic data are insufficient, and if they have underestimated cumulative releases of nuclides by an unknown amount, then how can the same DOE assessors conclude that the "evidence indicates that the Yucca mountain repository site would be in compliance with regulatory requirements" (Sinnock et al., 1986, pp. 1–2)? Likewise, DOE researchers say that deep tests for mineral and petroleum potential at the Nevada site are not necessary. Yet, despite their ignorance in this area, the DOE scientists concluded that the potential for petroleum or mineral reserves on site is low (Zhang, 1989, Vol. 2).

Using value judgment AI to draw a positive conclusion about the safety and effectiveness of radwaste storage at repositories like Yucca Mountain, given a variety of unknowns regarding data and hydrogeological theory, is not only logically invalid but empirically questionable. It is empirically questionable because assessors know that after high-level waste or spent fuel is placed in a repository, the radioactive fuel rods will be breached and eventually most of the cladding will corrode, exposing the fuel to oxidation that will split the cladding and expose additional fuel. Assessors admit that they have an "underlying uncertainty" about the rate of oxidation, and that the oxidation data that they have gathered has an uncertainty (of which they know) between 15 and 20% (Einziger and Buchanan, 1988). Moreover, experiments on stress corrosion cracking of spent fuel cladding in a tuff repository environment indicate that the cladding C-rings broke after 25 to 64 days, when tested in water, and after about 75 to 192 days in air (Smith, 1988a, 1988b). Hence, assessors already know that the highly radioactive fuel is likely to be exposed, rather soon after emplacement, to the uncontained environment. Given such knowledge — as well as fundamental uncertainties about unsaturated fractured geological zones, about adsorption, about seismic activity, about volcanism, and about human amplification of risk at Yucca Mountain — it is questionable for anyone to claim that the radioisotopes at the site will be isolated from the environment for 10,000 years. Indeed, it seems questionable for anyone to guarantee the requisite integrity of any repository anywhere for more than a century or two, even though such integrity is a regulatory requirement of most governments pursuing permanent disposal. However, if one cannot provide compelling evidence for site integrity for thousands of years and merely argues that there is no evidence of serious site instability, then such an argument for site suitability presupposes an implicit and questionable value judgment AI, an appeal to ignorance.

A TWO-VALUE DECISION FRAMEWORK UNDERLIES VALUE JUDGMENT AI

Perhaps one of the reasons that many Yucca Mountain assessors have appealed to ignorance is that they have been forced to employ a questionable two-value framework for making decisions about site suitability. In the DOE's 1992 *Early Site Suitability Evaluation* (ESSE), for example, this two-value decision framework is quite obvious. The DOE decided that all judgments regarding Yucca Mountain site suitability could be formulated in terms of only two options: that the site is either "suitable" or "unsuitable." In so doing, they assumed that they did not need a three-value decision framework, or a third option such as "no decision," or "the data, at present, are inadequate to assess site suitability," or "the suitability decision, at present, is uncertain." As the ESSE report formulated this two-value framework:

> ...conclusions about the site can be either that current information supports an unsuitability finding or that current information supports a suitability finding.... If...current information does not indicate that the site is unsuitable, then the consensus position was that at least a lower-level suitability finding could be supported (Younker et al., 1992a, pp. E-5, E-11; Shrader-Frechette, 1993, pp. 114–121).

Using this two-value framework presupposes that our inability to falsify an hypothesis (e.g., that the site is suitable) is sufficient grounds for confirming it. This presupposition, "affirming the consequent," is a classic form of invalid inference (Popper, 1959, 1963; Hempel, 1966; Fetzer, 1991b). It is problematic because, even if one has not been able to falsify some hypothesis (e.g., "the site is suitable"), it might still be possible to falsify it in the future. Of course, failure to find a problem with an hypothesis may be sufficient grounds (in some nonscientific or pragmatic sense) for accepting it. Presumably, however, risk assessments of Yucca Mountain (such as the ESSE) are supposed to be accepted on scientific, rather than nonscientific or pragmatic grounds. And if so, then it is arguable that the ESSE and other evaluations of proposed repository sites ought to employ a three-value, rather than a two-value, framework for site/hypothesis assessment, just as scientific evaluation does. Moreover, by not following a three-value framework for repository evaluation, assessors run the risk of begging the question in favor of site suitability. By framing the site-suitability question in terms of only two values/options, they control many of the answers to that question. Those who frame the questions control the answers.

Very few scientific decisions can be assessed in terms of two alternatives — either suitable or unsuitable — because often we do not have complete knowledge. Whenever we have less than perfect knowledge, a particular decision about a site or hypothesis can never be reduced to the two options of suitability or unsuitability. Later events could show that a judgment of suitability, for instance, was unsuitable (Fetzer, 1991b, pp. 223, 227). Hence, many situations, because they are open-ended and imperfectly known, require a three-value decision framework reflects the category of uncertainty. For example, in earlier days, scientists might have claimed that vitrifying (incorporating within glass) highly radioactive liquids was suitable as a means of preventing them from escaping into the environment (Younker et al., 1992a, p. 1–31). In 1992, however, scientists at Argonne National Laboratory learned that, contrary to previous scientific opinion, radioactive wastes may escape from glass via a new route (Bates et al., 1992). They discovered a mechanism for directly generating colloids, particles too tiny to settle out of water (Raloff, 1992). By releasing only one drop of water per week over an inch-long, half-inch diameter, glassy cylinder containing neptunium, americium, and plutonium scientists showed that exposure to slow dripping of water can change the largely nonreactive borosilicate glass into a form that facilitates the flaking of mineralized shards containing radionuclides. Hence, any claims about the suitability or unsuitability of vitrification for controlling radwastes depend on whether we have gained closure on the problems associated with vitrification. Likewise, in the absence of complete knowledge of, and closure on, numerous problems at Yucca Mountain, one can argue that scientists and policymakers ought to employ a three-value, rather than a two-value, framework for site evaluation (Ducasse, 1941; Rescher, 1969).

A three-value decision framework for risk-assessment decisions at Yucca Mountain is also more reasonable than the two-value framework, because it is more consistent with the ESSE peer reviewers' judgments about their level of scientific knowledge. In their "Consensus Position," the reviewers warned that the site was very poorly known. They said:

> ...many aspects of site suitability ... predictions involving future geologic activity, future value of mineral deposits and mineral occurrence models ... rates of tectonic activity and volcanism, as well as natural resource occurrence and value, will be fraught with substantial uncertainties that cannot be quantified using standard statistical methods (Younker et al., 1992b, p. B-2).

If many aspects of Yucca Mountain site suitability cannot be quantified and are uncertain, then the peer reviewers' own words appear to argue for a three-value framework and against the two-value decision framework that the DOE instructed them to use. Indeed, many of the ESSE peer reviewers such as D. K. Kreamer, M. T. Einaudi, and W. J. Arabasz, complained that they "were given" only the choices of site suitability or site unsuitability, despite the fact that "there is...currently not enough defensible, site-specific information available to warrant acceptance or rejection of this site" (Younker et al., 1992b, pp. 460, 257, 40–51). Hence, rejection of the two-value decision framework appears consistent with the "Consensus Position" of the 14 DOE peer reviewers for Yucca Mountain and with the comments of many individual reviewers of the ESSE.

However, if the Nevada site was poorly known and not amenable to quantitative risk assessment, why did assessors use the two-value decision framework (side suitable/site unsuitable), rather than the three-value decision framework (site suitable/site unsuitable/no decision)? The answer appears to be that, in the words of the ESSE Core Team:

> The DOE General Siting Guidelines (10 CFR Part 960) do not allow a "no decision" finding.... Thus the ESSE Core Team [of peer reviewers] followed the intent of the guidelines (Younker et al., 1992b, p. 460).

In other words, the DOE evaluation team and the DOE peer evaluators for Yucca Mountain appear to have answered the questions in the two-value decision framework that the DOE told them to use. They employed this framework, including the value judgment AI, despite the fact that their own words (see earlier paragraphs in this section) indicate that use of the two-value decision framework is questionable in the Yucca Mountain situation and in any risk assessments where there are significant or long-term uncertainties.

Like the value judgment AI, the two-value decision framework generally places the burden of proof on the disqualifier side. As we argued earlier, the ESSE appeals to ignorance and assumes that failure to disqualify a site counts as its being qualified. Hence, because of the value judgment AI, if assessors have only a short time period to disqualify a poorly known site, then it likely will be judged suitable or qualified. Such a situation obviously places a heavier evidentiary burden on the disqualifier side of the site controversy. This inequitable burden suggests additional problems with the ESSE value judgment AI and its associated two-value framework.

Many risk assessors might make a stronger case against AI. They might claim that arguments over the long-term stability of the Yucca Mountain repository, for thousands of years, are not based merely on a questionable appeal to a two-value decision framework and on attempting to draw a certain conclusion from partial ignorance. Rather, they might contend, Yucca Mountain conclusions relying on AI err because they draw a certain conclusion (about site suitability) from strong evidence to the contrary. This is part of the reason that some DOE evaluators have charged that the scientific integrity of the site-characterization at Yucca Mountain is not acceptable (Thompson, 1988). Such critics might charge that most existing U.S. nuclear-waste facilities have leaked in the past and continue to leak. Despite the improvements in the new high-level technologies, they might argue that past inductive evidence (about radwaste migration after only short periods) suggests that radwaste migration may also occur at Yucca Mountain, and facilities like it, for centuries. After all, all existing means of managing nuclear wastes have resulted in major leaks of radioactivity into the biosphere. Plutonium, for example, has travelled off-site from both high- and low-level storage facilities (Kiernan et al., 1977, pp. ix–17; U.S. ERDA, 1975, Vol. 2, pp. 11.1-H-1–11.1-H-4). At Hanford, the largest commercial storage site for high-level waste, over 500,000 gallons of high-level waste have leaked from their containers (Hart, 1978, p. 6). Officials from the U.S. Environmental Protection Agency (EPA) have indicated that, because of migration patterns of radioactivity released directly to the soil and water at Hanford, normal annual radiation releases from this facility "could result in a yearly impact of 580 man-rem total body exposure" (U.S.

ERDA, 1975, Vol. 1, p. X-74; Shrader-Frechette, 1983, ch. 2–3), in part because more than 50% of the radioactivity released directly to the soil at certain Hanford sites reaches the Columbia River (via groundwater) in 4 to 10 days (U.S. ERDA, 1975, Vol. 1, p. II., 1–57). Based on the great number of radioactive leaks from high-level storage tanks, government officials have stated: "extrapolation of past data would indicate that future leaks may occur at a rate of 2 to 3 per year" (U.S. ERDA, 1975, Vol. 1, p. III. 2–2). In addition to the Hanford leaks, extensive radioactive migration also has occurred at 1061 DOE sites, such as the waste disposal areas at Savannah River and at Idaho National Laboratory. These problems include widespread plutonium contamination in the groundwater (U.S. ERDA, 1977a, pp. II–20, IV–2; U.S. ERDA, 1977b, p. E–41; U.S. Congress, 1988, p. 46). If government officials did not prevent these leaks in the past, over a decade or two, and if they project them into the future (suggesting that not enough is being done to prevent them), then critics may be correct to question whether government officials can and will prevent leaks in the thousands of years to come at Yucca Mountain. This past inductive evidence about radioactive contamination may provide further reasons for criticizing risk assessors' appeals to ignorance, especially their value judgment AI, in evaluating future potential risks at sites such as Yucca Mountain. Indeed, U.S. EPA researchers have warned that it is in practice impossible to predict, beyond the next 100 years, what the institutional conditions and costs associated with such storage or management will be, or whether storage or management will even be possible (U.S. EPA, 1978, p. 26).

VALUE JUDGMENT CVV: USING COMPUTER VERIFICATION AND VALIDATION

Precise predictions about Yucca Mountain, over the next 10,000 years, likewise are difficult because of the long time frame. Incomplete data, especially for the long term, have forced assessors to make value judgment CVV: if data are lacking, computer models are adequate to verify or validate the required 10,000-year safety of the site. In subscribing to CVV, however, assessors must affirm the consequent, an invalid inference already discussed. The inference occurs whenever one postulates that a hypothesis is true or accurate (e.g., that the site will be suitable for 10,000 years), merely because some test result predicted to follow from the hypothesis actually occurs. In fact, however, although failure of predictions can falsify theories, success of predictions can never verify them. Of course, it is important to test one's hypotheses in order to determine whether the data falsify them or tend to confirm them. Moreover, the greater the number of tests, and the more representative they are, the greater is the assurance that the data are consistent with the hypotheses. Indeed, one of the repeatedly acknowledged failures of the Yucca Mountain assessments is that the models often are not tested (Loux, 1989, Vol. 1, p. 3;

Vol. 2, p. 2; Shrader-Frechette, 1993, pp. 121ff, 1994; Oreskes et al., 1994). If the models do turn out to be consistent with the data, however, it is wrong to assume that they have been absolutely "verified" or "validated" because, short of affirming the consequent, it is impossible to verify or validate any model. It is possible merely to achieve higher degrees of confirmation or probability through testing that the hypothesis or model has been confirmed to some degree.

At the proposed Yucca Mountain repository, risk assessors have repeatedly made incorrect "verification" or "validation" claims when they proposed to test some h, some hypothesis, such as that the number of calculated groundwater travel times are less than 10,000 years. When the calculations, data, and models are shown to be *consistent* with the hypothesis, then the assessors have erroneously assumed that the hypothesis has predictive power or has been "verified." Exemplifying their reliance on value judgment CVV, one group of assessors, studying groundwater travel time, concluded: "this evidence indicates that the Yucca Mountain repository site would be in compliance with regulatory requirements" (Sinnock et al., 1986). Many other risk assessors speak of "verifying" their models and "validating" them. For instance, one group of assessors concluded that the tools they used demonstrated "verification of engineering software used to solve thermomechanical problems" at Yucca Mountain (Costin and Bauer, 1990, p. i; see Hayden, 1985, pp. 1–1, 1–2; see also Barnard and Dockery, 1991, pp. 1–3, 4; Hunter and Mann, 1989, p. 5; Campana, 1988, p. 51; Stephens et al., 1986, p. xvi).

Software and systems engineers, of course, do speak of computer models' being "validated" and "verified." Yet, such "validation" language obscures the fact that the alleged validation really only guarantees that specific test results are *consistent* with a model or hypothesis. The results do not validate or verify the model or hypothesis, because affirming the consequent prevents legitimate validation or verification. Hence, when computer scientists speak of "program verification" (Dijstra, 1976; Hoare, 1969, 1986), at best they are making a problematic inference in affirming the consequent. At worst, they are trading on an equivocation between "algorithms and "programs". As Fetzer argues (1988, 1989b, 1991a), algorithms, as logical structures, are appropriate subjects for deductive verification. As such, *algorithms* occur in pure mathematics and pure logic. They are subject to demonstration or verification because they characterize claims that are always true as a function of the meanings assigned to the specific symbols used to express them. *Programs*, however, as causal models of logical structures, are not verifiable because the premises are not true merely as a function of their meaning. *Encodings* of algorithms that can be compiled and/or executed by a machine cannot be conclusively verified and possess no significance for the performance of a physical system. As Einstein put it, insofar as the laws of mathematics refer to reality, they are not certain; insofar as they are certain, they do not refer to reality.

Because official DOE documents and individual risk assessments (for repositories like Yucca Mountain) use "verification" and "validation" language, they are systematically misleading as to whether the studies are reliable. For example, explicitly affirming the consequent and relying on value judgment CVV, the DOE affirmed:

> Validation ... is a demonstration that a model as embodied in a computer code is an adequate representation of the process or system for which it is intended. The most common method of validation involves a comparison of the measured response from in-situ testing, lab testing, or natural analogs with the results of computational models that embody the model assumptions that are being tested (U.S. DOE, 1990, p. 3–11).

Authors of the same official DOE document, used to provide standards for Yucca Mountain risk assessments, also talk about the need to verify computational models of the waste site. They say:

> Verification, according to the guidelines in NUREG-0856 ... is the provision of assurance that a code correctly performs the operations it specifies. A common method of verification is the comparison of a code's results with solutions obtained analytically.... Benchmarking is a useful method that consists of using two or more codes to solve related problems and then comparing the results (U.S. DOE, 1990, p. 3–7).

Although the term "verification" as used by DOE assessors suggests that the computer models or codes accurately *represent phenomena* they seek to predict, the term "verification" is merely a misleading euphemism for "benchmarking." Benchmarking consists merely of comparing the results of two different codes (computer models) for simulating an identical problem. On the DOE scheme, one "verifies" a model of Yucca Mountain by comparing it to another model. What is required in the real world, however, is validating a model against reality. Assessors can accomplish this validation or confirmation only by repeated testing of the code or model against the real world, against field conditions.

Because tests can only falsify or confirm a hypothesis, not validate it, to assume otherwise is to affirm the consequent. Hence, every conclusion of compliance with government regulations, or every conclusion of repository safety, on the basis of "verified" or "validated" test or simulation results, is an example of affirming the consequent. Program *verification*, in other words, "is not even a theoretical possibility" (Fetzer, 1989a). One cannot *prove* safety. One can only demonstrate that one has attempted to falsify one's results and either has failed in doing so or has done so. In other words, there are at least three reasons that both the DOE risk documents and risk assessors at Yucca Mountain are misleading in speaking of "validation" and "verification" at Yucca Mountain. Real validation and verification are impossible, because of the problems of induction and affirming the consequent. Only falsification of an hypothesis, or determining that the data are consistent with it, is possible. In the latter case, when one obtains repeated results indicating that the data are consistent with the model or hypothesis, one is able merely to increase the probability that the model or hypothesis has been confirmed. Moreover, the DOE's and assessors' use of the terms "verification" and "validation" misleads the public about the reliability of studies allegedly guaranteeing repository safety. Finally, use of the term "verification" by DOE assessors is, in particular, misleading because they typically only compare different computer codes or models, with no reference to the real world. Any model can be tuned or calibrated to fit any pattern of data, even when the model is not well confirmed.

The arguments of this section show that there are both prudential and ethical problems with risk assessors' continuing to use the value judgment CVV and the language of "verification" in long-term, real-world phenomena. The *prudential* problem is that aiming at "verification" does not tell us what we most want to know — about complex relationships in the physical world. The more complex and the longer-lived the system, the less likely it is to perform as desired and the less reliable is short-term inductive testing of it. By emphasizing verification, at best risk assessors in such situations exhibit a "misplaced advocacy of formal analysis" (Nelson, 1989). At worst, they mislead others about the reliability of their conclusions. The *ethical* problem with risk assessors' using "verification" language is that, by encouraging confidence in the operational performance of a complex causal system, claims of "verification" oversell the reliability of software and undersell the importance of design failures in safety-critical applications like waste repositories. Such overselling and underselling exposes the safety of the public to the dangerous consequences of risk assessors' "groupthink" (Dobson and Randell, 1989). It also risks misunderstanding of software in cases where the risks are greatest. To avoid affirming the consequent, the invalid inference that repository safety models can be "verified," repository risk assessors need to refrain from the claim that their results "indicate" or "show" or "prove" compliance with government regulations or with some standard of safety. They need to avoid value judgment CVV. That is, when they have not checked the models against field data, they need to avoid misleading claims that they have "verified" or "validated" the mathematical models at Yucca Mountain and elsewhere (Hopkins, 1990, p. 1). Such terms suggest a level of reliability and predictive power which, because of the long time periods involved, is impossible in practice at Yucca Mountain. Instead, assessors might do better to speak in terms of *probabilities* that a given model or hypothesis has been *confirmed* and to avoid misleading claims (like CVV) about verification.

CONCLUSION

Science has much to teach us about risk assessment and evaluation, even though the work of assessors and evaluators relies on both scientific and policy decisions. Although risk evaluation must often move beyond science, evaluations ought not violate clear norms of science regarding the reliability of their claims. When risk assessors appeal to value judgments AI or CVV, the lesson of Yucca Mountain is that they do so at their own risk. If risk assessors "oversell" the reliability of their conclusions and decisions, they threaten not only projects like Yucca Mountain but also the credibility of future risk studies.

REFERENCES

Barnard, R., and H. Dockery. "Technical Summary of the Performance Assessment Calculational Exercises for 1990 (PACE-90) Volume 1: 'Nominal Configuration' Hydrogeologic Parameters and Calculation Results," Nuclear Waste Repository Technology Department, Sandia National Laboratories, U.S. DOE, SAND90-2727 (1991).

Bates, J., J. Bradley, A. Teetsov, C. Bradley, and M. Buchholtz-tenBrink. "Colloid Formation During Waste Form Reaction: Implications for Nuclear Waste Disposal," *Science* 256:649–651 (May 1992).

Borg, I. Y., R. Stone, H. B. Levy, and L. D. Ramspott. "Information Pertinent to the Migration of Radionuclides in Ground Water at the Nevada Test Site. Part 1. Review and Analysis of Existing Information," Lawrence Livermore Laboratory, U.S. DOE, UCRL-52078 (Pt.1) (1976). (Item 53 in U.S. DOE, DE89005394).

Campana, M. E. "Quantitative Analysis of Groundwater Flow Systems Using Environmental Isotope-Calibrated Flow Models," in *Yucca Mountain Program Summary of Research, Site Monitoring and Technical Review Activities (January 1987–June 1988)*, Desert Research Institute, Ed. (Carson City, NV: State of Nevada, Nuclear Waste Project Office, Agency for Nuclear Projects, December 1988).

Costin, L., and S. Bauer, "Thermal and Mechanical Codes First Benchmark Exercise, Part I: Thermal Analysis," Sandia National Laboratories, U.S. DOE, SAND88-1221 UC-814 (1990).

Dobson, J., and B. Randell, "Program Verification," *Communications of the ACM* 32(4):422 (April 1989).

Dijstra, E. *A Discipline of Programming* (Prentice-Hall: Englewood Cliffs, 1976).

Ducasse, C. "Truth, Verifiability, and Propositions about the Future," *Philos. Sci.* 8:329–337 (1941).

Einziger, R. E., and H. C. Buchanan, "Long-Term, Low-Temperature Oxidation of PWR Spent Fuel: Interim Transition Report," Westinghouse Hanford Company, U.S. DOE, WHC-EP-0070 (1988). (Item 26 in U.S. DOE, DE89005394).

Fetzer, J. H. "Program Verification: The Very Idea," *Commun. ACM* 31(9):1048–1063 (September 1988).

Fetzer, J. H. "Another Point of View," *Commun. ACM* 32(8):921 (August 1989a).

Fetzer, J. H. "Mathematical Proofs of Computer System Correctness," *Not. Am. Mat. Soc.* 36(10):1352–1353 (December 1989b).

Fetzer, J. H. "Philosophical Aspects of Program Verification," *Minds and Machines* 1:197–216 (1991a).

Fetzer, J. H. "The Frame Problem: Artificial Intelligence Meets David Hume," *Int. J. Exp. Syst.* 3(3):219–232 (1991b).

Hart, G. "Address to the Forum," in *Proc. Public Forum on Environmental Protection Criteria for Radioactive Wastes*, ORP/CSD-78-2 (Washington, DC: U.S. Government Printing Office, 1978).

Hayden, N. "Benchmarking: NNMSI Flow and Transport Codes: Cove 1 Results," Sandia National Laboratories, U.S. DOE, SAND84-0996 (1985).

Hempel, C. *Philosophy of Natural Science* (Prentice-Hall: Englewood Cliffs, NJ, 1966).

Hoare, C. "An Axiomatic Basis for Computer Programming," *Commun. ACM* 12:576–580, 583 (1969).

Hoare, C. "Mathematics of Programming," *BYTE*:115–149 (August 1986).

Hopkins, P. "Cone 2A Benchmarking Calculations Using LLUVIA," Sandia National Laboratories, U.S. DOE, SAND88-2511-UC-814 (1990).

Hunter, R., and C. Mann, "Techniques for Determining Probabilities of Events and Processes Affecting the Performance of Geologic Repositories," Division of High-Level Waste Management, Office of Nuclear Material Safety and Safeguards, U.S. NRC, NUREG/CR-3964 SAND86 0196 (June 1989), Vol. 1.

Jantzen, C. M., J. A. Stone, and R. C. Ewing. *Scientific Basis for Nuclear Waste Management VIII. Vol. 44* (Materials Research Society: Pittsburgh, PA, 1989). (Item 403 in U.S. DOE, DE90006793).

Kiernan, B., et al. *Report of the Special Advisory Committee on Nuclear Waste Disposal*, no. 142 (Legislative Research Commission: Frankfort, KY, 1977).

Lin, Y. "Sparton — A Simple Performance Assessment Code for the Nevada Nuclear Waste Storage Investigations Project," Sandia National Laboratories, U.S. DOE, SAND85-0602 (1985).

Loux, R. "State of Nevada Comments on the U.S. Department of Energy Site Characterization Plan, Yucca Mountain Site, Nevada," Nevada Agency for Nuclear Projects/Nuclear Waste Project Office, U.S. DOE, DOE/NV/10461-T40 (1989), 4 Vols.

Meyer, G. "Maxey Flats Radioactive Waste Burial Site: Status Report," Advanced Science and Technology Branch, U.S. EPA, unpublished report (1975).

Nelson, D. "Letters," *Commun. ACM*, 32(7):792 (July 1989).

Nevada Nuclear Waste Project Office (NWPO). "State of Nevada Comments on the U.S. Department of Energy Consultation Draft Site Characterization Plan, Yucca Mountain Site, Nevada Research and Development Area, Nevada: Volume 1," Nevada NWPO, U.S. DOE, DOE/NV/10461-T39-Vol.1 1988a). (Item 333 in U.S. DOE, DE90006793).

Nevada Nuclear Waste Project Office (NWPO). "State of Nevada Comments on the U.S. Department of Energy Consultation Draft Site Characterization Plan, Yucca Mountain Site, Nevada Research and Development Area, Nevada: Volume 2," Nevada NWPO, U.S. DOE, DOE/NV/10461-T39-Vol. 2 (1988b). (Item 334 in U.S. DOE, DE90006793).

Oreskes, N., K. S. Shrader-Frechette, and K. Belitz. "Verification, Validation, and Confirmation of Numerical Models in the Earth Sciences," *Science* 263:641-646 (February 1994).

Pacific Northwest Laboratory, et al. "Research Program at Maxey Flats and Consideration of Other Shallow Land Burial Sites," Pacific Northwest Laboratory, U.S. NRC, NUREG/CR-1832 (1980).

Pitman, S. G., R. E. Westerman, and J. H. Haberman. "Corrosion and Slow-Strain-Rate Testing of Type 304L Stainless in Tuff Groundwater Environments," Pacific Northwest Laboratory, U.S. DOE, PNL-SA-14396 (1986). (Item 172 in U.S. DOE, DE88004834).

Popper, K. R. *The Logic of Scientific Discovery* (Basic Books: New York, 1959).

Popper, K. R. *Conjectures and Refutations* (Routledge and Kegan Paul: London, 1963).

Pruess, K., J. S. Y. Wang, and Y. W. Tsang. "Effective Continuum Approximation for Modeling Fluid and Heat Flow in Fractured Porous Tuff: Nevada Nuclear Waste Storage Investigations Project," Lawrence Berkeley Laboratory, U.S. DOE, SAND-86-7000 (1988). (Item 42 in U.S. DOE, DE89005394).

Raloff, J. "Radwastes May Escape Glass via New Route," *Sci. News* 141(18):141 (1992).

Rescher, N. *Many-Valued Logic* (McGraw-Hill: New York, 1969).

Sastre, C. et al. "Waste Package Reliability," U.S. NRC, NUREG/CR-4509 (1986).

Shrader-Frechette, K. S. *Nuclear Power and Public Policy* (Reidel: Boston, MA, 1983).

Shrader-Frechette, K. S. *Risk and Rationality* (University of California Press: Berkeley, 1991).

Shrader-Frechette, K. S. *Burying Uncertainty: Risk and the Case Against Geological Disposal of Nuclear Waste* (University of California Press: Berkeley, 1993).

Shrader-Frechette, K. S. "Science, Risk Assessment, and the Frame Problem," *BioScience* 44(8):548–551 (September 1994).

Sinnock, S., and T. Lin. "Preliminary Bounds on the Expected Postclosure Performance of the Yucca Mountain Repository Site, Southern Nevada," Sandia National Laboratories, U.S. DOE, SAND84-1492 (1984).

Sinnock, S., Y. T. Lin, and M. S. Tierney. "Preliminary Estimates of Groundwater Travel Time and Radionuclide Transport at the Yucca Mountain Repository Site," Sandia National Laboratories, U.S. DOE, SAND85-2701 (1986).

Smith, H. D. "Electrochemical Corrosion-Scoping Experiments: An Evaluation of the Results," Westinghouse Hanford Company, U.S. DOE, WHC-EP-0065 (1988a). (Item 151 in U.S. DOE, DE90006793).

Smith, H. D. "Initial Report on Stress-Corrosion-Cracking Experiments Using Zircaloy-4 Spent Fuel Cladding C-Rings," Westinghouse Hanford Company, U.S. DOE, WHC-EP-0096 (1988b). (Item 153 in U.S. DOE, DE90006793).

Stephens, K. et al. "Methodologies for Assessing Long-Term Performance of High-Level Radioactive Waste Packages," Division of Waste Management, Office of Nuclear Material Safety and Safeguards, U.S. NRC, NUREG/CR-4477 ATR-85 (5810-01)1ND (January 1986).

St. John, C. "Thermal Analysis of Spent Fuel Disposal in Vertical Emplacement Boreholes in a Welded Tuff Repository," Sandia National Laboratories, U.S. DOE, SAND84-7207 (1985).

Thompson, F. et al. "Preliminary Upper-Bound Consequence Analysis for a Waste Repository at Yucca Mountain, Nevada," Sandia National Laboratories, U.S. DOE, SAND83-7475 (1984).

Thompson, J. L. "Laboratory and Field Studies Related to the Radionuclide Migration Project: Progress Report, October 1, 1986-September 30, 1987," Los Alamos National Laboratory, U.S. DOE, LA-11223-PR (1988). (Item 358 in U.S. DOE, DE90006793).

U.S. Congress. Nuclear Waste Policy Act, Hearings Before the Subcommittee on Energy and Environment of the Committee on Interior and Insular Affairs, House of Representatives, 100th Congress, September 18, 1987 (U.S. Government Printing Office: Washington, DC, 1988).

U.S. Department of Energy (DOE). "Nuclear Waste Policy Act, Environmental Assessment, Reference Repository Location, Hanford Site, Washington," U.S. DOE, DOE/RW-0070 (1986), 3 Vols.

U.S. Department of Energy (DOE). "Nuclear Waste Policy Act, Environmental Assessment, Yucca Mountain Site, Nevada Research and Development Area, Nevada," U.S. DOE, DOE/RW-0073 (1986), 3 Vols.

U.S. Department of Energy (DOE). "Performance Assessment Strategy Plan for the Geologic Repository Program," Office of Civilian Radioactive Waste Management, U.S. DOE, DOE/RW-0266P (January 1990).

U.S. Energy Research and Development Administration (ERDA), "Final Environmental Statement: Waste Management Operations, Hanford Reservation, Richland, Washington," National Technical Information Service, U.S. ERDA, ERDA-1538 (1975).

U.S. Energy Research and Development Administration (ERDA). "Waste Management Operations: Savannah River Plant, Aiken, South Carolina," National Technical Information Service, U.S. ERDA, ERDA-1537 (UC-2-11-70) (September 1977a).

U.S. Energy Research and Development Administration (ERDA). "Waste Management Operations, Idaho National Engineering Laboratory, Idaho," National Technical Information Service, U.S. ERDA, ERDA-1536 (September 1977b).

U.S. Environmental Protection Agency (EPA), "Considerations of Environmental Protection Criteria for Radioactive Waste," U.S. EPA (1978).

Wang, J. S. Y., and T. N. Narasimhan. "Hydrologic Mechanisms Governing Fluid Flow in Partially Saturated, Fractured, Porous Tuff at Yucca Mountain," Lawrence Berkeley Laboratory, U.S. DOE, LBL-18473 (1984). (Item 105 in U.S. DOE, DE88004834).

Wang, J. S. Y., and T. N. Narasimhan. "Hydrologic Mechanisms Governing Partially Saturated Fluid Flow in Fractured Welded Units and Porous Non-Welded Units at Yucca Mountain," Lawrence Berkeley Laboratory, U.S. DOE, SAND-85-7114 (1986). (Item 282 in U.S. DOE, DE88004834).

Younker, J. L. et al. "Report of Early Site Suitability Evaluation of the Potential Repository Site at Yucca Mountain, Nevada," U.S. DOE, SAIC-91/8000 (1992a).

Younker, J. L. et al. "Report of the Peer Review Panel on the Early Site Suitability Evaluation of the Potential Repository Site at Yucca Mountain, Nevada," U.S. DOE, SAIC-91/8001 (1992b).

Zhang, P. "Evaluation of the Geologic Relations and Seismotectonic Stability of the Yucca Mountain Area, Nevada Waste Site Investigation," Center for Neotectonic Studies, Mackay School of Mines, University of Nevada, (NNWSI) (1989), 2 Vols.

22 THE STEWARDSHIP ETHIC — RESOLVING THE ENVIRONMENTAL DILEMMA

David W. Schnare[1]

CONTENTS

The Dilemma — Prelude to an Ethic 312
  Basic Environmental Reality............................... 312
    What Is the Environment? 313
    What Is Man's Place in the Environment? 314
    What Is the Nature of Threats to the Environment? 316
    What Are Normal Responses to Environmental Threats?........ 317
  Rights and Responsibilities — The Environmentalist's Dilemma.... 321
    Rights and Values...................................... 321
    Responsibilities and Values 322
The Stewardship Ethic 323
Some Final Thoughts on Stewardship and Estoban 327
Endnotes and References 328

Estoban led the old man to the rock outcropping overlooking a lush valley pasture. On the sloping hillsides below they watched his brother and two border collies move 400 sheep to a new meadow near the ridge line. Resting 500 yards up the ridge, a graying wolf did not escape the notice of Estoban or the old man. Freshly graduated from an unencumbered childhood that ranged throughout these valleys, the young man counted the wolf as one of his former fellow travelers. Explaining to his older companion, "He took a ewe last night and two lambs last week. But, if I kill him, a piece of my soul will die. Yet if I protect him, the sheep will die and my family will suffer. What is the right thing to do?" They sat in their aerie contemplating this problem as the morning sun warmed their bones. It is a dilemma mankind has long confronted. It is time the long-known ethic of multivalued stewardship is remembered.

THE DILEMMA — PRELUDE TO AN ETHIC

This conflict between the use of resources for human purposes and the desire to preserve resources has been termed the environmentalist's dilemma.[2] It is the central challenge confronting stewardship. At its heart is the conflict between the most basic of competing human values. As well, this dilemma is colored by questions about man and his place in the universe. The discussion which follows is an old-time view of how to examine and resolve this dilemma.[3] The actual advice given to Estoban is not important. What is important is that a reasonable answer can evolve — that the dilemma can be resolved.

The individual who would craft a good solution to the environmentalist's dilemma is the steward. The steward, and his craft, reflects an ethic that has long existed. It is the application of that ethic to resolve the conflict of values that results in a resolution of the dilemma. Thus, we visit the dilemma, the stewardship ethic, and the application of stewardship. In so doing, several dialectics arise. Is there an anthropocentric right that is greater than a biocentric one — does man have rights that other species do not? Is man within or apart from the environment? Is extremism justifiable on moral grounds? Is there a way to answer these questions that fairly considers the values of all interested parties?

The last question can be answered with a simple yes. Application of stewardship principles is a means to reasonable, acceptable and moral resolutions.

The environmentalist's dilemma is set against a landscape of rights, responsibilities and reality. Rights and responsibilities are value-laden and when in conflict are the source of the dilemma. In some cases the conflicts are simple — values are opposed. In other cases, values are in conflict with reality. For example, demands for a risk-free life are unrealistic in the real world.

It is the real world that often defines the limits of rights and responsibilities. The rights of a species to survive are moot if they cannot be guaranteed by a real force active in the real world. Likewise, the ecological system that is "chaotic" is beyond responsible management if no one can control the sensitive initial conditions on which the system's behavior depends. Thus, reality bounds the arena within which values can be applied. As a prelude to examining values and ethics, it will pay handsomely to refamiliarize ourselves with the reality boundary so that arguments remain within the confines of the real world.

Basic Environmental Reality

Four ecological questions frame environmental reality. What is the environment? What is man's place within it? What is the nature of threats to the environment? What are normal responses to environmental threats? The answers to these questions are basic environmental truisms.[4]

What Is the Environment?

The planetary environment is made up of small, interconnected systems. Nature is not defined by specific species but by the niches and ecosystems within which species are found. As Leopold describes, nature is "a tangle of chains so complex as to seem disorderly, but when carefully examined the tangle is seen to be a highly organized structure. Its functioning depends on the cooperation and competition of all its diverse links."[5]

As well, the environment is everywhere. Attempts have been made to distinguish between the ecology and the environment, the former free of human niches, the latter which includes them. Such a division, value based as it is, does not comport with reality. Ecological boundaries are fuzzy and a variety of ecosystems may overlay each other. The most mobile and adaptable species will travel over many local ecosystems and these adaptable species have ecosystems that are extremely large. In the human case the ecosystem extends off the planet. When an environmental system is defined too narrowly, the inputs and outputs of the system overwhelm the internal workings of the system itself. An environmentalist's dilemma may arise because the system under study is too circumscribed.[6]

The environment is also resilient. Nature is commonly and usually able to assimilate insults and return to a robust state, even if the insults appear to be devastating. Lamenting the harvest of timber in the east, Udall wrote, "There was enough wood for a thousand years, but the lumbermen leveled most of the forests in a hundred."[7] Yet I often roam the sentinel woods surrounding my home. Here my dogs and I walk among the mature second growth forest, dawdling amongst iris planted by long dead settlers in a clearing nearly overgrown by oak and hickory.

The residents on these lands, some still my neighbors, clear-cut the forest from the eastern seaboard, through my property and on to the Appalachian mountains. By 1910 there were only a few patches of mature hardwoods in all that stretch of countryside. Today, 80 years later, the hardwood stands that have grown are routinely being harvested. The deer herds are larger than before the arrival of Europeans. Travel by airplane over this land in summer provides an almost seamless view of forest canopy throughout Pennsylvania, Maryland, the Virginias and North Carolina, leaving the viewer amazed that the signs of the millions of residents in these area are so difficult to identify. If this past provides any measure of the future, Secretary Udall should take comfort that these woods will last for a thousand years.

Nature is hardy and limits to the assimilative capacity of nature are only understood after reflection on the period required to establish climax conditions, in addition to consideration of the absolute limits of short term ecosystem survival. This is true for every ecosystem on the globe.[8]

Wastes are usually integral to the environment. Waste products are not the same thing as pollution. All species produce waste. Wastes that fill an ecosystem need are merely inputs to the ecosystem. It is the abundance of the unusable portion of waste that constitutes pollution or an environmental threat. A dynamic environment will find uses for waste products. Pollution is defined by the inability of the waste to be recycled or made benign.

What Is Man's Place in the Environment?

No one has ever refuted the axiom that humans are a species living within the environment. We are a top predator species with many identifiable subpopulations and a variety of ecosystems that are uniquely human. Our human ecosystems[9] are defined by political boundaries and contain or cross the boundaries of other ecosystems. They are politically bound only in that politics is the means humans have used since aboriginal times to define areas of conquest, control, and responsibility.

It is a false road that leads to consideration of whether man stands outside the environment (in God-given dominion over the earth); as a current top predator-actor whose conspicuous success defines his ascendancy over other species; or as an equal among species with the moral or ethical responsibility to grant all other species equal rights of survival. Long debates on this question are available,[10] but they are moot. We have long recognized that humans are within the system. Consider, for example, Chief Arapooish, titular head of the Crow nation:

> The Crow country is exactly in the right place. It has snowy mountains and sunny plains; all kinds of climates and good things for every season. When the summer heat scorches the prairies, you can draw up under the mountains, where the air is sweet and cool, the grass fresh, and the bright streams come tumbling out of the snow banks. There you can hunt the elk, the deer, and the antelope, wherein their skins are fit for dressing; there you will find plenty of white bears and mountain sheep.
>
> In the autumn, when your horses are fat and strong from the mountain pastures, you can go down into the plains and hunt the buffalo, or trap beaver in the streams. And when winter comes on, you can take shelter in the woody bottoms along the rivers; there you will find for you horses; or you may winter in the Wind River Valley where there is salt weed in abundance.
>
> The Crow country is exactly in the right place. Everything good is to be found there. There is no country like the Crow country.[11]

Investigation shows that every significant environmental leader has acknowledged that man is an integral part of the environment. More important, while most of these leaders are deeply committed to preserving nature, they all indicate that it is essentially for the sake of mankind, as well as for other species.

Stewart Udall: "[W]e have slowly come back to some of the truths that the Indians knew from the beginning: that men need to learn from nature, to keep an ear to the earth, and to replenish their spirits in frequent contacts with animals and wild land. We seek to provide a habitat that will, each day, renew the meaning of the human enterprise."[12]

Haitian-born Jean Rabin, who later changed his name to John James Audubon, killed many fowl and painted likenesses of the best. He once boasted of shooting enough birds to make a feathered pile the size of a small haycock in a single day. Later, when concerned about the large kill of mink and marten by one fur company, he asked, "Where can I go now, and visit nature undisturbed?" His wish was not to preserve nature undisturbed, but to preserve it so that he could disturb it with his own hunting and fishing.[13]

George Perkins Marsh: "Men now begin to realize what as wandering shepherds they had before dimly suspected, that man has a right to the use, not the abuse of the products of nature."[14]

Overton Price and Gifford Pinchot: "Conservation [was coined to reflect how stewardship would deal with the] central problem of the use of the earth for the good of man."[15]

The Sierra Club: [As stated in its constitution and bylaws] "to explore, enjoy and protect the nation's scenic resources." This is *human* exploration and enjoyment.

George Bird Grinnell, Teddy Roosevelt and the other 98 men who formed the exclusive Boone and Crockett Club, and the subset who later established the National Audubon Society, protested the commercial hunting of birds and the senseless slaughter of wildlife, but for highly personal reasons. These 100 men were big game hunters and wanted to preserve lands and species so they could hunt them through the years ahead.[16]

Aldo Leopold: "When we see land as a community to which we belong, we may begin to use it with love and respect.[17]

Wallace Stegner: "We simply need that wild country available to us, even if we never do more than drive to its edge and look in. For it can be a means of measuring ourselves and our sanity as creatures, a part of the geography of hope."[18]

And finally, there was Benton MacKay, Harvey Broome, Bernard Frank, and Robert Marshall who sat on a riverbank in the Great Smoky Mountains in the fall of 1934 and conceived of the Wilderness Society that would save remnants of the nation's virgin lands. They did this for the traditional reason of wanting it maintained for human entertainment. Not for fishing and hunting, but merely for observation — a new time forest religion.[19]

The basic human concern for the environment is about man's use of nature. When someone argues "Nature for nature's sake" they lose touch with human society, human political institutions and the chance to foster human investment in nature, regardless of how they define nature.

The question of man's place in nature is related to the size and type of change man permits himself to impose on existing conditions. The conflict is between material use of the environment and aesthetic and spiritual uses. As Udall admits, "Drawing a line between the workshop and the temple was, and still is today, the most sensitive assignment for conservation planners."[20] This is what divided Muir and Pinchot. It is at the core of arguments against the so-called 'wild lands' proposals of today.

The reality of the world is that humans have multiple needs the environment must be able to supply. The status of the human ecosystem is defined by its members' health, cultural, economic, political, aesthetic, and spiritual quality of life. Joseph Campbell has argued that myths merely reflect the greater truths of human life. As such, the mythical Anaeus was invincible as long as he was able to touch the earth. This seems true for mankind. We are linked to the earth but not bound by it or to it. Man's place within the environment is the human ecosystem and it is wherever man can survive and chooses to go.

What Is The Nature of Threats to the Environment?

A hiker coming from the forest next to my home noted on her second visit that Pine Siskins, small sparrow-like birds found in mature pine ecologies, no longer seemed to be found in our woods. As we sat in the cool evening air on the forest's edge, I lamented this loss. In keeping with the gentle cooing of the doves feeding at the fringe underbrush, this friend, a forest ecologist, quietly suggested that she never spent much time lamenting change that was inevitable. The Pine Siskin needs a sufficient pine stand in which to live. As this Virginia forest grows into a climax hardwood stand, the pines die out. The Pine Siskin moves on and other birds move in. Nature is dynamic. When judging threats to the status quo environment, they must always be cast against the inevitable changes normally expected.

The concept of preservation, as opposed to conservation, brooks no change. This is not realistic except under very narrow conditions, or for relatively short time horizons. Take, for example, the climax condition of the prairies or steppes of North America or Asia. The prairie conditions that favor the wild flowers, the seemingly endless meadows on which the bison grazed, the wide expanses of gently undulating grass lands could only remain clear from encroachment of trees through the irregular, but frequent wild fires that scoured the land and destroyed the saplings. As mankind extinguished these fires, the prairie ecology he meant to preserve was lost. When fires were reintroduced and natural fires were permitted to burn, the prairie returned.

To understand what is a threat to the environment, first it is necessary to understand what is a desirable environment. This is a human function, regardless of one's position on biocentric or anthropocentric philosophies. The reality is, preservation, conservation, exploitation, and benign neglect all reflect human decisions. Preservation and conservation require selection of a desired ecology. Exploitation of the environment reflects a selection of a specific human ecology, often mated with the explicit decision to devalue nonhuman ecologies. Even the advocates of so-called "wild lands", who would remove all traces of human intervention and eliminate the human footprint from these lands argue from their very human perspectives that whatever results therein is desirable and to be protected from human activity.

The reality is, we decide what is desirable in the environment affected by humans. Only when that is decided is it possible to determine what is a threat.[21]

Threats to ecosystems are graduated. The radiological harm of the Chernobyl reactor explosion is dramatically different than the controlled releases at the Three Mile Island reactor. In the Ukraine, up to 120 tons of transuranic long-lasting radioactive materials were explosively discharged into the surrounding countryside and if not removed from the surrounding area, will present a life-threatening risk to the higher order populations for tens of thousands of years. The American release was less than one-millionth that size, was in the form of short-lived materials, and stopped presenting a risk outside the plant within weeks of the final containment.

A realistic picture of an environmental threat must describe the specific attributes of the ecology at risk after disturbance and relate it to the frequency, duration, timing, and spacial character of the potential effect. In addition, indirect effects on economic (human) and ecological competition, disease, trophic-level relationships like predation, resource utilization, effect severity, the number and life stages of organisms affected, the role of those organisms in the community or ecosystem, and ecological compensatory mechanisms need to be described.[22]

These are the many ecological characteristics on which humans place value. Any description of an environmental threat that reflects the disruption of only a single or small number of human values is a misrepresentation of reality. As we have found when trying to resolve the conflicts over old growth forests, there is no innately correct balance between man and nature (the ecological setting). There is only a range within which the human being is comfortable. It is a very big range, perhaps with no center. To assume that there is a perfect balance is to ignore the dynamic nature of humans and nonhuman species. The dynamic nature of the environment is a basic axiom. That humans are elements of nature is a basic axiom. Thus, the dynamic nature of humans is the basis for the axiom that there is no perfect balance, only a progression of movements toward improved conditions, based on the transient sense of what is good. Environmental threats are those things that upset this balance beyond what the people are willing to accept.

What Are Normal Responses to Environmental Threats?

We now come to the crucial point that precedes a clear exposition of the environmentalist dilemma. What are the real world boundaries within which we can act to maintain or reach our environmental goals? Whenever this question arises, Joseph LeConte comes to mind. Co-founder of the Sierra Club, he wrote, "It is true that the trees are for human use. But there are aesthetic uses as well as commercial uses — uses for the spiritual wealth of all, as well as the material wealth of some."[23]

In the most stark terms, the direct implication of LeConte's realization is that, regardless of what value system applies or to what use they are put, the trees belong to someone. To restrict their use, whatever the use, is a taking. If the taking restricts the planned uses of the owner, the owner must be paid. The user must pay. End of argument.[24]

The reality of environmental protection is that mankind always pays one way or another. The environment presents material, emotional, intellectual, aesthetic, and spiritual opportunities to mankind. To some degree, opportunities are not independent of each other. If one opportunity is exploited, another opportunity is diminished. Thus, our response to environmental threats is related to the wealth of mankind. And wealth is directly related to productivity.

A basic tenet of societal and cultural maturation is that people can become sufficiently productive to permit leisure hours. In those hours they can devote time to emotional, intellectual, aesthetic, and spiritual growth. In the most basic terms, people use the time provided by increased productivity to enhance the human ecosystem. That productivity may not be applied to all sectors of mankind. Consider the exchange between President James Monroe in response to a spokesman for the Ottawa, Sioux, Iowa, and Winnebago, made at their appeal at the Council of Drummond Island in 1816: "The hunter or savage state requires a greater extent of territory to sustain it than is compatible with the progress and just claims of civilized life and must yield to it."[25] The predator human that could be described by Darwin as the successful competitor, chose first to preserve the European human ecology as it provided for greater species growth in both quantity and quality, based on the values of the top predator. The Native American lifestyle was not as productive and could not compete.

In time, however, the Europeans became so productive that they could chose to pay the opportunity costs of maintaining the less productive Native American culture, and many other low productivity cultures as well.[26] Udall, reflecting on the profligate waste of the early American plunderers of the American countryside, did not grasp the essential axiom that a species' success is a direct function of its productivity. On the other hand, Emerson, the first important American philosopher and a touchstone for the early environmentalists, did understand. He viewed exploitation of resources by a developing nation without much concern. He argued that pirates and rebels were the real fathers of colonial settlement and men would adopt sound policies once the frontier was settled and the ennobling influence of nature took effect.[27]

The axiom that productivity is linked to environmental protection has been found true even in times of economic hardship. FDR realized the unemployed, more productive when working than when begging, were a direct means of producing the wealth the society needed to pay for environmental goals. In an interesting twist to current concerns about the rights of future generations, FDR looked to the future generation to be responsible for immediate needs. The

President said: "It is clear that economic foresight and immediate employment march hand in hand in the call for reforestation of the vast [clear cut] areas."[28] He then put the unemployed to the task of reforestation through the work programs of his administration. The payment for this effort was made by the children of the workers through federal debt. These children would be the same group that would reap the rewards. In this manner, the wealth of the society was in the unemployed people and the productivity was realized well before the monetary exchange which completed the transaction.

Government is generally not the source of increased productivity. The economic engine of a society, when made more efficient, provides the resources and time necessary to protect and preserve environmental goods. The donations that lead to the Acadia National Park, the Hudson River Palisades parks, the Great Smokey mountains and Shenandoah National Parks and the Jackson Hole facilities within the Grand Teton National Park came directly from the wealth generated by a society moving from wood to oil for heat and energy, a quantum leap in productivity.

There is, however, a limit to productivity increases, at least in the short term. Thus, real world limits on our responses to environmental threats take the form of trade-offs between investments in the human ecological system. As Thomas E. Lovejoy explains, "Any preservationist, however militant about maintaining wild places, knows full well that their protection rests on a stable human population/resources equation elsewhere."[29] Lovejoy's admission reflects a realistic understanding that more and more environmentalists have come to share. In the real world, it has been recognized that humans are not going to (nor should be expected to) sacrifice human material goals on the alter of nonhuman species survival.[30] Rather, a balancing between human material goods and human emotional, intellectual, aesthetic, and spiritual goods will be necessary to ensure environmental protection.

An understanding of the need for trade-offs has lead to accusations that unrepentant environmentalists reflect an elitist view which does not bode well for the less wealthy within the society who suffer when nonhuman species take precedence over humans. A shrinking number of environmentalists dispute the Emersonian view of trickle-down environmental benefits in an increasingly productive society. Their approach is to demand reliance on values requiring social justice and environmental stability over unrestrained economic growth. These values may reflect legitimate human desires, but when used as some form of trump card over trading among social goals, they fall well outside the real world. As discussed later, there is no one value that is a stable trump card, nor should one be expected in a dynamic human ecosystem.

If one is to care about social equity, one must examine the wants and needs of the economically disadvantaged. They are not the ones asking for more spotted owls or unspoiled wild lands. They are the ones asking for better jobs and more consumer goods than they can now afford.[31] Real world governments and leaders have increasingly looked to the needs of the disadvantaged and have not permitted the more wealthy classes (and their aesthetic and spiritual needs) to bar resource uses needed to improve the condition of the poor. On the other hand, economically motivated leadership, from the Roosevelts to the Rockefellers, have been cognizant of the disadvantaged classes' emotional, intellectual, aesthetic, and spiritual needs that are satisfied by nature.

The fact remains, the productivity (not growth) of the economy determines the amount of additional time and energy available for higher order survival desires like aesthetic and spiritual requests on behalf of the environment. The individual that wants a density of no less than 20 acres per human, to allow sufficient inculcation of his spiritual and aesthetic needs, will find himself in a bidding war with others of like mind, because of the limits of land. The bidding may or may not be in money, but it will always favor the wealthy. As the greatest pool of wealth is held not by the upper income earners, but by the large masses of middle income people, the desires of the wealthy middle class will be the desires reflected in the political will of the majority. This is a real world boundary on the nature of our response that must be kept firmly in mind.

What then is the normal response to environmental threats? It is to assess the competing human needs and select responses based on dominant human values. By so doing, for example, we develop subsidies for small family farms instead of returning part of those marginal farms to nature and the rest to large scale, high productivity farming conglomerates. In so doing, we place a subset of humans over greater productivity and thus over the environmental good that increased productivity would have wrought.

In the real world of federal budgets, responses to environmental threats are assessed in comparison with the needs of veterans, the national science agenda and the space program. By far, it is the space program that dominates federal spending competition with the U.S. Environmental Protection Agency (EPA).[32] We assess the need to expand the human frontier and compare it to the need to expand control over private use of ecological opportunities.

The reality of our response to environmental threats, thus, reflects the simple truths about how we view the environment and ourselves. In the most basic terms, the human species, and its various subpopulations, always act to survive and seek to improve their quality of life. All people display environmental stewardship within their personal ecosystems. As a result, the wealth of a community defines its capacity to protect the environment.

Finally, beyond the desire and productive capacity to act on threats to the environment is a practical aspect to stewardship. One cannot manage the unknown, but perfect knowledge is unnecessary. Species within healthy dynamic ecosystems are successful as long as they have some basic information. Thus, reality does not require omniscience. However, species and the human subset of stewards fail in the absence of basic information.

With this review of reality in mind, it is possible to examine the basics of the environmental dilemma, the questions of rights and responsibilities in conflict.

Rights and Responsibilities — The Environmentalist's Dilemma

At the core of many an environmental dilemma is the question of the rights of man and other species, as well as the responsibilities of man to other species. While the values of various environmental and human goals are many in number, reflecting the many who hold those values, the issue of rights and responsibilities is more simple.

Rights and Values

The most important element defining the clash of values leading to the environmentalist's dilemma is that the conflict is a disagreement between humans about the value of things from the human perspective. This dilemma is of human construction — not because of human actions, but because of human interests and values.

The second most important element in the dilemma is that any rights claimed on behalf of other species or ecosystems are claims for rights within the human community. Thus, in truth, there is no actual claim that a snail darter has the "right" to survive. Rather those humans who care about the snail darter claim the right to prevent human activity from eradicating the snail darter. The term "rights" has meaning only within the human political system.

Keeping these two elements in mind, the concept of the environmentalist's dilemma begins to dissolve into a tractable issue. It also becomes obvious that the original specification of this so-called dilemma ignored these underlying principles.

Specifically, Norton introduced the concept of the environmentalist's dilemma in his 1991 work. He describes the dilemma as the difficulty of balancing the use of resources for human purposes against the moral desire to preserve resources, perhaps for future human use, but mostly because those resources have innate value that is lost if they are used in any way.

It is the possible existence of this innate value that is a red herring with regard to social decision making on environmental protection. Regardless of whether a nonhuman species (or the Grand Canyon, for that matter) has innate value, no degree of innate value implies or requires "rights" within the human social and political system, the only system available to protect and preserve the environment.

It is important to understand the system under which "rights" themselves exist. Rights are imbued by a political system. They are given, they are not innate, unless one takes the view that God imbues them. In the absence of an enforceable spiritual mandate, all rights stem from the system that grants them. Thus, other species have rights equal to man only if man grants them or if God begins to discipline mankind for some human violation of other species. As there is no credible example of the latter, then it is the law of man which must be examined to find the basis for plant and animal rights.

To be protected under human law, human law must establish these rights and in the U.S., there must be a plank in the federal constitution that empowers the society to make such a federal law. It is an open question as to whether this case has been fully made. Further, we can certainly turn to many cases where mankind has refused to grant rights to individual animals, and to a few cases where entire species are given no political or legal truck.

Thus, the dilemma is first and last a question of balancing between human desires, some of which may appear as actual rights (e.g., property rights) and others which merely masquerade as rights.

Responsibilities and Values

While rights reflect social values, responsibilities reflect individual values. A society may choose not to guarantee the survival of all ponds, but individuals can choose to guarantee the survival of any particular pond. This, of course, is a luxury only available in a relatively productive society that has guarantees of individual freedom, including property rights.

The individual in the United States has the requisite freedoms and guarantees. He or she can build a minority consensus within the society to protect or preserve a resource. This group, while unable to build a majority political position based on their personal morals and values, can generate sufficient capital to purchase the right from the majority.[33]

The remarkable success of the "purchase and preserve" community demonstrates the power of moral suasion combined with economic strength. It is also a testament to the underlying concern shared by nearly all people that some degree of attention must be placed on rebuilding and maintaining environmental resources, especially those near where they live, or of special import to them as individuals.

Thus, we recognize the personal and individual nature of environmental responsibility as a moral imperative rather than a political one. This allows a reformulation of the environmentalist's dilemma into a question that, while value based, is tractable and available for resolution.

In short, the dilemma can now be seen as no more than the challenge of balancing human resources among diverse and occasionally mutually exclusive human wants and needs (that reflect conflicting or competing human values). The central question then becomes, is it possible to relate these conflicting and competing human values to a commonly or universally acceptable ethic. If so, then we would have a balancing mechanism by which to meet the challenge heretofore called the environmentalist's dilemma.

THE STEWARDSHIP ETHIC

Regardless of whether the body politic chooses to resolve its environmental dilemmas within legislative, judicial, or executive arenas, it is common that balancing efforts are inefficient because of the typical two-valued adversarial nature of conflicts. Over the last four decades it has come to pass that there is usually no neutral third party who embraces the values of the various sides to the conflict. If there were, there would also be a moral perspective that has already integrated the divergent values, at least to some degree, and that can then serve to inform the decision process.

Historically, stewardship has been the ethic that integrated the divergent values. Stewart Udall, writing 30 years ago, noted this ethic when describing the outlook of George Perkins Marsh — an ethic dating from 100 years ago:

> Science, guided by a new land conscience, should have an opportunity to give primacy to the needs of coming generations; the success of our stewardship would be measured by the extent to which the land was redeemed and enriched for those to follow. Man was part of the cycle of nature, and the fall of a sparrow or the felling of a tree should be studied in the context of the total environment. These concepts were Marsh's most valuable contributions, and in time they became part of a saving American creed.[34]

The steward, however, is not a value-free scientist or the perfect embodiment of societal values. No one could be that. Rather, the steward is a means to transfer the values of the society onto the economic, political, and scientific arguments, thus ascribing humanistic solutions that represent the best knowledge about the protection or enhancement of nature and man's place within it. The steward is responsible for articulating the "contextual pluralism" that Norton describes in his chapter.

While certainly not exhaustive, there are 12 aspects of stewardship that describe the ethic that integrates concern for man with concern for the rest of the environment. Application of this ethic results in the melding of society's values into balanced decisions and actions.

1. The Responsibility for Stewardship Falls to Humans Alone and to All Humans to Some Degree.

There is no stewardship without a steward. Because man has been a steward of his own environment throughout time, this is a comfortable ethic embraced openly by most.

2. The Steward Represents Human Concerns for the Environment, Remembering that Humans Are Part of the Environment Too.

The environmental advocate spins his or her own single thread. The steward weaves these many environmental threads into the multi-hued fabric of human society. There is no cloth woven directly from the spindles of the advocates alone. Cloth with an environmental component comes only from the looms of a multi-valued human body politic.

3. The Purpose of Stewardship Is to Promote as Much Environmental Quality as Humans Want and Can Afford.

Environmental quality is neither a duty nor a right, it is a public good. Because man's purse is not a cornucopia, no citizen or citizenry can afford to purchase an unlimited basket of public and private goods. Just as we must decide how much human health care we can afford, we must also decide how much environmental quality we can afford. When we buy environmental protection, we cannot use the same dollars to buy assistance to children, the poor, the hungry, and the sick. Citizens must chose which public goods to buy. Environmental quality, the common defense, good roads and bridges, subsistence for those in need — all are things we want. The steward identifies those among the environmental goods we most need and want, and works to ensure those needs are understood in the context of all human needs.

4. The Essence of Stewardship Is to Promote a Thoughtful Balancing Among Investments in Human Needs, Including Environmental Quality.

As discussed early in the chapter, there is no innately correct balance between man and nature, only a progression of movements toward improved conditions, based on the transient sense of what is good.[35] This is akin to the merchant who must add or remove weights as the customer places or removes fruit onto the merchant's scale — an ever-changing balancing act.

More important than this balancing act, keep in mind, it is not the steward who does the actual balancing. That is a political and economic process. While debate rages on how best to make detailed decisions, there is no question that the means of melding conflicting values is a political process.[36]

5. Stewardship Has Always Been a Politically Assigned Function Grounded in Science and Humanity.

It should not be a surprise, therefore, to find Marsh, Powell, Pinchot, Muir, Leopold, Carson, and a legion of current stewards with government background and political experience. The environmental advocate that does not reflect a politically assigned function is unlikely to display a robust stewardship ethic.

6. Stewardship Existed Before Formation of the State and Is Recognized by the State, Not Authorized Thereby.

There is no better example of this than that provided by Theodore Roosevelt. As Udall describes: "He was the servant of the people, not of the Congress, and the charter he looked to for his power was the Constitution itself. Furthermore, he proposed to function fully and affirmatively as the nation's landlord and chief husbandman wherever and whenever his sense of stewardship told him action was needed."[37]

7. Economic and Political Processes Define the Scope of a Steward's Control, but Not the Limit of a Steward's Concern.

The steward recognizes the dynamic and inter-related nature of the environment. Thus, the steward's concerns lie with the whole system, not only that part to which mankind has access and can dominate. Regardless, while concern for the whole will color his or her actions, those actions have moment only within the bounds placed by economics and politics.[38]

Economics is a means to understand the relative value of goods. In the case of those goods not traded on a market, economics falters. However, the political system values goods which the economic system cannot. Between the two systems, it is always possible to understand the relative value of one human's environmental desire compared with that of another.

Environmental advocates eschew economics because they claim economics does not value the contextual nature of a species living within a greater whole. In general, this is not true. Economics and politics place higher values on rare goods. When a desired good is threatened, its value goes up. Thus, a species, living within an ecosystem, has a value derived by the degree of threat to the species. That is a function of the status of the entire system. It is the steward that ensures the public and decision makers understand the relationship of the species to its ecosystem. The steward provides a balanced view of the threat to a species. The public incorporates this information and places a relative value on the species. The economist, explaining the meaning of alternative public investments, places a value on the species that is derived from the public interest. The steward's role is to bridge the information gap between the environmental scientist, the public, and the economist — a well-known and long-standing responsibility.[39]

The validity of this seventh element is also made plain when political and economic systems dissolve, for example, during human conflict. Then, stewardship is not only very difficult, but may lose some relevance. Concern for threatened species and niches are irrelevant if the productivity of the human niche is lost, as it is the surpluses of the productive human niche that allows stewardship. In the absence of the productive surpluses, the threatened niches are condemned anon. Thus, in times of political crisis, the steward's first concern may well end up being for the maintenance of human productivity. In essence, there are no clean bombs, only the potential for a clean peace.[40]

8. Natural and Social Sciences Define the Breadth of Knowledge Available to the Steward.

Perfect knowledge is unnecessary, but a minimum amount is essential. The deer does not need to see, or even smell the fire before it starts to move out of an area threatened by flame. More sensitive species alert the deer. They do not yell fire, they merely break their normal routines, giving the deer a warning. In like measure, a modicum of information can lead to telling conclusions, even when the data in use are highly uncertain. As presented in the chapter on Risk and Rationality (Schnare), modern analytical techniques permit the steward to know when he is on uncertain ground, when that is important, and when it is not.

The abiding concentration of the private and government sectors to better understand ecologies reflects this information need. However, data alone are not sufficient. Misrepresentations of data, or the failure to explain data in the context of society's values, is tantamount to no data or wrong data. The steward must not only amass sufficient data, but they must be in a form that is fair to the body politic. This is another reason the environmental advocate is rarely willing to assume stewardship values, even though the steward integrates the advocates' values into the full picture presented.

9. The Needs and Desires of the Human Species Define the Scope of a Steward's Responsibility.

This also means environmentalism is grassroots oriented. As Pinchot pointed out, "The public good comes first. Local questions will be decided by local officers on local grounds." Thus, stewardship is a local responsibility and the scope of that responsibility is defined by the human system that exercises the stewardship role. When the needs and wants of the locale are continent wide, then so too is stewardship. When there is no need or want, there is no stewardship, in part because there is no interest, mostly because there is no investment of the human species.

10. Together, the Degree of Knowledge, Responsibility and Control Available to the Steward Defines the Degree of Environmental Quality Achievable Within the Steward's Ecosystem.

It has been found that the individual with the greatest knowledge and responsibility will also have the greatest control, and that these three elements work together.[41] As one increases, so do the others — and vice versa. The clever and experienced bureaucrat will quickly realize that the steward may become the decision maker, insinuating himself between knowledge and the body politic. However, the reservoir of responsibility and control lies with the body politic, and the steward cannot win out in the end. Thus, the steward must recognize his or her own role and play it. However, in so doing, he must realize the limits of control and responsibility assumed and available to the body politic, and not expect more than is reasonable from the society, as it is at that time.

11. As the Steward Represents All People and All Their Values, Stewardship Requires Multi-Value Logic.

Muir, who created the concept of single-value political action to achieve environmental goals, was the first to use extremist approaches. Udall refers to him as a "zealot who preached a mountain gospel with John the Baptist fervor".[42] Muir described his political success in Washington and Sacramento on behalf of Yosemite Part as follows: "I am now an experienced lobbyist; my political education is complete. I have attended Legislature, made speeches, explained, exhorted, persuaded every mother's son of the legislators, newspaper reporters, and everybody else who would listen to me."

Norton, describing Muir, notes the weakness of this approach. "That rhetoric left no room for integrating legitimate uses of nature." Single value advocates cannot recognize or craft a balanced solution. Thus, anyone that is an "advocate" would likely make for a poor steward.

The steward, however, recognizes the breadth of human values and seeks to integrate them all. Thus, the steward pays attention to issues of human equity, as well as the desires of the public for employment, good health, and opportunity. The steward realizes that placing one value over another remains a political concern.

12. The Steward Dispatches Responsibility by Informing the Political Process and Executing the Political Decision.

The steward does not replace the political will with some other value system, but educates the body politic in a manner that will allow them to make full use of environmental knowledge. Hays and Norton recognize the steward must confront conflicting values, but they focus on being a "broker among conflicting demands".[43] They miss the lessons of Leopold, who had to operate within government. The steward, even the bureaucrat who has this role, is a framer of values. The role is not one of an arbiter, but as an explainer. Otherwise, the steward and decision maker is subject to finding a compromise among extreme alternatives, reflecting conflicting values. Rather, the job is to develop shared values, or at least an appreciation of the values of the others involved, and hence the willingness to agree on a mixed-value solution — the hallmark of the steward and the product of the political system.[44]

SOME FINAL THOUGHTS ON STEWARDSHIP AND ESTOBAN

If there is one value required of the steward it must be a logical respect for the whole. There must be a place in the heart of the steward for both humans and nonhumans and this heart must be firmly connected to the steward's head.

The steward must have, in his or her soul, a love for the land that can propel him or her past the human penchant for rules designed for mankind alone. It must push mankind past greed — the excessive desire of men for creature comforts. Inexorably, when productivity is stalled, the boundary of human comfort is the carrying capacity of the land and the marginal value of nature over humans. We have a penchant for nature as one of our creature comforts. Thus, we should have confidence that there dwells within the heart of mankind a desire for balance. That desire alone is not enough.

The society needs a corps of stewards who have a sufficient breadth of knowledge. They will not be wrought solely from some transcendental visit to a wild place. It will have to be the crucible of human existence and a sympathy for survival of the body and the spirit that makes the heart of the steward.

With regard to the young boy's dilemma which opened this chapter, a balancing between values is not impossible. His companion is a weathered steward who knows the wolf is outside his range. He explains to Estoban, "This wolf is an opportunist. He comes for your sheep because it is easier than to compete within his old system. If you shoot this wolf, he will die a quick and merciful death compared to the slow starvation and disease he faces if he were captured, moved and forced to compete in his home range. You can also see he is an old wolf. He has lived his life and left his sons to carry on. You must now look to your own."[45]

We find there is no black and white. There is no simple rule to apply. People tend to want an easy rule — a simple philosophic concept that would embrace all values and provide a simple tool for making the hard choices. That can never be the way. Even my own penchant for dependence on dynamical ethics and an interdependent reliance on knowledge, responsibility and control, demand a complex tool. We must recognize that complex is far more common than simple and while we may desire a simple "elegant" approach, we are not God and have no infinite capacity to devise, much less recognize, perfection when making hard choices. That is what makes these choices "hard". The steward must have the tenacity to apply the complex tools of our society to the complex issues of the environment and let a complex society decide.

ENDNOTES AND REFERENCES

1. Dr. Schnare is a former Branch Chief of Economic, Legislative, and Policy Analysis in the Office of Water at the U.S. Environmental Protection Agency. He holds degrees in chemistry (B.A.), public health (MSPH) and environmental management (Ph.D.). The opinions expressed in this chapter are his own and do not necessarily reflect those of the U.S. EPA.
2. Norton, B. G., *Toward Unity Among Environmentalists*, Oxford University Press, New York, 1991.
3. There is a long tradition of governmental environmentalists providing thoughtful contributions on the large issues of the day. Muir, Pinchot, Leopold, Marsh, and Carson quickly come to mind. This chapter is in the mold of those writers. As such, although this essay is heavily footnoted, the reader cannot help but recognize the personal nature of this presentation. In it are a few opinions that reflect my 24 years of government service and academic enterprise. I have carefully restricted those opinions to subjects on which I have a modicum of expertise. I note this for the reader (on advice of a reviewer) because of the penchant of a few academicians to discount experience, regardless of where it is accumulated. I am happy to accede to this request if it will make the chapter more acceptable to this small group. In the larger sense, however, the purpose of this chapter is to present an argument that is generally self-evident and based on observations that anyone could make. That I have made them only gives me confidence that others will do so as well.

4. A colleague questions whether this formulation of reality excludes the perceptions and values of those who believe in God, and/or an afterlife in heaven or
   on earth. As is discussed later, stewardship reflects the entire breadth of values.
   The role of God on environmental decision making is usually related to the
   rights of animals and plants. That too is addressed elsewhere. However, because
   this question often comes up as readers commence this essay, I want to assure
   you, these values and concerns are not to be ignored.
5. Leopold, A., A Biotic View of Land, *J. For.*, 37, 727, 1939. See also: Leopold,
   A., *Sand County Almanac*, Oxford University Press, London, 1949; Marsh,
   G. P., *Man and Nature*, 1864; and Elton, C., *Animal Ecology*, 1926.
6. Norton, B. G., *Toward Unity Among Environmentalists*, p. 176. Pinchot also
   makes this point with regard to planning and conservation. Pinchot, G., *Breaking New Ground*, Island Press, Washington D.C., 1987 (originally published in
   1947).
7. Udall, S. J., *The Quiet Crisis*, Holt, Reinhart and Winston, Salt Lake City, 1963,
   p. 56. Udall updated this work, expanding it some and appending to the title
   "The Next Generation" (Peregrine Smith, 1988).
8. A recent example of the assimilative capacity of nature is the Yellowstone
   forest. Burned 6 years ago to less than 80% of pre-fire size, this forest is
   regrowing over much, but not all of the burned areas, with more diversity that
   its previous climax condition. The wildlife have also reemerged in a very strong
   condition (perhaps too strong).

   Recovery from the Yellowstone fire also provides an important lesson on
   stewardship. The Park Service, required to "let nature take its course" is
   condeming the park to an extremely slow (200 year) recovery and possibly a
   permanent change in ecology. With an elk herd that is more than 20 times the
   size that the land can carry, reemerging aspen and cottonwood is being so
   overgrazed during the winter that these species probably will disappear, leaving
   the elk to starve or move onto private lands where they will demolish the hay
   stores of the farming community. Further, lodge pole pine is continuing its
   encroachment onto the park, especially due to the fire, and the two thirds
   reduction in white bark pine will mean the grizzly bear population will shrink
   to the same degree. Reflecting discussions presented later, real stewardship
   would require action by humans, not benign neglect, and if done would reduce
   the time needed for restoration of the mature ecology by a century, if not more.
   Real stewardship would mean drastically reducing the elk herd, at least until the
   newly released wolf population can produce a more self-regulating balance. It
   would also mean planting aspen and cottonwood as well as a mix of pines. It
   may also mean culling the grizzly population, if warranted.
9. There is a propensity among some to apply the term "ecosystem" exclusively
   to the dwelling place of nonhuman species. The term is a relatively old one,
   dating to 1935. The Webster's Ninth New Collegiate Dictionary, Merriam-
   Webster, Springfield, MA, 1984, reflecting its original use, provides a single
   definition: "the complex of a community and its environment functioning as an
   ecological unit in nature." This applies to humans as well as it does to any other
   species. What may be different is that we define our system in subjective
   cultural and spiritual terms as well as practical objective physical ones.

Nevertheless, the term applies well to us and gives us a very workable tool to understand our relationship to the land and the other species within the community in which we live.

10. Norton, B. G., *Toward Unity Among Environmentalists*, p. 66. He provides a detailed examination of this subject. Several entirely new vocabularies have been created to attempt to define the differences between what most view as either dominance given by God or success achieved by survival of the fittest. In the final call, the question is moot and the vocabularies are unnecessary because humans have dominion over human political systems and it is those systems that generate the power and energy to massively alter the landscape for good or evil (regardless of how you define either).

11. Udall, S. J., *The Quiet Crisis*, Holt, Reinhart and Winston, Salt Lake City, 1963, p. 5.

12. Udall, S. J., *The Quiet Crisis*, Holt, Reinhart and Winston, Salt Lake City, 1963, p. 12, 175.

13. Udall, S. J., *The Quiet Crisis*, Holt, Reinhart and Winston, Salt Lake City, 1963, p. 44.

14. Marsh, G.P., *Man and Nature*.

15. Udall, S. J., *The Quiet Crisis*, Holt, Reinhart and Winston, Salt Lake City, 1963, p. 106.

16. Udall, S. J., *The Quiet Crisis*, Holt, Reinhart and Winston, Salt Lake City, 1963, p. 148.

17. Leopold, A., *Sand County Almanac,* Forward, p. xviii.

18. Stegner, W., *The Sound of Mountain Water*, 1961.

19. Udall, S. J., *The Quiet Crisis*, Holt, Reinhart and Winston, Salt Lake City, 1963, p. 212.

20. Udall, S. J., *The Quiet Crisis*, Holt, Reinhart and Winston, Salt Lake City, 1963, p. 119.

21. Rowe, W. C., *An Anatomy of Risk*, John Wiley & Sons, New York, 1977, p. 66, for a discussion of the hierarchy of human values reflected in the decision about the desirable environmental condition. These values have been directly related to human functions and ecosystem utility by Manning E. W., and Sweet, M.F., *Environmental Evaluation Guide Book: A Practical Means of Relating Biophysical Functions to Socioeconomic Values*, Centre for a Sustainable Future, Foundation for International Training, Toronto, 1993.

22. There is a growing literature on how to examine ecological threats that is only beginning to catch up with the mature subject of human risk assessment. Extremely influential are two National Research Council publications: *Risk Assessment in the Federal Government: Managing the Process*, National Academy Press, Washington, DC, 1983, pp. 254; and *Ecological Knowledge and Environmental Problem-solving: Concepts and Case Studies*, National Academy Press, Washington, DC, 1986. The first of these two is now undergoing revision.

23. Quoted in: Stephen Fox, *John Muir and His Legacy*, Little, Brown, Boston, 1981, pp. 115.

24. There is no value in arguing about whether man can own nature or whether he has the right to impose on other species. This has been discussed earlier in the paper, and deserves reiteration to those offended by this fact. The rights of humans devolve from political systems. The right to use the land, air, and water

is, for the most part, a matter of settled law. Even where there is argument about who owns the right to control uses of the environment, the default condition is that mankind owns it, at least to the degree that mankind must decide whether and how to invest or disinvest in the resource. There is always an opportunity cost to using or not using a resource. Mankind always pays this cost. It is the nature of choice within the human ecosystem. Thus, the issue is not whether man can, should, or does own the land, air, and water. The issue is how mankind will use its limited resources to satisfy its very human wants and needs.

25. Udall, S. J., *The Quiet Crisis*, Holt, Reinhart and Winston, Salt Lake City, 1963, p. 10.

26. Other examples are family-owned small farms and counter-culture artists.

27. Udall, S. J., *The Quiet Crisis*, Holt, Reinhart and Winston, Salt Lake City, 1963, p. 48.

28. Roosevelt, F. D., Democratic Convention Presidential Nominations Acceptance Speech, 1932.

29. Lovejoy, T. E., in the preface to Norton, B. G., *Toward Unity Among Environmentalists*, Oxford University Press, New York, 1991.

30. Norton, B. G., *Toward Unity Among Environmentalists,* Oxford University Press, New York, 1991, p. 139.

31. Graham, F., *Since Silent Spring*, Houghton Mifflin, Boston, 1970, p. 75; to wit, "In a 1985 agenda document, the Group of Ten said: 'Past environmental gains will be maintained and new ones made more easily in a healthy economy than in a stagnant one with continued high unemployment.'"

32. Appropriations for the U.S. Environmental Protection Agency fall within the House and Senate Appropriations Subcommittee on HUD, VA, and Independent Agencies. Despite the work of the Budget Committee, the full Appropriations Committee redivides the overall budget into pieces, giving each subcommittee a ceiling. Within that ceiling, the agencies and departments compete for appropriations. EPA competes with HUD, the Veterans' Administration, the National Science Administration, NASA, and a handful of very small agencies. NASA is a giant among these competitors.

33. Keep in mind that many minority concerns have been turned into majority political positions. In this way, concerns about "environmental justice" are beginning to emerge as a means to protect minorities and their values. The interesting political question that is becoming a significant current issue is how to reimburse those members of the society that suffer as a result of privilege being given to a minority. The typical approach is to spread the cost widely, where no payment can be easily made, or to make the payment and transfer the rights to the minority.

34. Udall, S. J., *The Quiet Crisis*, Holt, Reinhart and Winston, Salt Lake City, 1963, p. 81.

35. For a deeper discussion of this concept, see: Rawls, J., *A Theory of Justice*, Harvard University Press, Cambridge, 1971, p. 19.

36. Norton, B. G., *Toward Unity Among Environmentalists*, p. 123. There, Norton, Portney, and Ayers discuss whether detailed decisions should be done through legislation or rulemaking. In either case, the actual balancing is done in a manner reflecting the political will of the public. This is also true for local decisions, or even ones made by a family. The means mankind uses to order his society, regardless of its size, is the means used to combine values into decisions about how to balance competing needs.

37. Udall, S. J., *The Quiet Crisis*, Holt, Reinhart and Winston, Salt Lake City, 1963, p. 131.

38. Horton goes beyond this to argue even more strongly: "To preserve the full diversity of such connections into the next century will require a broader view of environmental protection than is likely to evolve through our legal and political systems alone. Horton, T., *Bay Country*, Johns Hopkins University Press, Baltimore, 1987, p. 35.

39. Norton recognizes this linkage to some degree as well: "The emerging consensus among environmentalists regarding biological complexity therefore rests not on nonanthropocentrism, but on a growing recognition of the systematic nature of our biological context and an associated realization that the good life must be an ecologically informed life." Norton, B. G., *Toward Unity Among Environmentalists*, Oxford University Press, 1991, p. 154.

40. When faced with protecting humans and protecting the environment, we have long recognized the primacy of the human. General Sheridan speaking to the Texas legislature said: "Let them kill, skin and sell until the buffalo is exterminated, as it is the only way to bring about lasting peace and allow civilization to advance." (Udall, *The Quiet Crisis*, p. 65) In 1875, President Grant vetoed the buffalo-protection act, the first bill passed by Congress to protect an endangered species, for the same reason — peace on the plains. Only when the nation became productive enough was it possible to reestablish the bison herds, as we have now done.

41. Pineo, C., Miller, G. W., and Schnare, D. W., *Environmental Health and Integrated Health Delivery Programs*, Agency for International Development, Washington, DC, 1981.

42. Udall, S. J., *The Quiet Crisis*, Holt, Reinhart and Winston, Salt Lake City, 1963, p. 117.

43. Norton, B. G., *Toward Unity Among Environmentalists*, p. 105. Also see: Hays, S. P., *Beauty, Health, and Permanence: Environmental Politics in the United States, 1955–1985*, Cambridge University Press, London, 1987, p. 128.

44. Many forget that Leopold always begins and ends with humanity. He saw conservation biology and environmental management as a value-laden search for a culturally adequate conception of man's ethical relation to land, discussed in: Norton, B. G., *Toward Unity Among Environmentalists*, p. 107. It is not the culturally adequate conception formed in the mind of a muskrat — it is the one formed in the mind of man.

45. Although wolves are protected by the Endangered Species Act in the United States, this wolf, having killed stock, is not protected and can be destroyed.

SECTION IV
Commentary

23

INTRODUCTION
TO THE COMMENTARY SECTION

C. Richard Cothern

It is important to consider the areas of values, perceptions, and ethics in environmental risk decision making from all the different vantage points of disciplines, specialties, and biases in our society. The complexity of the range of inputs to decision making is reflective of our society and also in this volume. This section showcases how this problem is viewed from three particular vantage points. The three particular areas presented here are a provider of financial and moral support for research, a former Congressman representing the political and public views, and a long-time observer of our society from academe and originator of the concept of bioethics.

There are several emerging social science areas involving research that involve values and value judgments. These include, among others: risks due to natural or physical phenomena; risks created by social systems, e.g., due to human error or mismanagement; and risks of destruction of disease to animals, plants, or ecosystems. A further dimension of research in this area is that applied or practical ethics is becoming part of the federal discussion of risk in research and development (R&D) budget discussions. Recent work in biotechnology and risk assessment in particular have been involving more aspects of social science and ethics.

Scientists tend to isolate themselves from the world of civic and political activity. Further, there is a gulf of ignorance between scientific community on the one hand and public officials, the media, and public, on the other hand. Some perceptions of a scientist are: evil genius, absent-minded, unable to cope with the real world and politics, corrupt sex fiend, power hungry, or willing to sell his grandmother to accomplish his goals. These perceptions prevent effective communication of the real values and value judgments underlying decisions in the environmental area and perhaps in other areas as well.

1-56670-131-7/96/$0.00+$.50
© 1996 by CRC Press, Inc.

The concept of bioethics was introduced over two decades ago as a combination of biological science and knowledge from the humanities. Global Bioethics calls for environmental ethics and medical ethics to look at each others' problems. The integration of these concepts is an important contribution to our understanding of values, perceptions, and ethics as they apply to environmental risk decision making.

AWAKENINGS TO RISK IN THE FEDERAL RESEARCH AND DEVELOPMENT ESTABLISHMENT

24

Rachelle D. Hollander*

CONTENTS

Background: Constructing Risk 338
   Risk as a Contested Domain 338
   Normative Classification Schemes for Risk 338
   Risk Acceptance and Risk Rejection 340
   Normative Classification Schemes for Risk Policy 340
   Decisions When Process and Character Matter 342
The Federal Context for Risk Research 343
   The National Science and Technology Council 344
   The Biotechnology Research Subcommittee 345
   Subcommittee on Risk Assessment 346
References 348

     This chapter has two parts. The first part provides an orientation to some ideas about risk from the fields of science and technology studies and ethics. The second part describes some current federal activities that give priority to research on risk that incorporates approaches from social sciences and ethics.

     The thesis of the chapter is that these activities demonstrate characteristics that would be expected if ideas from science and technology studies are true. They show how federal activities are faced with and trying to cope with conflicts about what should be counted as "risky", and how and who should be involved in these determinations. Part of this engagement involves what counts

* Dr. Hollander directs the Ethics and Values Studies Program at the U.S. National Science Foundation. This chapter presents her own views and does not represent those of the NSF.

as science and what as values (or policy), and demonstrates how science and values (or policy) intertwine.

BACKGROUND: CONSTRUCTING RISK

Risk as a Contested Domain

The field of science and technology studies is concerned to examine how knowledge gets legitimated and socially appropriated. While other disciplines approach this question also, science studies has a distinctive view involving the notion of intellectual boundaries.[1] The notion of boundaries can be usefully applied in thinking about different groups — professional, disciplinary, geographical, political, social — in relation to risk.

To start, assume a knowledge field, one which is socially recognized as such. Knowledge workers define problems, develop and defend approaches to solve them, test their approaches, and present findings. Others, both working within the field and outside of it, and perhaps belonging to factions within and outside at the same time, challenge these activities and findings. Factions within and without defend or challenge the challenges, and so on. Fences are raised and lowered, defining something as science, something else as not; or maybe so or maybe not. This creates change both within the field and in the larger world. Any change incorporates prior findings, modifies, discards, or transforms them. The change may bring outsiders from the field inside, or push insiders out. This interactive process constitutes legitimation and appropriation. It is multidirectional. So risk analysis develops and incorporates constructs in social science and ethics, while having to take care to maintain its boundaries as science.

The concepts of risk, risk assessment, risk management, and other risk terms-of-art fall within such a contested domain. The contest involves who legitimately speaks about risk and how. This chapter identifies some positions in this contest, and how the social sciences and the area of ethics sometimes called practical or applied ethics are beginning to be recognized as legitimate actors.

Normative Classification Schemes for Risk

One important recognition for examining risk issues is the recognition that any talk about them involves normative matters. Even actuarial accounting, say for mortality figures, will find at times that the assignment of a death to one cause rather than another will be contested, and there will not be an unambiguous scientific answer. Assigning the benefit of the doubt to one cause rather than another will not have a univocal scientific justification. Certainly, where public policies or court claims are involved, normative matters are unavoidable.

Besides these internal normative components to risk analysis, there are external, or worldview, orientations to risk that the different actors bring to its discussion. William Aiken has classified normative views he heard expressed concerning agricultural research priorities into four types: top priority, trade-offs, constraints, and holism.[2] Scientists, wishing to defend one priority or another for agricultural research, would articulate these positions. Similarly, discussions about what is risky often contain these four views.

One example of the top priority point of view, for instance, is productivity in agricultural research. Zero exposure to risk might be the equivalent in risk assessment, or perhaps de minimus risk, although both are quite controversial notions.

In contrast to the top priority view, persons concerned about the impacts of agriculture, for instance on land and water, might take a trade-offs view. This view would accept lower productivity, for instance, if it has a better balance overall. This trade-offs view is quite common in risk assessment and risk management. It is very important, particularly when societies have limited resources to invest in preventing or ameliorating risk problems.

A third view is the rights or constraints view. It would maintain that human rights, (for example, not to be exposed to pesticides without voluntary, informed consent) must "trump" outcome-oriented top priority or trade-offs views. This, too, is a common view of persons concerned with risk, and captures some human concerns about freedoms and having a voice in decisions that affect them.

Aiken calls the fourth view holistic or systemic. This view attends to the crucial element of interconnectedness that is left out of the other views. Interconnectedness means that a negative cannot be simply traded off against a positive; it may be necessary to the maintenance of a desirable whole. This view is often used to justify preservation of small farms and ecosystems, although they may be uneconomical.

The fourth view allows us to recognize positive features of risk. We often do so, for instance, when we would allow people to accept the risks, for example, of skiing. It is not just that people accept the trade-off, thus satisfying the constraints point of view; but the pleasure may require the risk, even be heightened by it. The trade-off might, could one measure it without artificially weighting the measures, come out negative.

On a grander level, life as we know it requires predation and death. Evolution is risky, and not reversible. These features are intrinsic and not adequately understood in a trade-offs point of view.

Which kinds of risks we desire to diminish, prevent, or control, and which kinds we accept will express Aiken's four views. All of these views are normative or value laden. In addition, attempts to justify research priorities for risk assessment and the ways in which the assessments are done will incorporate these views. If objectivity requires value-free justifications, then exercises to establish and justify research priorities cannot be objective.

Risk Acceptance and Risk Rejection

It is also important to recognize that accepting a risk, in common parlance, does not mean that one expects or should expect to fall prey to it. Nor need it mean that it is morally acceptable. For instance, persons accept the risk of being accosted in walking on certain streets of the city at certain times. However, some may be known to be experts at self-defense or have an evil eye, and be unlikely to be harmed. If some are accosted or harmed, the moral onus remains on their assailants.

One problem with the scientific and engineering construction of risk as currently practiced is that it often seems to assume that risk is impervious to human influence. Additionally, it does not seem to recognize the extent to which its groups include disparate kinds of individuals whose individual risks are different from each other. Also, it seems sometimes to assume that acceptance equals moral acceptability; that is, that moral acceptability requires only voluntary informed consent. This is incorrect, as is the reverse view that voluntary informed consent is required to make a risk morally acceptable.

Normative Classification Schemes for Risk Policy

There is growing recognition that adequate answers to questions of acceptable risk and acceptable evidence of risk will have important social and ethical dimensions. Answering these questions requires acknowledging the different positions groups take about what is risky and what to do about it. These positions contain social and ethical dimensions and have consequences, which themselves affect the risk. The different groups involved and affected include scientific and nonscientific ones, in roles ranging from undertaking risk assessments, to attempting to bring different dimensions of risk assessments to the attention of relevant scientists and policymakers in order to make them part of the formal process, to disputing their results, to adopting their results in policy or practice.

When risk assessors refuse to incorporate the positions different groups take about what is risky, the conclusions of risk assessments may be irrelevant, invalidated, or harmful. Constituents and stakeholders ignore, modify, or overthrow the results. This invalidation marks, to use the term of Roger and Jeanne Kasperson, a hidden hazard or risk (that of being wrong because of overlooking relevant factors, including social response) to risk assessments and risk assessors.[3] Approaches from the social sciences and ethics can help overcome this hidden hazard.

Paul Thompson develops a classification scheme that is useful in understanding the different kinds of components that are important to people in assessing risk policy. He points out that ethical discourse can focus on outcomes, structures, and conduct, but that policy discourse has been usually limited to talk about outcomes.

Thompson develops this classification scheme in a recent discussion paper on food labeling policy.[4] He points out that it is not sufficient to limit policy discourse to outcomes talk. The languages of social and scientific discourse about risk and risk assessment can and should incorporate attention to all three elements. They will need to do so to develop not just a predictive understanding of societal response to risk, but of elements to consider in decision making that can change the nature and extent of future risks. Policies can affect future structures and norms, and future human conduct, after all, as well as direct outcomes. These effects can help to improve workplace and environmental safety, or worsen them. They can create better norms and structures; and they can shape human conduct in ways we would applaud or condemn. Furthermore, "people's attitudes and judgments about the alternatives" will change over time and with the process of decision making and its outcomes.[5]

In the contested domain we are considering, combatants quarrel about what outcomes should be included as risky. Some believe that the players' concerns should be limited to the outcomes of morbidity and mortality. Some bring in issues of their distribution, raising the issue of fairness of outcomes, or equity. These parties often behave as if the "real" risks are those posed by the natural or physical phenomena under consideration; problems created by the social systems within which they reside are somehow less real and not to be granted legitimate status as a risk or as part of risk assessment. But of course these are sociotechnical systems and this exclusionary posture seems arbitrary. Risks of morbidity and mortality created by mismanagement and human error, or modified by good management and careful practice, need to be factored into this equation. Otherwise the answer is wrong. Also, risks of destruction and disease to nonpersons (animals, plants, ecosystems), need to be considered, it seems. Questions of economic risks, amenity risks, and aesthetic risks; risks to social structures and processes; and to social and ethical behaviors are all relevant.

These kinds of questions are raised not just in the context of risk outcomes, but also in the context of concerns about the structures by which risk decisions are made, as well as the structures and behaviors to which they may lead. These are not concerns about the outcomes from exposures to putative hazardous substances; they are concerns for the laws, norms, procedures, rules, for process and for fairness in process; they ask about such things as protection of human rights, and of the integrity and public confidence in social systems. They can be found in Aiken's constraints and interconnectedness categories identified above.

Also as indicated above, various parties dispute whether or not such concerns belong in legitimate processes of risk assessment. While they might see them as legitimate to risk management, they do not view risk management as subject to what they would call scientific or objective approaches. Surely this is wrong. How can risk assessment help the risk management process if the latter is not subject to rational or reasonable approaches for improvement?

Why should risk assessment be studied scientifically, if risk management cannot be?

Concerns about conduct focus on another dimension. They ask such questions as: What does it do to people making these assessments and those affected by them, that the decisions are made in these ways? How does this way of doing things affect their behaviors? To what habits of character does it lead? Will it result in more care, or more carelessness? To efforts to improve in the future, or complacency? Should we, or when should we try to quantify the value of an individual human life and then use that as a basis for making social decisions? We may be concerned not just about the influence on outcomes or structures of doing so, but about its influence on human beings' regard for each other. If we refuse to place monetary values on individual human lives, it does not mean that we cannot justify decisions about scarce resources. It means that we refuse to do so by a consequentialist procedure that assigns monetary values to individual lives. At least, we recognize that questions about norms and structures, and about human character and conduct need to be incorporated into the decision procedures.

Decisions When Process and Character Matter

Both Aiken's and Thompson's classification schemes provide an interesting matrix with which to analyse the recent decision of the New York Police Department to equip regular park users with cellular telephones and bright blue vests marked "Safe Parks". The impetus for this idea was several incidents in Prospect Park. While police report the city's parks are safer than the blocks surrounding them, they say that crime is "more offensive" to people in parks, who "do not want to always have to look over their shoulders". The extra benefit is "to reassure people that other people in there are their friends and neighbors".[6]

How does this example relate to those from environmental policy? Think about the enormous technical expenditures to clean up toxic sites near areas where children might play. Suppose people from nearby neighborhoods were hired to be sure that they did not? This is a very low-tech solution. It is also one that could provide useful jobs to people whose skills may not be of high value in the market otherwise. Why is it that such an idea has not found a voice in the public agenda or decision making processes about this issue?

There are a number of good reasons. Toxins migrate. The problems in environmental clean-up do not involve protecting a valued resource, but improving a degraded one. They do not involve deterioration in which the affected communities play an active part; rather they are or are perceived as problems kept secret from those affected, and perpetrated by big business and government. Nonetheless, an approach requiring active engagement may be one way to help to overcome this unfortunate past legacy, responsive as it would be to concerns for structures and conduct, as well as outcomes.

This discussion is not just fanciful or theoretical. Approaches responsive to issues of process and character or conduct may be essential to overcoming major policy problems, such as those surrounding the selection of Yucca Mountain in Nevada as the site for a high-level radioactive waste repository.[7] The risks there were and are, in substantial part, risks of and to democratic processes. Adequate delineation of the risks requires attention to these historical and current processes and characterization of those kinds of risks.

The importance of these considerations for this chapter is not whether the delineations that some scholars in ethics have developed are correct. In fact, they are only beginnings. Their importance lies in having such considerations recognized as legitimate concerns for studies in risk and risk assessment. Some scholars in the field of science studies could identify this recognition as anti-democratic and another manifestation of bureaucratic and expert attempts to wrest control of politics from the hands of citizens. The opposite interpretation is that this represents a necessary broadening and deepening of the process that is occurring because democracy in the late 20th century U.S. demands it.

The policy discourse currently focuses primarily on probabilities and consequences with respect to harms to health and environment. It is outcomes oriented. However, the probability and consequence of harms to social structures and processes is a risk issue as well. The probability and consequence of harms affecting how human beings behave towards each other, their organizations, and environments is also a risk issue. People pay attention to all of these kinds of risk issues for good reason: because of the influences of these latter two on the first, as narrowly defined, and on social outcomes as defined by the second and third categories.

The contest is occurring because democracy demands it. It is an expression of the interconnectedness view. Also, the contest improves the processes of risk assessment and management and demonstrates the value of science studies approaches to understanding the social construction of risk. The contest about risk involves what normative dimensions are legitimate to discuss for policy purposes. Here is where social science and ethics join the fray, with some of the contest concerning not what view of risk is correct, but whether the voices are recognized. Since part of the recognition involves what aspects can be called science, it is important to find ways to incorporate these normative dimensions into scientific assessments and into scientific and policy discussions of risk.

THE FEDERAL CONTEXT FOR RISK RESEARCH

Recently, a number of phenomena indicate that the social sciences and applied or practical ethics have arrived as legitimate actors in federal discussions of risk. One is the 1994 symposium at the American Chemical Society meeting on values, perceptions, and ethics in environmental decision making, for which this chapter was prepared. Another is occurring in the federal interagency process for developing and budgeting for strategic research and development activities.

The National Science and Technology Council

Budget setting for research and development has traditionally been accomplished separately in the numerous federal agencies with research and development (R&D) responsibilities. Only in the later years of the Reagan administration (around 1986) did attention begin to be paid in the budget setting and, of course, following in the wake, the research priority setting process, to cross-cutting agency initiatives. This interagency process, which began to get underway around 1986, was called the FCCSET, the Federal Coordinating Council on Science, Engineering and Technology; it was comprised of federal bureaucrats from R&D agencies, coordinated through the Office of Science and Technology Policy (OSTP).

With the Clinton administration, this effort has increased visibility. It is now formalized as the National Science and Technology Council, chaired by the President, established by Executive Order, November 23, 1993.[8] OSTP and its head, who also has elevated status in the domestic Cabinet, manage the effort. The formal structure is still evolving, but a serious attempt is being made to organize the federal R&D budget in terms of cross-cutting areas. Now there is considerably increased attention to what are labeled strategic research priorities and to the relevance of research in certain areas to policy making.

The process by which areas of research get labeled "strategic" must involve an intersection of, and contest among, scientific and social interests of the kind identified above. It is a political process as well. We can see this process recently in increasing attention to needs for research in the social and behavioral sciences devoted to improving education and workplace productivity, and strengthening families and neighborhoods.[9] We will see it as the new Congress and the Administration grapple with regulatory policy in 1995. However, this is not the subject of this chapter.

This chapter accepts the current strategically defined areas and the NSTC committees as given. There are nine committees with assignments for fostering interagency cooperation in such areas as high performance computing and the national information infrastructure (the superhighway), environment, biotechnology, global change and human and economic dimensions of global change, and advanced manufacturing processes.

The notion of risk is relevant to all of these areas, and it is likely to be relevant to new areas in the future. This chapter describes briefly the status of efforts in subcommittees of two NSTC committees, the Committee on Environment and Natural Resources Research, and the Committee on Fundamental Science.

The efforts show how different parties inside and outside of this process try to bring to bear knowledge in the social sciences and ethics. Of considerable importance is the ability of these parties to gain acceptance for their approaches as scientific. For the parties in the process with expertise in the social sciences and ethics, this activity enables them to act as knowledge workers, to interact with other experts and policymakers, to help to change the concept of risk and the dimensions of risk assessment that are part of the process. Outsiders to the activity, including academic, industrial, and nongovernmental organizations also influence and are influenced by it.

The Biotechnology Research Subcommittee

The effort involving biotechnology and social science and ethics is part of the development of the third report on research priorities for biotechnology. This report started development under the old FCCSET process. The group with that responsibility has now become the Biotechnology Research Subcommittee of the Fundamental Science Committee, National Science and Technology Council. Representatives from FDA, NIH, EPA, NSF, DOD, DOC, USDA, NIST, State, and probably a few other agencies as well comprise the membership of the Biotechnology Research Subcommittee, which has issued reports since 1992.[10] In prior versions, the reports have paid little attention to research on social and ethical issues.

In 1994, the federal actors involved in drafting the report decided to concentrate on four specific priorities for biotechnology research: marine, environmental, agricultural, and bioprocessing-manufacturing. However, as they were preparing the report, they realized that outsiders question and challenge much of the research they favored. They realized that many questions would arise for which their expertise would be irrelevant. As these concerns began to be voiced, a member of a working group concerned with research priorities in biotechnology at the National Science Foundation agreed to work with interested individuals in other agencies to draft a section on research on social and ethical dimensions of biotechnology.

The development of the section foundered, however, for a number of reasons. There was little time left and it was difficult to recruit representatives from the various agencies to work on this section of the report. The Fundamental Sciences Committee and its subcommittees include few people with expertise in social science and ethics, and do not consider those areas of priority for their attention. Additionally, members of the Biotechnology Research Subcommittee and its representative from OSTP have different views about what kinds of research, for what purposes, should and will have priority in the subcommittee and its parent committee. Unless it could be seen as promoting or neutral with respect to biotechnology research and its applications, a section on social and economic dimensions would not be likely to garner unconditionally positive responses.

The subcommittee tried to develop a section recommending priorities for research on social and ethical dimensions of biotechnology. However, members were not able to come to agreement nor to persuade the representative from OSTP as to what this section should contain. While the subcommittee continues to regard research on social and ethical dimensions to be important, and individual agencies are pursuing research projects in this area, the new report will be limited to outlining biotechnology research priorities in areas of natural and physical sciences and engineering.

While there will be no section on social and ethical dimensions of biotechnology in the upcoming report of the Biotechnology Research Subcommittee, subcommittee members did seem to agree:

- That ethical and social factors shape organizational and public responses to new developments in biotechnology,
- That biotechnology has ethical and social impacts and presents ethical and social opportunities,
- That there are methods of inquiry in social science and other disciplines that can help to discover and examine what these factors and impacts and opportunities are, and
- That these discoveries can assist in the development of responsible polices and organizational, group, and individual behavior.

Acceptance of this language indicates how social science and ethics disciplines are beginning to be influential, shaping the norms and conduct and outcomes in risk assessment and risk management. It provides evidence for the view that knowledge develops and is legitimated and appropriated in an interactive process involving social as well as scientific actors.

Unless the subcommittee can develop a statement of research priorities and a plan to implement them, however, there will be little systematic attention to these areas in the context of biotechnology R&D. It is likely, however, that attention to social and ethical dimensions in environmental risk research, described below, will be useful in considering biotechnology policy.

Subcommittee on Risk Assessment

There has been a more extensive development within the Committee on Environmental and Natural Resources, which now has both a Risk Assessment Subcommittee and a Subcommittee on Social and Economic Sciences attached to it. The Risk Assessment Subcommittee views the development of a predictive understanding of societal response to risk to be part of its research endeavor. Note that a societal response to risk will be to what society perceives or decides to be risky, not to what any particular group of experts or other stakeholders define as risk.

The discussions in the Risk Assessment Subcommittee indicate two areas for high-priority research on social and behavioral elements of risk: one focused on individual and group responses to risk and one focused on institutional responses. Note that both of these areas do not require "risk" to meet some kind of objective standard, nor a rigid distinction between risk assessment and risk management.

Under the first area, research topics range from the need to develop better techniques, paradigms, and integration between fields studying individual responses and social influences, to the need to understand risk taking and avoiding behaviors and the influence of such factors as trust and justice. In institutional responses to risk there is similar emphasis on improving tools and understanding of such issues as the interrelationships of management practices, human error, and sources of risk; the influences of organizational, institutional, and social factors on risk behaviors and communication; and the feedback from policy options to behavioral responses of individuals, groups, and institutions.[11]

Further discussions in the Subcommittee are adding another priority for risk assessment in the context of science policy, indicating "Research should focus on improving the value and effectiveness of scientific information for decision making about risks by emphasizing (1) methods for characterizing uncertainty and default assumptions; (2) methods for discriminant analysis and for determining the adequacy of these methods in the context of risk decision making; (3) the relationships among social concerns and decisions, risk assessment methods, and data needs; and (4) methods to evaluate, weigh and compare different endpoints, time frames, and populations."

The language in which these research priorities is cast is acceptably scientific. Nonetheless, these approaches are not limited to a concern for probabilities or consequences associated with morbidity and mortality outcomes (including ecological morbidity and mortality). They are open to considerations of risk framing, risk response, and issues of decision making and management. They are interested in individual, group, and institutional responses. They have an interdisciplinary and cross-cultural agenda: to examine such issues as risk communication and the role of risk assessment in policy analysis and to require improved theory, methods, and data. These statements allow social science and ethics into risk assessment and risk management.

The next step in this process requires the development of interagency statements of priorities as goals and objectives, and milestones and outcomes for implementation. As these federal interagency efforts continue, and work to shape the research agenda on risk, they will shape norms and conduct, as well as outcomes in risk assessment and risk management. They will incorporate top priority, trade-offs, constraints, and holistic views. They will be evidence for the view that knowledge develops and is legitimated and appropriated in an interactive process involving social as well as scientific actors. What is risky is a blend of the objective and the subjective, as is what is real.

REFERENCES

1. See, e.g., Gieryn, T.F., 1983. "Boundary-Work and the Demarcation of Science from Non-Science: Strains and Interests in Professional Ideologies of Scientists," *Am. Sociol. Rev.,* 48, pp 781–795. Also, Jasanoff, S.S., 1990. *The Fifth Branch: Science Advisors as Policymakers.* Harvard University Press. Cambridge, MA.

2. Aiken, W.H., 1986. "On Evaluating Agricultural Research," in *New Directions for Agriculture and Agricultural Research,* K.A. Dahlberg, Ed., Rowman & Allanheld, Totowa, NJ, pp. 31–41.

3. Kasperson, R.E. and J.X. Kasperson, 1991. "Hidden Hazards," in *Acceptable Evidence, Science and Values in Risk Management,* D.G. Mayo and R.D. Hollander, Eds., Oxford University Press. New York, pp. 9–28.

4. Thompson, P.B., 1993. Food Labels and Biotechnology: The Ethics of Safety and Consent. Discussion paper for the Center for Biotechnology Policy and Ethics, Texas A&M University, unpublished.

5. Ianonne, A.P., 1994. *Philosophy as Diplomacy.* Humanities Press International, Inc., Atlantic Highlands, NJ, p. 24.

6. Martin, D., 1994. "Police Enlist Park Users in Safety Drive," *New York Times,* Sunday, August 21 (Metro Section) pp. 45–46.

7. Colglazier, E.W. 1991. "Evidential, Ethical, and Policy Disputes: Admissible Evidence in Radioactive Waste Management," in *Acceptable Evidence: Science and Values in Risk Management,* D.G. Mayo and R.D. Hollander, Eds., Oxford University Press. New York, pp. 137–159.

8. Sclove, R.E., "Report to the General Program, J.D. and C.T. MacArthur Foundation, May 2, 1994, unpublished.

9. Investing in Human Resources, A Strategic Plan for the Human Capital Initiative. Report prepared for the National Science Foundation, from a Workshop held March 17–18, 1994. Printed for NSF in 1994. 18pp. The Senate Appropriation Bill language for fiscal year 1995 contains an allocation for this effort.

10. Biotechnology for the 21st Century; Report of the FCCSET Committee on Life Sciences and Health, Washington, D.C. 1993. Biotechnology for the 21st Century: Realizing the Promise; Report of the FCCSET Committee on Life Sciences and Health, Washington, D.C., 1994.

11. Robin Cantor, unpublished document for the Subcommittee on Risk Assessment of the Committee on Environment and Natural Resources Research of the NSTC, discussing social and behavioral elements in risk assessment, table on high priority research needs appended, 7/6/94.

25 THE CITIZENSHIP RESPONSIBILITIES OF CHEMISTS

Hon. Mike McCormack

CONTENTS

Chemists as Good Citizens 349
Ethics, Ignorance, and Public Service 350
Understanding Public Officials — And Ourselves 352
Getting Informed and Involved 352
Chemists and Environmental Concerns 353
The Hazards and Benefits of Emotionalism 354
Common Sense Environmentalism 355
Being a True Environmentalist 356
Chemists — Environmentalists and Earth Day 357
Citizenship, Ethics, and Commitment 358

I am pleased to have this opportunity to present some thoughts about the citizenship responsibilities of chemists; about how we can, and should, undertake to be of still greater service to the people of our nation and the world.

CHEMISTS AS GOOD CITIZENS

Of course, we already — and justifiably — think of ourselves as good citizens. We vote, pay our taxes, obey the laws and serve our country. We contribute to the social welfare through our professional activities. However, I think the concept of good citizenship for us in today's world calls for much more than such worthy attributes and activities. It involves a sincere and intelligent commitment to participation in public life, to getting out of the laboratory or the classroom, and bringing our professional skills to the service of society through political and civic involvement.

1-56670-131-7/96/$0.00+$.50
© 1996 by CRC Press, Inc.

This is a responsibility of citizenship that frequently is not appreciated (and just as frequently avoided) by too many chemists. While you contribute so much to society through your professional research, study and teaching, most of you, as chemists, have been trained in a culture that considers public activity outside one's discipline (and certainly outside scientific research and study) as being unprofessional; and that getting involved in politics, of all things, is something like intellectual prostitution.

The tendency to isolate oneself from the world of civic and political activities is not unique to, nor universal among chemists. It has long been a traditional practice among most scientists and engineers. Without attempting to analyze it further, we must recognize that it is substantially of our making; and, more importantly, that it is a major factor contributing to the gulf of ignorance that exists today between the scientific community, on the one hand, and most public officials, the news and entertainment media, and the public, on the other. This constitutes a serious handicap for the people of this country who are faced with a multitude of public problems and issues that relate to science and technology.

ETHICS, IGNORANCE, AND PUBLIC SERVICE

We should think of this as an ethical issue with which we must deal: ignorance has always been the fundamental and most dangerous enemy facing mankind. It always spawns fear, superstition, bigotry, repression, and hatred. Today, we are witness to a society so susceptible to unreasoned fear of science and technology, especially "hazardous chemicals", that the general public is often unable rationally to consider important issues involving such subjects.

Unfortunately, most scientists are just as ignorant of the realities of the political world as the average citizen is of the basic concepts of the physical sciences. Thus, we have a gulf of mutual ignorance that must be bridged if we, as chemists, are to make a meaningful contribution to future public policy decisions involving science and technology. This is our obligation. Many of our fellow citizens outside the scientific community do not understand that a scientifically illiterate public can, and frequently does, support the creation of unwise, and even self-defeating policies. The Delaney Clause, the ban on cyclamates, and the EPA's exaggeration of the hazards of radon and asbestos are examples.

Scientists are, in many cases, the only ones who recognize and appreciate the seriousness of potential problems associated with the enactment of scientifically unsound legislation. Today, this is especially true for chemists, because public concerns involving environmental protection and human health and safety relate primarily to chemicals and chemistry. We are the ones who understand. There is no one else to send. The bell tolls for us. The Greek word "ethos", from which we derive the word ethical", referred to that self-imposed obligation recognized by responsible citizens "to do what they ought to do anyway" as an obligation of citizenship, but which the law did not necessarily require.

If we assert that there is a special ethical obligation on chemists, as chemists and citizens, then we must ask ourselves if it is ethical for us to remain silent and hide behind our professional dignity, and refrain from speaking out on public issues involving science and technology. With the question framed in this context, most of us, I believe, would respond that we should speak out.

Why, then, do we not? I think that at least part of the answer originates with the images and illusions with which most Americans, including scientists and politicians, continually live.

On most any day we can find movies or television programs about scientists or public officials. More often than not, the scientist is either an evil genius, madly scheming to wipe out half the earth, rule the rest, and enslave the beautiful girl — or he is a once-brilliant scientist, but now an absent-minded, shuffling old fuddy-duddy, unable to cope with the real world, and about to lose his beautiful daughter to the evil genius.

On the other hand, the politician is usually a high ranking office holder, or a candidate for such office — a corrupt sex fiend, scheming to become the next president and rule the world, and willing to sell his grandmother to accomplish his goals. He may be found at exclusive cocktail parties, accepting bribes under the table, or being lured into side rooms by slinky prostitutes.

It may not have occurred to you, but these totally slanderous caricatures of scientists and politicians are brought to us and all the other citizens of this country by the same unconcerned peddlers of sensationalist fiction: political cartoonists and movie and TV script writers.

Of course, we scientists know that at least part of that message is untrue. We know that most scientists are really intelligent, hard working, patriotic, responsible family men and women of high integrity, attempting to make an honest living, and dedicated to serving our fellow men. Naturally, we are not so sure about the politicians. After all, we have seen the movies and cartoons.

However, we politicians know that most public officials are hard working, intelligent, patriotic, responsible family men and women of high integrity, attempting to make an honest living by serving our fellow men — but we are not so sure about those scientists. We have read that they have sold out to corrupt corporations that are only interested in making a profit and have no concern for the damage being done to the environment or the threat to public health and safety from their actions.

Deplorably, the average citizen (and all too many members of the news and entertainment media) appear to think that both groups: scientists and politicians, are correct in their impressions of each other, but not of themselves.

The damage to our country from such attitudes is extreme. Lampooning public officials or scientists for fun is perfectly acceptable in a democracy, but it is unacceptably damaging when otherwise thoughtful citizens take such caricatures seriously. As with flattery, lampooning is okay — as long as one does not inhale.

UNDERSTANDING PUBLIC OFFICIALS — AND OURSELVES

As a matter of fact, our legislators and Congressmen are, allowing for a few conspicuous exceptions, a distinct cut above the average citizen in intelligence, understanding, dedication, and integrity — the same as scientists. However, legislators and Congressmen have one of the toughest and most pressured jobs imaginable. Having worked for 20 years as a research scientist and then 10 years as a Member of Congress, I can assure you that the workload of Members of Congress and the pressures on them are orders of magnitude heavier those that come from working in the laboratory or classroom.

Dedicated and hard working as most public officials are, however, there is another phenomenon of great importance at work. One may think of it as an aberrant case of natural selection. The sad fact is that the tendency toward isolation of most scientists from society is fairly well institutionalized.

As a result, remarkably few scientists, engineers, or mathematicians even consider running for a significant partisan office. I estimate that only about 1% of our state or federal lawmakers have had any advanced study in any scientific discipline. Not only does this make legislative bodies unrepresentative, it dramatically reduces their ability to handle issues involving science or technology.

Probably one of the reasons chemists avoid public life is that becoming involved might be frowned upon by an employer and/or peers and might threaten one's professional stature and/or income and retirement security. The insidious effect of lampooning caricatures creates a barrier to rational thinking, to say nothing of the acceptance of civic responsibility. This is a disgrace, and the country suffers grievously from it.

In general, scientists understand little more about what is involved in being elected and serving as a state legislator or Member of Congress than most public officials understand of the more sophisticated aspects of physical or biological sciences.

GETTING INFORMED AND INVOLVED

Lawmakers need to understand the concept of the scientific approach to resolving political issues involving science or technology. At the same time, scientists need to learn about the realities of serving in elective office. Working within the political community, scientists can come to appreciate its realities, and, at the same time, may have an opportunity to make a meaningful contribution to legislation as it is being considered. There is a great need for this type of assistance, because much of the legislation enacted at the state and federal level today relates directly to science or technology.

Examples include laws providing support for and/or regulation of education, basic research, energy research and development, environmental protection, space projects, astronomy, health care, biotechnology, safety, risk assessment, agriculture, reclamation and conservation, and others. Much of this legislation is enacted in an emotional atmosphere which tends to preclude rational consideration of scientific information, even when it has been made available to lawmakers.

There are effective mechanisms through which chemists can become involved in political activity. To start, I suggest that each of you take the initiative to form a committee of about a dozen scientists and engineers in your local community. Draw upon the membership of other scientific and engineering societies and key civic leaders, and arrange for your committee to meet individually with your local legislators and congressmen several times a year, when the lawmaker is at home and can meet with you, on neutral turf, in a relaxed atmosphere.

Such meetings must serve the mutual interests of all participants. The scientists and engineers, taking the initiative, can provide accurate and understandable information on current issues, and, at the same time, gain an understanding of the political realities at work, and the pressures under which public officials must operate.

It is important to earn the confidence of your legislators and Congressmen. The first rule is to remember that absolute integrity must be observed in all instances, because this is the standard by which elected officials relate successfully to each other. The next rule is that you must be good teachers, recognizing that most public officials do not have the background that makes the scientific approach to problem solving second nature to you. In addition, you should try to understand the political restraints under which they must function. Beyond this, you can become active in the political party of your choice, or the campaign committee of your favorite candidate. This is certain to be one of the more educational experiences of your life. Best of all, after you gain some experience, you can run for office. Start modestly, but plan to move up. You, and the country will be better for it.

CHEMISTS AND ENVIRONMENTAL CONCERNS

There is a special reason for public involvement by chemists today. All chemists should be deeply concerned about the unreasoned fear of man-made chemicals that is expressed almost daily by the news and entertainment media, and in the halls of our legislative bodies. This is especially true because chemists are frequently cast in a "bad guy" context, while self-appointed "environmentalists" cast themselves as the "good guys", and claim the moral high ground.

Unfortunately, many chemists accept this characterization without reflecting on what its implications may be. It may be a good exercise, therefore, for each of us to stop and ask, "Am I not also an environmentalist?" If we do, I am confident that our answer will be a resounding "yes!" If we have any courage, we will dare anyone to demonstrate that he or she has any greater claim to the title "environmentalist" than we have. As an environmentalist, and a chemist who understands the technical aspects of many environmental issues, your voice can contribute to their rational consideration.

There are few challenges facing the people of America today that are more important than the need for such a rational approach to pollution control and environmental protection; and the need for a serious and reasoned consideration of the impact of human activities on ecological relationships within the various environments which we humans occasionally visit or in which we live every day.

THE HAZARDS AND BENEFITS OF EMOTIONALISM

I think it is extremely unfortunate that a discussion of these subjects has become charged with emotion, because most Americans really want about the same standards of environmental protection. This disruptive atmosphere, and the uncertainties it breeds, is inhibiting the development of environmentally attractive, job-producing industries and the generation of energy from benign domestic sources, both of which the people of this country desperately need.

We must remember that we live in a period of extremely rapid transition. Attitudes and standards related to environmental protection are no exception. It is not unfair to observe that only a few decades ago, the prevailing attitude with respect to the environment in a society dominated by laissez-faire and frontier philosophies was ignorance and indifference. These attitudes prevailed until the 1950s, when our understanding of their consequences began to overtake us. It was not until the late 1960s that we began to institute policies and programs to protect the air, the water, the land, and our wild ecosystems. Serious implementation did not occur until the 1970s and 1980s, and changes in related regulations are being made almost daily.

Unfortunately, our guilt and our good intentions caused many sincere Americans to overreact, and to sweep aside rational consideration of specific issues with one generalization: "The 'Environment' (with a capital E) must be protected at any cost!" Moreover, for many persons this "Environment" came to mean primarily air, water, and wilderness areas. This disregarded the environments of our homes, work places, cities, or neighborhoods. Most of us will agree that these environments are equally sacred and important. All too often, however, they have not even been considered.

As might be expected, there are those who, for their own purposes or their own benefit, are exploiting societal susceptibility to fear mongering with regard to environmental and health issues. Fanning the flames of thoughtless emotionalism, they attempt to discredit scientists who assert that we should engage in a rational deliberation of these issues, give serious consideration to all the environments in which we live, to national strength and security, and to the individual welfare of all concerned.

In asserting that most Americans really are sincere environmentalists, we recognize that some think more about the environment than others; and some are thoughtless about their own individual contribution to pollution. For instance, some persons still smoke in the presence of non-smokers, dogs are walked without consideration, and litter and garbage are often scattered about. Fossil fuels, including coal, with its polluting contaminants, are burned with reckless abandon. Such insults to the environment make me angry, and they should make you angry also. Thus, we have an emotional reaction to a problem, and this emotionalism is quite healthy — when directed toward rational solutions. Emotionalism can, therefore, be beneficial.

Most Americans are probably emotionally committed to the premise that we should, as individuals, and as a society, leave this world a cleaner, better place for our children than we found it. This is, perhaps, the most important criterion for sincere environmental concern, but we must be certain that common sense is paramount in arriving at solutions to it.

COMMON SENSE ENVIRONMENTALISM

Most Americans know that common sense must play a major role in dealing with environmental issues. They recognize (if it is called to their attention) that chlorine, which we add to our drinking water to kill germs; and fluorides, which we add to protect our teeth, are both deadly poisons in higher concentrations, but certainly safe and generally accepted as beneficial in the concentrations used. Most Americans are rational about pollution and environmental protection. They know that our air and rivers and lakes can be made much cleaner, but never absolutely pure; that pollution from industrial activity can be significantly reduced but never totally eliminated; that there will always be some pollution of the environment from human activity; that there are inevitable economic trade-offs that will always affect the degree of pollution control and environmental protection; that wilderness classification is good for some land but not best for all; that nature frequently pollutes offensively, and often changes the environment of a given area, without the consent of any activist organization. They recognize that one person's concept of environmental enhancement may be another's concept of environmental degradation: such as a multi-purpose dam on a mountain stream, or, closer to home, somebody's perfume or loud music.

Some of our fellow citizens will recognize, that if one thinks of the earth as a living entity, then the exponential growth of the human race is, in a very real sense, like a cancer that is destroying its host. Sincere concern for environmental protection must begin with an intense campaign, at home and worldwide, for zero population growth. There is no general awareness of the fact that the earth cannot continue the exponential population growth it is experiencing in the 20th century. Americans must understand that there are limits to the earth's resources. Moreover, science is not magic, and scientific research cannot produce miracles. It cannot provide a decent standard of living, a clean environment, and adequate water and resources for the 10 billion or more people who will populate the earth by the middle of the 21st century if present population growth rates continue.

Most Americans do not know that ski buffs in Colorado receive about 100 times as much extra ionizing radiation as a person living next door to a nuclear power plant, or that the crews on most long-distance commercial air flights receive more extra ionizing radiation per year because of their job than the average amount received per year by employees in our nuclear power plants. If more Americans were aware of these facts, they would probably put into better perspective their concerns about the hazards of low-level radiation and nuclear power; but few of them would suggest making Vail or Aspen, Colorado off limits, or forbidding high altitude flights of commercial jets.

In other words, most Americans would agree that being a responsible environmentalist involves the use of common sense. One of **our** primary responsibilities as scientists, is to help them understand that common sense must be based on as much scientific information as reasonably can be obtained.

In addition, we may have a good reason, on some occasions, to deliberately alter or destroy an existing ecosystem. Each time we plant a garden or a field of corn or wheat, or plant or cut down a tree, or build a dam, we are modifying the existing ecosystem at that location. Sometimes we totally eradicate one ecosystem and replace it with another, as, for instance, the construction of an irrigation project to convert the desert to productive farmland. Such activities are generally considered to be constructive, and consistent with our ideas of making the world a better place in which to live. Of course, we have an absolute obligation to always conduct ourselves so as to reduce all pollution to the lowest practical limit, and to consider seriously the impact of our intended actions on the ecological systems at any location. In short, we must protect any environment over which we have control, but this may legitimately involve altering it in a constructive manner, understanding what we are doing.

BEING A TRUE ENVIRONMENTALIST

Now I take these assumptions, and my own personal, sincere dedication to realistic environmental protection, just as much for granted as I do my obligation to be considerate of others, to maintain appropriate standards of personal hygiene, to pay my bills, to keep the commitments I make, and respect the scientific approach to problem solving. In fact, I take my concern for the environment so much for granted that I do not go around talking about it all the time. I do not wear it on my sleeve. It is simply a part of my life, and I assume most other Americans feel the same. So I am an environmentalist, essentially no more or no less than anyone else. I think each of you is, too.

Accordingly, I object to any claim by anyone that he or she is an "Environmentalist", and therefore possessed of some special higher right to judge the actions or attitudes of others on this subject. I respect those who are legitimate students or professionals in environmental sciences, and I look to them for guidance and factual information. However, there is no elite priesthood which can claim that its members occupy a special position in society as "Environmentalists".

I believe that all responsible Americans should reject the activities of anti-energy, anti-growth, anti-technology, or anti-chemicals activists, especially when they falsely wrap themselves in the banner of environmentalism, attempting to hide behind a worthy concept in their efforts to introduce emotionalism to obstruct realistic programs that provide the employment, the materials, and the electricity the people of this country need.

As soon as we recognize that there is a positive benefit to mankind in planting a garden, we cannot avoid the obvious conclusion that there must be thousands of rational acts by human beings to change their environments; acts that are thoroughly justified in terms of improved living conditions, safety, food supply, housing, etc. Ever since our ancestors evolved from wandering hunters and foragers, adopted agriculture and developed organized communities, their attempts to protect their children, to provide shelter, avoid disease and to raise enough food for winter, they, and we, have impacted our natural environment — the wilderness around us.

This is one of the hallmarks of civilization: the conversion of wilderness into productive land to provide higher standards of living. The march of civilization is catalogued by the development of our natural resources to provide the energy and materials needed to free men and women from the physical tyranny that survival without them imposes. Today such thoughts come almost as a surprise to some who have been conditioned during recent years to think of the earth as one grand, vague, pure "Environment", somewhat like an imaginary wilderness that is sacrosanct, and must never be altered in any way. Those who set such emotional traps in the public mind or the political community are guilty of dangerous mischief.

Emotional exhortations to oppose rational development — to "Protect the 'Environment' At All Cost" — frequently mean that those of another environment will pay the penalty. For example, preventing the construction of a power plant, pipeline, or transmission line in the guise of environmental protection in a rural area often means that unemployment will increase, energy will be more expensive in a nearby urban environment, and vulnerable groups of our population such as senior citizens and the poor may be deprived of adequate heat in the winter or air conditioning in the summer. It may also result in the cutting of forests — just for wood to burn for heat.

CHEMISTS — ENVIRONMENTALISTS AND EARTH DAY

We annually celebrate Earth Day. I hope that each one of us has participated in this celebration in some constructive way. I have always believed, and I have always tried to live by my belief, that we should appreciate and respect the beauty of Earth and sky around us and do everything we can to protect that beauty from unnecessary degradation. However, I hope that we will reject the thoughtless advocacy of unrealistic causes that gain undeserved attention on occasions such as this. All of us who are true environmentalists have an obligation to speak out in support of common sense, and scientific rationality as we observe Earth Day.

CITIZENSHIP, ETHICS, AND COMMITMENT

Martin Luther King emphasized that, in the struggle to overcome racism, the sins of commission by a few were of little consequence as compared to the sins of omission by the many. This suggests what the chemist's ethical position must be with respect to public service. The issues that have come to us in the form of concerns about health, safety, and environmental protection cannot be ignored, nor can we ignore our unique responsibility with respect to helping our fellow citizens and our public officials understand the facts that we understand that are associated with these issues. Remember, no matter how emotional the reaction to any environmental or safety problem may be, *the solution is always — and exclusively — cold-blooded science, engineering, logistics and economics,* which is another way of saying "common sense".

I hope that you will, as individuals, and especially as chemists, become involved in public service, in politics, as aggressively as is practical. Do not stand back and say you will not dirty your hands. Remember that adage about nature and a vacuum. If you do not do it, someone else will, and he (or she) may not understand — or may even be hostile to — a rational, scientific approach to solving modern societal problems.

The Greek word from which politician is derived means "servant of the people". The Greek word "idios," from which we derive our word idiot, means "those deprived of the vote". In a sense, we are all continually choosing between those roles. Citizenship in a democracy is a privilege, but for the ethical citizen, it carries with it the obligation of public service. To you, who are already doing so much in your professional fields, I must emphasize that the depth of any individual's commitment to being worthy of his or her citizenship must be a function of what that person has received from society, and what that individual's skills and wisdom may be. Each of us has received a splendid education, and much more — substantially at public expense or as a legacy from our forefathers. We are the beneficiaries of the investment that society has made in us, and our ethical obligation is to return that investment to society.

The time has come for each of us to look beyond the limits of our professional activity, and into the public arena. Your country needs you also as a public servant, and it has never needed you more.

26

GLOBAL BIOETHICS: ORIGIN AND DEVELOPMENT

Van R. Potter

CONTENTS

The Contemporary Scene...................................... 359
Tracing the Origin of Global Bioethics......................... 365
References 372

THE CONTEMPORARY SCENE

Since my introduction of the word *bioethics* (Potter, 1970) I have never deviated from an emphasis on long-term "acceptable" survival of the human species. My bioethics has always been defined as an attempt to combine biological science and knowledge from the humanities, stressing that ethics cannot be separated from biological facts. *Global Bioethics* (Potter, 1988) maintained that medical ethics and environmental ethics should not be separated from each other. Only by a philosophy of individual health for all the world's people, and not for just a chosen few, can we achieve the goal of survival and improvement of living conditions for the human race (Jameton, 1994). "Acceptable" survival must be understood as possible only in this context (Potter, 1988, 1990). Environmental ethics needs to accept the goal of health for the human species as part of health for the biosphere. Global Bioethics calls for environmental ethics and medical ethics to look at each others' problems, along with the others' insights, to achieve the goal. Environmental ethics should be part of the mission of professional public health officers as much as anyone, for it is their work that crosses international borders in both environmental and medical domains (Potter, 1993).

The cross-currents between medical bioethics and environmental bioethics can best be understood if we realize that they are two streams of conflicting values flowing in a river of reality. In each case, a set of *quality* values is

1-56670-131-7/96/$0.00+$.50

challenging a set of *sanctity* values, where the word sanctity implies a value that cannot be challenged. (One definition of "sanctity" is "inviolable".) In both the medical and the environmental conflicts of interest we embrace a wealth of professional information and ethical opinion in the words *quality* and *sanctity*. The vision of Global Bioethics calls for an ethic of responsibility (cf. Jonas, 1984) for long-term acceptable survival based on Health Care and Earth Care worldwide. The conflicts of interest and the failures to reach consensus (Bayertz, 1994) are truly colossal.

In *Health Care*

Quality of Life challenges Sanctity of Human Life.

While in *Earth Care*

Quality of the Environment challenges Sanctity of the Dollar.

The fact that interaction occurs between all four components should be obvious to everyone, but the two fields remain separated, probably because the details in each are so complicated. Yet no one can deny that "sanctity of human life" impacts "quality of the environment" and "sanctity of the dollar" impacts "quality of life". What is clear in principle is that too much emphasis on the *sanctity* side of the balance is, in each case, damaging to the *quality* component. Perhaps the vision of Global Bioethics is naive and the reality is the problem of getting the dominant culture to willingly accept the idea that the masses of people in poverty are *persons* (cf. Engelhardt, 1986, p. 109). Global Bioethics calls for on-going discussion of what is required to *permit* human survival and to make it *acceptable* and *deserved* by the dominant culture.

In the development of Global Bioethics, the idea of bioethics moved from a broad concept of integration of biology and the humanities to a narrower integration of medical and environmental bioethics. Then it became apparent that medical and environmental ethics had proceeded down separate paths. Global Bioethics was proposed as an attempt to integrate the medical and environmental branches. In my vision, the evolution of bioethics as a discipline moved from humility, responsibility, and competence (Potter, 1975) to encompass five cardinal virtues: (1) humility, (2) responsibility, (3) interdisciplinary competence, (4) intercultural competence, and (5) compassion, but always with acceptable long-term survival of the human species in a worldwide civil society as the goal (Potter, 1988, 1990).

As of the summer of 1995, Global Bioethics is proposed as an idea whose time has come. The concept calls for a coalition for all the efforts to bring science, religion, the humanities, governments, business, industry, and people together in interdisciplinary groups that can agree on the five cardinal virtues and a goal that goes beyond "stewardship" to embrace "Acceptable Survival".

It is mandatory that agreement be reached on the necessity for limiting human reproduction to levels that are compatible with long-term acceptable survival in a "civil society". Indeed, the concept of Global Bioethics calls for development of the whole idea of what constitutes a civil society on a global basis.

In the book, *The Idea of Civil Society* (Seligman, 1992), we find the author commenting, as many recent scholars have, that neither religion nor philosophy can come up with ethical rules for moral action that can provide the guidelines for a civil society in today's world. He remarks:

> With the loss of these foundations in Reason and Revelation, the idea of civil society itself becomes the problem rather than the solution of modern existence. And while it is certainly true that this realization would seem to leave us less than sanguine about the possibility of reconstructing civil society, as idea or ideal, it will, I would hope, make any move in this direction more realistic.[198]

Seligman does not even consider the possible role of science in reconstructing civil society, probably because it is widely agreed that Science cannot in itself provide ethical rules for moral action. Recently, it has been proposed that science and religion (Potter, 1994), and much earlier that science and philosophy (Potter, 1962), must share the quest for global survival. Now, the insidious development of postmodernism (Himmelfarb, 1994) seems to propose that neither religion nor philosophy, nor science, or any combination of the three, could provide the guidelines for an acceptable global ethic. On the other hand, Global Bioethics is a long-range intuition that leads to the idea that a concern for future generations can lend meaning to life today. This long-range intuition, choosing "Acceptable Survival" as a goal, is an idea that cannot be proved by religion, philosophy, or science but *having chosen the goal*, Global Bioethics challenges the cynicism aspect of postmodern deconstructionism and chooses to seek truth, reason, morality, civil society, and realism using the values and resources of all three: religion, philosophy, and science. Global Bioethics is proposed as the stage after the "death" of postmodernism that Himmelfarb looks forward to (Himmelfarb, 1994, p. 161).

It is considered appropriate at this point to take notice of a recent address given by Vaclav Havel, President of the Czech Republic, when he was awarded the Philadelphia Liberty Medal at Independence Hall on July 4, 1994. His remarks were presented as excerpts by the *New York Times* (Havel, 1994) and quoted selectively here. While Seligman ignored science completely, Havel was primarily concerned with the state of science in the present world. He noted that we are in a time,

> ...when all consistent value systems collapse ... Today, this state of mind, or of the human world, is called post-modernism....our civilization does not have its own spirit, its own esthetic.

This is related to the crisis, or to the transformation, of science as the basis of the modern conception of the world. The dizzying development of science, with its unconditional faith in objective reality and complete dependency on general and rationally knowable laws, led to the birth of modern technological civilization. It is the first civilization that spans the entire globe and binds together all societies, submitting them to a common destiny.

At the same time, the relationship to the world that modern science fostered and shaped appears to have exhausted its potential ...

Politicians are rightly worried by *the problem of finding the key to insure the survival of a civilization that is global and multicultural*; how respected mechanisms of peaceful coexistence can be set up and on what set of principles they are to be established. ... (italics added)

The moment it begins to appear that we are deeply connected to the entire universe, science reaches the outer limits of its powers. ...

The only real hope of people today is probably a renewal of our certainty that we are rooted in the Earth and, at the same time, the cosmos. ...

Only someone who submits to the authority of the universal order and of creation, who values the right to be a part of it and a participant in it, can genuinely value himself and his neighbors and thus honor their rights as well (Havel, 1994).

In a final remark Havel turned to the Declaration of Independence which

...states that the Creator gave man the right to liberty. It seems man can realize that liberty only if he does not forget the One who endowed him with it (Havel, 1994).

Thus in the end, having accepted the postmodern challenges to philosophy and science by drawing on the Declaration of Independence, it appears that Havel chose to end with words suitable to the occasion — his acceptance of the Philadelphia Liberty Medal. In taking a course that is too facile, Havel ignores the daunting fact that millions of people have widely divergent views of how to honor and obey the *"One"* they all believe in. It is the divergent views of duties to the *One* that have resulted in endless conflict and killing. Going beyond remembering the unknowable *One*, Global Bioethics places ethics and morality in the context of an acceptable survival to the year 3000 and beyond in tolerant coexisting civil societies that are worldwide, despite the cynicism of postmodernism.

Himmelfarb (1994), with Seligman (1992), sees postmodernism as the deconstruction of religion, philosophy, and history, and neglects the role of science following the Enlightenment. Now we find Jon Franklin (1994), a professor of journalism, agreeing with Havel (1994) that postmodernism has led to a loss of faith in science. Franklin, however, brings in a new perspective not found in any sense in the words of the other three scholars. Speaking as the keynote speaker at the annual meeting of the Chemical Industry Institute of Toxicology (CIIT) on May 10, 1994 on "Poisons of the Mind", he gave a talk that must have given considerable comfort to his hosts. Although not once did Professor Franklin mention "ethics" or "environmental risk assessment", the relation between ethics and the reporting of environmental risk is really what his talk was all about. Franklin's core message was that it is unethical for journalists, the media, or anyone, to propagate "poisons of the mind" by exaggerating the dangers of deliberately disseminated biologically active chemicals. He cited his own extensive investigative journalism in the case of "Agent Orange". Out of all this, Franklin came up with some very sound philosophical reflections that are highly relevant to global bioethics and my references to Seligman, Havel, and Himmelfarb. He opened by relating his main topic to his philosophy:

> Here we are at the dawn of the what some call the post-modern and others are beginning to think of as the neo-Medieval — and poisons are back in the news (p. 1).

Getting at the core of the scientific ethic, he noted

> As faith had been the heart of Medieval consciousness, so now truth was the touchstone [new faith] of the modern. The idea that the truth could be known replaced faith in the unknowable as the basis of the social contract; the prevalent faith was that truth would always win out, and that when it did it would be visible to all (p. 2).

Franklin has come to believe that the truth will *not* always win out and to doubt that once attained it would be visible to all.

[Perhaps naively, the basic bioethic has always been "Ethics cannot be separated from biological facts" (Potter, 1988, p. 75). Global Bioethics continues this claim.]

Coming to his main thesis, criticism of scientists, journalists, and lawyers (omitting postmodern historians and philosophers scorned by Himmelfarb, 1994) Franklin pinpointed the problem:

> As more and more of us draw sustenance from propositions that we know to be false, if only in their disproportion, so we devalue the respect for truth that is the foundation of our civilization. Finally it comes down — it has come down — to a corruption of the faith that once underlay the modern age (p. 5).

That is, as noted earlier, what underlay the modern age was belief in the idea "that the truth could be known". This idea, then, "replaced faith in the unknowable" referring, in a sense, to the *One* of Havel (1994). Focusing on his fellow journalists he then commented in conclusion:

> What we are seeing, in the press and in our society, is nothing less than the deconstruction of the Enlightenment and its principle (sic) institution, which is science ... And I would remind you as well that human history admits to greater dangers than you can titrate in your laboratories ... we must never forget for an instant that there are poisons, too, of the mind (p. 6).

"Greater dangers", indeed. Are poisons of the mind the only threat to "acceptable survival" to the year 3000 and beyond? Perhaps, as an umbrella term, if we include the political assassination broadcast daily, or the failure to see ethics requiring a knowledge of biological truths, or the biological fatal flaw (Potter, 1988, 1990) that causes cultural evolution to seek short-time gain that ignores long-term acceptable survival.

It can be suggested that, in a grim metaphor, we are all passengers on the "unsinkable" Titanic and the captain (the dominant culture) has ordered full speed ahead ignoring possible icebergs (greater dangers) while the crew (all of us professionals) are simply arguing about who gets the deck chairs and where they are to be placed ("liberty" and "freedom"). "In general it is worth taking action in advance to deal with disasters" (Watt, 1974, p. 7).

In contrast to Watt's thesis, and as a classic example of the Titanic phenomenon, the People's Republic of China is going full speed ahead with developing industry, superhighways, and cars for private owners (Shenon, 1994; Tyler, 1994). Vaclav Smil, a professor of ecology at the University of Manitoba in Winnipeg, is an advocate of interdisciplinary programs. He is probably the world's leading expert on China's economy (Smil, 1984, 1993). On the environmental impact of decisions in China's automobile industry he commented "That is an insane route" (Tyler, 1994). However, could not the same be said for the rest of the world?

Although the goal of Global Bioethics cannot be *proved* to be right, reasonable, desirable, or ethical, it is proposed as a long-range intuition, a vision, or an opinion. What is demanded of anyone who accepts the concept of Global Bioethics is to *communicate convincingly the vision to enough people to create a following*. A "following" may be gained in various ways that overlap in some respects: In science, one "communicates convincingly" by having an intuition as to the meaning of some result, and then inventing an experiment that will test the idea in a way that can be repeated at another time and place by any independent operator. In religion one "communicates convincingly" by claiming authority from God, by force of argument, and by charisma. In philosophy, it is a matter of forceful logic. In the arts, acceptance by critics and the public is required. Global Bioethics must convincingly communicate ends and means: the *end* as "acceptable survival" and the *means* as interdisciplinary effort that includes willing religious leaders.

TRACING THE ORIGIN OF GLOBAL BIOETHICS

The term *Global Bioethics* is an outgrowth of the purely personal use of the word that began with my original coining of "bioethics" in 1970. The triggering event in my epiphany occurred in 1957, when I was 46 years old. The late Margaret Mead was a well-known anthropologist. She was associate curator of Anthropology at the American Museum of Natural History in New York. She presented the Phi Beta Kappa Lecture to the meeting of the American Association for the Advancement of Science in New York, December, 1956, later published with the title "Toward More Vivid Utopias" (Mead, 1957). In her closing paragraph, she presented an idea that struck me as a proposal that had never before been activated, or, indeed, visualized. It was not her reference to *survival*, which concerned everyone in those days because of the atom bomb that had been constructed and used over Hiroshima and Nagasaki. Rather, it was because of her vision of a new kind of university professor. While her inspiration was survival in the nuclear age, her words are equally applicable to my idea of "acceptable survival" in the Third Millennium, beginning with the 21st century. She said:

> Finally: It seems to me, in this age when the very survival of the human race and possibly of all living creatures depends upon our having *a vision of the future for others which will command our deepest commitment*, [she continued] we need in our universities, which must change and grow with the world, not only chairs of history and comparative linguistics, of literature and art — but *we also need Chairs of the Future, chairs for those who will devote themselves*, with all the necessary scholarship and attention *to developing science to the full extent of its possibilities for the future* ... (italics added) (Mead, 1957).

Of course, I had read *The Challenge of Man's Future* by Harrison Brown (1954) soon after I had returned from a leave of absence that took me to Peru for studies on adaptation to high altitude for the U.S. Air Force (during the Korean War). Now the vision seen by Margaret Mead inspired me to act on the side in a new direction, while devoting full time to my role as a Professor of Oncology.

My first thoughts along the new line were expressed in 1962 when I was invited to help celebrate the 100th Anniversary of the creation of the Land Grant College system at my alma mater, South Dakota State College, in Brookings. (Abraham Lincoln had signed the Morrill Act in 1862). I conceived a presentation on an occasion that offered a real opportunity as well as an obligation to depart from my usual topic as a generalist on "the Biochemistry of Cancer." For the first time I developed a talk that would ultimately launch *bioethics* and now Global Bioethics, as "The Science of Survival." My title was "Bridge to the Future: The Concept of Human Progress" (Potter, 1962). After categorizing three divisions of the concept of progress — the religious, the material or consumption lifestyle, and the scientific-philosophic, I concluded by declaring that:

> The scientific-philosophic concept of progress which places its emphasis on *long-range wisdom* is the only kind of progress *that can lead to survival* [italics added]. It is a concept that places the destiny of [humankind] in the hands of [humans] and charges them with the responsibility of examining the feedback mechanisms and short-sighted processes of natural selection at biological and cultural levels [later to be characterized as the "fatal flaw,"], and of deciding how to circumvent the natural processes that have led to the fall of every past civilization... Let us use our tremendous capacity for production to produce the things that make us wiser, rather than the things that make us weaker... The addition of curricula that will enable students to achieve a balance between the three concepts of progress and to translate them into action will, in my opinion, give the universities a role of leadership in this complex world of today. Only by combining a knowledge of the sciences and of the humanities in the minds of individual [persons] can we hope to build a "Bridge to the Future" (Potter, 1962).

I had read *The Step to Man* by John R. Platt (1966), and *New Views on the Nature of Man* (1965), which he edited, so I was alerted to his article on "What We Must Do" (1969) and his book, *Perception and Change: Projections for Survival* (1970). "What We Must Do" was rather apocalyptic, but its final conclusion struck a chord with me. His subtitle was, "A large-scale mobilization of scientists may be the only way to solve our crisis problems." He concluded by declaring:

> The task is clear. The task is huge. The time is horribly short. In the past we have had science for intellectual pleasure, and science for the control of nature. We have had science for war. *But today the whole human experiment may hang on the question of how fast we now press the development of science for survival* (Platt, 1969) (italics added; "today" was 1969; the proposition is still true in 1994).

Platt called for a "Science of Survival" in 1969 just as I had called for a scientific-philosophic approach "that can lead to survival" in 1962. Platt declared "The only possible conclusion is a call to action." He asked:

> Who will commit themselves to this kind of search for more ingenious and fundamental solutions? Who will begin to assemble the research teams and the funds? Who will begin to create those full-time interdisciplinary centers that will be necessary for testing detailed designs and turning them into effective applications?" (Platt, 1969).

I had responded to Margaret Mead's call by setting up an informal and unofficial "Interdisciplinary Seminar on the Future of Man" that included Merle Curti of the History Department, Willard Hurst of the Law School, and

Farrington Daniels of the Chemistry Department, among others. By 1962, President Fred Harrington of the University of Wisconsin had made my committee official as the Interdisciplinary Studies Committee on the Future of Man [i.e., humankind] and told me to get in touch with Professor Reid Bryson, a professor of meteorological science, who was up to that time unknown to me. Together, we enlarged the interdisciplinary mode and each of us served as chairman on occasion. When I was chairman I drafted a paper that responded to a widely distributed request by the Board of Regents asking about the purpose of the University. This was a golden opportunity to pursue the agendas I had vocalized in the 1962 talk on "The Concept of Human Progress." The opportunity also permitted a review of the agendas published by Margaret Mead and by John R. Platt. My draft of the paper was circulated to each member of the committee and was discussed and revised during several meetings. In its final form it was published. It was also presented to the faculty and unanimously approved on December 1, 1969 as "an appropriate and timely supplement to previous statements of University purpose and function." The faculty specifically endorsed the statement of primary purpose. The title of the report was "Purpose and Function of the University" with a subheading "University scholars have a major responsibility for survival and quality of life in the future." After describing the mission we stated:

We affirm the views

- that the survival of civilized [society] is not something to be taken for granted.
- that governments throughout the world are experiencing great difficulty in planning for the future while trying to cope with the present.
- and finally, that the university is one of the institutions that has a major responsibility for the survival and improvement of life for civilized [society] (Potter et al., 1970).

Our report highlighted some of the dangers in the ambivalence in previous reports in claiming priority for the "search for truth" on the one hand, and the trend to assume responsibility for finding solutions to problems of the *immediate present* on the other. We pointed out the danger that in the latter case universities could become merely "public utilities". While the entire report remains in itself very close to a statement of Global Bioethics, only the blank verse form of the statement is given here:

The primary purpose of the University
Is to provide an environment
In which faculty and students
Can discover, examine critically,
Preserve and transmit
The knowledge, wisdom, and values
That will help ensure the survival

Of the present and future generations
With improvement in the quality of life.
(Potter et al., 1970).

It may be noted that this statement and, indeed, the whole report was issued in a time of great unrest on every university campus. Commenting on the above statement of purpose we noted that:

Ways should be found to allow students and faculty to engage in interdisciplinary efforts that are implied by the statement of purpose. Such an orientation might help to close the "relevance gap" that now exists between faculty and students (Potter et al., 1970).

It will be recalled that the first Earth Day was held on April 22, 1970. It initiated environmental teach-ins on nearly every campus in the country, and has been repeated every year since that time.

Unfortunately, the committee that President Harrington authorized and placed in the hands of Potter and Bryson ceased operations shortly after the above report was finalized. The Regents authorized the formation of the Institute for Environmental Studies with Professor Reid Bryson as Director and Potter included on the Executive Committee. Potter continued to focus on cancer research while on the side pursuing the scientific–philosophic mission described in 1962 and formalized in a series of lectures and articles. This effort led to the coining of the word *bioethics* which first appeared in print in two articles in 1970: "Bioethics: The Science of Survival" (Potter, 1970) and "Biocybernetics and Survival" (Potter, 1970). The substance of these articles and others appeared in a book in January 1971 with a title that included the key word: *Bioethics, Bridge to the Future* (Potter, 1971). The book was adopted, mainly by biology departments in over 1500 colleges and universities, responding to the first Earth Day (April 22, 1970).

Meanwhile, the University continued along traditional lines and the Institute for Environmental Studies incorporated several new technological advances in environmental assessment with many faculty members having outside departmental affiliations. Many courses were assembled into a teaching program including a course on Environmental Ethics. However, the goals of the old committee on the future of the human species were not pursued. Environmental ethics was not bioethics in the sense of acceptable survival in the long term based on integrating health care (medical bioethics) and earth care (environmental ethics).

Shortly after *Bioethics, Bridge to the Future* appeared (January 1971), a special section of *Time* magazine was devoted to the new dilemmas in biology:

Cancer Researcher Van Rensselaer Potter of the University of Wisconsin has suggested in a new book, *Bioethics*, that the U.S. create a fourth branch of

Government, a Council for the Future, to consider scientific developments and recommend appropriate legislation.

Indeed some form of super-agency may be the only solution to the formidable legal problems sure to arise [from new developments in biology] ... (*Time*, 1971)

After quoting from the chapter on Teilhard de Chardin who found "fathoming everything, trying everything, extending everything" on the way "to our ultimate Omega Point of shared godhood", *Time* referred to "the religious community, especially Roman Catholics [who] warn that man must not tinker with such sacred values or life and the family for fear of disturbing the natural order of things." (Here was the epitome of the late André Helleger's agenda in the Georgetown program). The *Time* article continued:

Those in the scientific world, more pragmatically, tend to mirror Potter's warning about "dangerous knowledge" — knowledge that accumulates faster than the wisdom to manage it (*Time*, 1971).

While the *Time* publication and the 1970 articles should have alerted Dr. Helleger and colleagues of the prior introduction of the word *bioethics* in the broad sense, they incorporated the word into their newly funded Institute for Human Reproduction and Bioethics, omitting any mention of the books by Potter or his references to the "Land Ethic" of Aldo Leopold (1949). These developments are described by Warren Reich in "The Word 'Bioethics': Its Birth and the Legacies of Those Who Shaped It" (1994). Thus, bioethics as a scientific–philosophic emphasis on human survival in the long-term became an orphan, incorporated into neither environmental ethics nor medical ethics. However, the original effort was continued with the publication of *Global Bioethics: Building on the Leopold Legacy* (Potter, 1988). Here it may be noted that *Bioethics, Bridge to the Future* was dedicated to Aldo Leopold, and that the original title of the 1988 manuscript was *Global Bioethics for Human Survival: Aldo Leopold's Land Ethic Revisited.*

In the 1990's, the problem for Global Bioethics is how to relate to religion. In "Global Bioethics Defined" (Potter, 1988, p. 151) the ethic was defined as "a secular program of evolving a morality that calls for decisions in health care and in the preservation of the natural environment. It is a morality of responsibility. Although described as a secular program, it is not to be confused with *secular humanism*. On the other hand, it was noted that "a global bioethic cannot be based on any single religious dogma; and even if there were no other reason it must be secular...". While in 1988 it was argued that conflicting religious factions, "must be persuaded that mutual respect and tolerance for other groups is part of a viable global bioethic", there was no mention of how global bioethics might proceed to develop a relationship with such groups.

Neither was that issue approached in a key article on evolution's fatal flaw in relation to survival "Getting to the Year 3000: Can Global Bioethics Overcome Evolution's Fatal Flaw" (Potter, 1990). The neglect of religion was continued when an article on "Scientists' Responsibility for Survival of the Human Species" was authored by Potter and Richard Grantham (1992), an emeritus professor at Université Claude Bernard in Lyons, France. We listed the following seven principles in a "Declaration for Geotherapy and Global Bioethics" reproduced from Grantham's report on his conference on "Modeling and Geotherapy for Global Changes". A draft Declaration was circulated to the conferees and debated in the final session to produce the final version:

1. Accelerating environmental degradation threatens the habitability of the biosphere. We believe that corrective action is possible and urgent.
2. Our goal is long-term survival in an acceptably maintained global ecosystem.
3. We as human beings need to take full responsibility for our actions by not sacrificing natural resources for short-term gains and by working to make the world a better living place.
4. This choice will influence our future biological and cultural evolution; we cannot avoid it without grave consequences.
5. A global bioethic should be further developed to guide and motivate geotherapy and our cultural evolution.
6. A root problem is excessive demographic growth; the Earth's carrying capacity is being exceeded. With present lifestyles and patterns of development, pollution of all kinds will increase as long as the population increases.
7. We declare that scientists should adopt the aforementioned goals and participate in meetings at all levels to apply these principles (Grantham, 1992).

The breakthrough in the further evolution of the original bioethics occurred when Hans Küng published *Global Responsibility: A Search for a New World Ethic* (1993). As Director of the Institute for Ecumenical Research at the University of Tübingen, Germany, he had never read or referred to the word bioethics or the books or articles dealing with it. However, his ecumenism went beyond the usual narrow definition to call for a "coalition of believers and non-believers ... in mutual respect ... for a common world ethic" that would lead to survival. Thus, his opening section carried the title "No Survival Without a World Ethic. Why We Need A Global Ethic." (Potter, 1994b). Another article also brought religion into play in the global bioethic: "Religion, Science Must Share Quest for Global Survival" (Potter, 1994a). This title is in contrast to the 1992 title "Scientists' Responsibility..." in which a bond between science and ethics was urged but religion was not mentioned (Potter and Grantham, 1992). It was proposed (Potter, 1994a) that the National Academy of Science might forge a beginning in the U.S. by bringing together leading scientists and willing religious leaders.

Then I received a number of responses that were forwarded to the Editor at his request and published in *The Scientist* (August 22, 1994, p. 12) with permission from the persons involved. One writer called my attention to the fact that religion and science were already together in the "Joint Appeal by Religion and Science for the Environment" from 1990 to 1993, when the name was changed to the "National Religious Partnership for the Environment." Another called my attention to the book, *Earth Keeping: Christian Steward-ship of Natural Resources,* edited by Loren Wilkinson (1980), and a revised edition with a new title, *Earth Keeping in the 90's: Stewardship of Creation* (Wilkinson, 1991). The various main line religions in the U.S. are working closely with the Union of Concerned Scientists, and will probably join in expressing concern about what many call "overpopulation". No doubt they will use the word "stewardship." The two books cited above have clearly expressed concern about overpopulation, along with stewardship.

Not cited here are a number of books dealing with the need for religion to become involved in "Care for Creation", but to my knowledge they have not included a concern for long-term survival as proposed here. Two authors who deal with survival in secular terms need to be cited however. The late Hans Jonas (1984) speaks to the "new millennialism of the postreligious age" (p. 177) and of "the long-range responsibility toward the future" that must be "included in the ethical theory and become the cause of a new principle ... that the prophecy of doom is to be given greater heed than the prophecy of bliss" (p. 31). Coming from a different direction, Manfred Stanley, a sociologist, expresses similar concerns for survival. These two books deal with survival realistically in contrast to a collection of essays by 27 authors who try to answer philosophical questions (Partridge, 1981).

This concludes the history of the events that have led to the concept of bioethics for the 1990's as Global Bioethics. Global Bioethics maintains the original 1970 interest in "Bioethics: The Science of Survival" and credits the role of Aldo Leopold (1949) in formulating the "Land Ethic". Global Bioethics now recognizes the need for ecumenism in the sense advocated by Hans Küng (1993), and agrees with Küng in advocating human survival in the long term (Potter, 1994a). This account is not a history of bioethics as launched by André Hellegers at Georgetown University (cf. Reich, 1994). The Georgetown concept of bioethics has spawned literally thousands of articles, books, and conferences, so many that a comprehensive overview may never be written. In the words of H. Tristam Engelhardt, Jr., the word bioethics

> ...is like a child who left home, renouncing the disciplines of its father but with substantial talents and capacities of its own. It has willfully chartered its own successful but narrower destiny by spawning an *Encyclopedia of Bio-ethics* [Warren Reich, Editor] and a large number of volumes and essays on bioethics, as well as a journal by that name. For the most part the term bioethics has been taken to identify the disciplined analysis of the moral and conceptual assumptions of medicine, the biomedical sciences, and the allied health professions (Engelhardt, Foreword, from Potter 1988).

In contrast to the narrow view that bioethics is concerned only with the moral and conceptual assumptions of the health care professions, Professor Andrew Jameton (1994) has accepted the vision of Global Bioethics and concluded that major changes will be required in how we practice healthcare in the context of environmnetal, population, and poverty problems worldwide.

REFERENCES

K. Bayertz, Introduction: Moral Consensus as a Social and Philosophical Problem. In *The Concept of Moral Consensus. The Case of Technological Interventions in Human Reproduction.* (K. Bayertz, Ed.). Kluwer Academic Publishers, Dordrecht, Germany. 1994.

H. Brown, *The Challenge of Man's Future.* Viking, New York, 1954.

H. Tristram Engelhardt, Jr., *The Foundations of Bioethics.* Oxford University Press, New York. 1986.

J. Franklin, Poisons of the Mind. *CIIT Activities,* 14:1–6, 1994.

R. Grantham, Declaration for Geotherapy and Global Bioethics, *Global Environmental Change,* pp. 66–67, March 1992.

V. Havel, The New Measure of Man. Excerpts from *New York Times,* OP-ED page. Friday, July 8, 1994.

G. Himmelfarb, *On Looking Into the Abyss, Untimely Thoughts on Culture and Society,* Alfred A. Knopf, New York. 1994.

A. Jameton, Global Bioethics. Casuist or Cassandra? Two Conceptions of the Bioethicist's Role. *Cambridge Quarterly of Healthcare Ethics,* 3:449–466, 1994.

H. Jonas, *The Imperative of Responsibility. In Search of an Ethics for the Technological Age.* University of Chicago Press, Chicago. 1984.

H. Küng, *Global Responsibility: A Search for a New World Ethic.* Continuum Publishing Company, New York. 1993.

A. Leopold, *A Sand County Almanac and Sketches Here and There.* Oxford University Press, New York. 1949.

M. Mead, "Toward More Vivid Utopias," *Science,* 126:957–961, November 8, 1957.

E. Partridge, *Responsibilities to Future Generations. Environmental Ethics.* Prometheus Books, Buffalo, NY. 1981.

J.R. Platt, *The Step to Man,* Wiley, New York, 1966.

J.R. Platt, *New Views on the Nature of Man,* University of Chicago Press, Chicago, 1965.

J.R. Platt, What We Must Do, *Science,* 166:1115–1121, 1969.

J.R. Platt, *Perception and Change: Projections for Survival,* University of Michigan Press, Ann Arbor. 1970.

V.R. Potter, Bridge to the Future: The Concept of Human Progress, *J. Land Econ.,* 38:1–8, February, 1962.

V.R. Potter, Bioethics: The Science of Survival, *Perspect. Biol. Med.,* 14:127–153, Autumn, 1970.

V.R. Potter, Biocybernetics and Survival, *ZYGON, Journal of Religion and Science,* 5:229–246, September, 1970.

V.R. Potter, *Bioethics, Bridge to the Future,* Prentice-Hall, Englewood Cliffs, NJ. 1971.

V.R. Potter, Humility with Responsibility — A Bioethic for Oncologists. Presidential Address. *Cancer Res.,* 35:2297–2306, 1975.

V.R. Potter, *Global Bioethics: Building on the Leopold Legacy,* Michigan State University Press, East Lansing. 1988.

V.R. Potter, Getting to the Year 3000: Can Global Bioethics Overcome Evolution's Fatal Flaw? *Perspec. Biol. Med.,* 34:89–98, 1990.

V.R. Potter, Bridging the Gap Between Medical Ethics and Environmental Ethics, *Global Bioethics,* 6(3):161–164, 1993.

V.R. Potter, Religion, Science Must Share Quest for Global Survival, *The Scientist,* p.12, May 16, 1994a.

V.R. Potter, An Essay Review of *Global Responsibility. In Search of a New World Ethic. Perspec. Biol. Med.,* 37:546–550, 1994b.

V.R. Potter, D.A. Baerreis, R.A. Bryson, J.W, Curvin, G. Johansen, J. McLeod, J. Rankin, and K R Symon, Purpose and Function of the University, *Science* 167:1590–1593, March 20, 1970.

V.R. Potter and R. Grantham, Scientists' Responsibility for Survival of the Human Species, *The Scientist,* pp. 10–11, May 25, 1992.

W. Reich, The Word "Bioethics": Its Birth and the Legacies of Those Who Shaped It, *Kennedy Inst. Ethics J.,* 4:319–335, 1994.

W. Reich, The Word "Bioethics": The Struggle Over Its Earliest Meanings, *Kennedy Inst. Ethics J.,* 5:19–34, 1995.

A. Seligman, The Idea of Civil Society, *The Free Press,* New York, 1992.

P. Shenon, Good Earth is Squandered. Who Will Feed China? *New York Times.* September 21, 1994.

V. Smil, *The Bad Earth. Environmental Degradation in China.* M.E. Sharpe, Armonk, NY. 1984.

V. Smil, *China's Environmental Crisis An Inquiry into the Limits of National Development.* M.E. Sharpe, Armonk, NY. 1993.

M. Stanley, *The Technological Conscience. Survival and Dignity in an Age of Expertise.* University of Chicago, Press, Chicago. 1981.

Time, "Man into Superman: The Promise and Peril of the New Genetics," pp. 33–52, April 19, 1971.

P.E. Tyler, China Planning People's Car to Put Masses Behind Wheel. *New York Times.* September 22, 1994, page A-1.

K.E. Watt, *The Titanic Effect. Planning for the Unthinkable.* Sinauer Associates, Stamford, CT, 1974.

L. Wilkinson, Ed., *Earthkeeping: Christian Stewardship of Natural Resources,* Wm. B. Eerdmans, Grand Rapids, MI, 1980.

L. Wilkinson, Ed., *Earthkeeping in the 90's: Stewardship of Creation,* Wm. B. Eerdmans, Grand Rapids, MI, 1991.

SECTION V
Summary

27

Bayard L. Catron

This summary highlights three central themes of the chapters presented in this volume. The preceeding chapters traverse a very wide range of issues, making a true synthesis quite impossible. The chapter concludes with four modest suggestions.

This summary is organized around three pairs of propositions. Each pair contains a "weak" version, which most people attending the symposium (though not necessarily in the broader risk community) would endorse, and a "strong" version which will be much more controversial, I think. The three sets are

1. (a) Risk assessment is value laden.
   (b) The whole enterprise of risk assessment is socially constructed — meaning that it has no independent validity or objectivity.
2. (a) Risk assessment is an appropriate and useful aid in environmental decision making, despite its deficiencies.
   (b) Risk assessment should be relied on more heavily in decision making (as has been proposed in recent bills before Congress).
3. (a) Public values should be taken into account in decision making and in setting risk reduction priorities.
   (b) Where there is persistent disagreement, public ("political") values should trump expert ("scientific") values.

PROPOSITION 1(A) RISK ASSESSMENT IS VALUE LADEN

Everyone who expressed a view on this point at the symposium agreed that values are present in risk assessment. For example, Schnare's model of risk analysis shows how values, often those of the assessor, were fundamental to the method used in the analysis. Nash argued not only that the notion of a

1-56670-131-7/96/$0.00+$.50
© 1996 by CRC Press, Inc.

scientifically pure analysis of risk is an illusion, but also suggests that the pretense of value neutrality itself poses a major danger to scientific integrity. There is also a practical challenge here, and Burt Hakkenen presented several examples of industry efforts to incorporate values into risk decision making.

There might not be such unanimity in the home disciplines represented at the symposium, and/or the professional communities and/or scientific organizations represented, but this proposition is rather widely accepted by this time. For example, as Nash notes, the EPA Science Advisory Board in *Reducing Risk*[1] speaks of "inevitable value judgments". When risks are borne differently by different groups of people, or cross generations are discussed by Catron et al., questions of fairness or justice arise.

Scott Baker pointed out that 1983 NAS risk assessment/risk management paradigm attempted to limit values to the risk management side, preserving risk assessment as value free. However, he says each step involves "best professional judgment" which is subjective as well as objective, as he argues in some detail. He suggests that subjective values are acceptable "as long as they do not introduce bias". (However, it might be argued that this is exactly what is at issue.)

Some, but not all, types of values create difficulties in the "scientific" status of particular claims. It is important to sort out the several kinds, such as moral, aesthetic, economic, and scientific values. In an unusual treatment, James Nash identified as *moral* values the following scientific values — honesty in selecting data, rationality, tolerance of diversity, freedom of inquiry, corrective dissent, cooperation, and open communications. Nash claims that moral values are present in all phases of risk assessment — motives, purposes, definitions, methods, and assumptions. However, granting that risk assessment is not value free, is it necessarily a *moral* enterprise in *all* these ways?

Doug MacLean's chapter on intrinsic vs. instrumental values illustrated the kind of careful analysis of particular values that is needed regardless of the method used to assess risks. Virginia Sharpe adopted a different strategy with respect to values. Beginning with a normative commitment to a particular value — sustainability — she explores the relation to ethical theory.

PROPOSITION 1(B) THE WHOLE ENTERPRISE OF RISK ASSESSMENT IS SOCIALLY CONSTRUCTED — MEANING THAT IT HAS NO INDEPENDENT VALIDITY OR OBJECTIVITY

The idea of social construction of reality was introduced by Bill Freudenberg. According to this epistemology or theory of knowledge, facts are not meaningful without human interpretation. There is no such thing as "brute facts" independent of a context, which is provided by the language and categories we use to understand anything at all. Facts do not "speak for themselves". So, for example, as William Cooper pointed out, in answering the question "How safe is a 10^{-5} risk?", we might conclude that it provides a generous level of protection or a license to kill, depending on the interpretation. Not only the concept of risk itself, but other basic concepts like fact, value, and objectivity, are socially constructed according to this theoretical framework. This orientation would not be accepted by many of those at the symposium. For example, it would undermine the distinction between subjective and objective that Scott Baker builds his paper around — what he calls objective is no less socially constructed, from this point of view, than what is acknowledged as subjective. However, contrary to the fears of many positive scientists, the orientation does not undermine the scientific enterprise — at least in some accounts of what that enterprise is essentially. According to philosophers and historians of science following in the tradition of Thomas Kuhn (*The Structure of Scientific Revolutions*)[2], superhuman objectivity has never been a requirement of scientific method. The values of the scientific community include those Nash listed (cited earlier) and, as Kristin Shrader-Frechette notes, testability and reliability. (This does not deny the usefulness of understanding the personal attributes of risk assessors, which Crawford-Brown and Arnold address in their chapter.)

Two other chapters seem relevant here, although neither explicitly mentions the social construction of reality. Don Brown argues that there is no neutral discourse — whether law, economics, or natural science. He might or might not agree with the further inference that science does not or should not have a privileged position — for example, as being "more rational" than other types of discourse. Rachelle Hollander focuses on the question: "How does knowledge get legitimated and socially appropriated?" The risk domain is contested — within/between different scientific fields, between experts and the public, between industry and government, etc. She says that the contest involves who legitimately (I would add "authoritatively") speaks about risk, and how. (This issue is important in the third set of propositions below.)

PROPOSITION 2(A) RISK ASSESSMENT IS AN APPROPRIATE AND USEFUL AID IN ENVIRONMENTAL DECISION MAKING, DESPITE ITS DEFICIENCIES

All authors who discuss risk assessment here seem to take this for granted, even as many of them acknowledged limitations and deficiencies of various sorts. No one in this group, at least, fundamentally challenged the utility of risk assessment.

PROPOSITION 2(B) RISK ASSESSMENT SHOULD BE RELIED ON MORE HEAVILY IN DECISION MAKING (AS HAS BEEN PROPOSED IN RECENT BILLS BEFORE CONGRESS)

To agree that risk assessment is useful, however, is not to commit oneself to using it to drive decision making. Of the authors in this volume, Don Brown would probably disagree most strongly with proposition 2(b). He would argue,

I think, that the deficiencies of risk assessment are intrinsic, like they are with cost-benefit analysis, and cannot be fixed. Therefore, we should use the methodology as a heuristic, not as a decision-driving method.

Bryan Norton would also disagree with this proposition. He argued that risk assessment as a methodology "will never be adequate" to deal with all we ordinarily think of as risk (which is more than the probability of an event occurring and an estimate of the magnitude of its consequences). He emphasized particularly our limitations in using *ecological* risk analysis (as William Cooper did as well), and our problem of understanding long-term and large-scale events.

PROPOSITION 3(A) PUBLIC VALUES SHOULD BE TAKEN INTO ACCOUNT IN DECISION MAKING AND SETTING RISK REDUCTION PRIORITIES

This is such a common litany these days that few people would contest it — at least in public. It is often used in advocacy, as when Tom Burke appealed to public values in arguing that we need to "rediscover public health". However, there is a question of how deeply and widely this proposition is held. Is it just a self-protective strategy, to coopt public opinion and avoid litigation? Or is it a basic principle, and a normative commitment?

Bert Hakkenen stated that Procter and Gamble has elevated to a principle the idea of involving the public or considering its values through two-way communication. Of course this is at least partly self-interested on the part of the company — every company wants its products to be "safe and perceived to be safe". How important is the "public values" principle in relation to other principles used in decision making (in the case of P&G efficiency and risk-based priority setting)?

PROPOSITION 3(B) IN CASES OF PERSISTENT DISAGREEMENT, PUBLIC ("POLITICAL") VALUES SHOULD TRUMP EXPERT ("SCIENTIFIC") VALUES

This is perhaps the real acid test. Scientists and technical experts of all sorts who value reason and evidence highly are not often comfortable seeing themselves as stakeholders in public debates, and certainly not as having vested interests. However, the public will inevitably see risk experts as wedded to their methodology and not as the final authority on what social risks should be accepted. This is not necessarily "irrational," as many experts would have it; as Freudenberg suggests, opponents of particular technologies are often as well informed as advocates.

At least some of the symposium presenters were not willing to put scientists in a privileged position with respect to value determinations. Scientists are "as competent" as others, according to Freudenberg, which suggests that their value judgments should not be given preference automatically. Nash has a more skeptical view: scientists are no more competent — and perhaps less so — than the public as a whole to make these value determinations.

There are many examples of disparity between public and expert estimates of the importance of particular risks. One of the most striking is nuclear waste, which the public ranked first in importance and the experts ranked 20th out of 30 in a recent survey. Freudenberg gives a useful example of differing values between scientists and the public. He says that scientists and engineers as a whole value efficiency and cost-effectiveness more highly than long-term safety, compared to the population as a whole. In that situation, should the scientists' view prevail? If so, why?

Of course there is not always disagreement between experts and the public. No one would prefer a lay opinion to an expert assessment of probabilities and consequences of a technical sort. However, risk is more than that, as Norton pointed out, and the circumstances under which risk should be undertaken or one risk preferred to another is not a technical question. Perhaps in addition to the categories of risk assessment, risk management, etc., we need the concept of a risk *judgment* which emanates from a social process involving all parties.

Ideally, it seems that we need better decision mechanisms that will avoid such polarization. How can public participation be made meaningful rather than cosmetic? Paterson and Andrews offered some excellent ideas in their chapter. They use state and local comparative risk projects to illustrate ways of involving the public meaningfully in establishing criteria for risk assessments and setting priorities, including recommending strategies to public officials. Technical advisory committees are used to bring science to bear, do first-phase assessment and ranking, and assist in generating and evaluating risk reduction strategies. In their example of the state of Washington, the public advisory committee seems to be in the driver's seat in at least some stages of the process. Paterson and Andrews' chapter provides a good basis for hoping that we will do better in blending public and expert judgments in the future.

CONCLUSION — FOUR MODEST SUGGESTIONS

Several recommendations can be made, partly derived from the symposium, but ultimately the personal suggestions of the author:

1. Pay more attention to the problems of definition and analysis in this arena of values and ethics in risk assessment. It may be true, as Baker said, that "God's favorite color is gray", but that does not excuse casual analysis and the loose, insensitive use of key terms.
2. Do not rely too heavily on risk assessment or claim more for it than it can deliver. Given our blind spots and pervasive overconfidence (and, as Freudenberg said, especially the *unknown* unknowns) and the limitations of all our techniques, humility is important.

3. Stop disguising value judgments as technical ones. Freudenberg was particularly forceful in emphasizing the seriousness of potential public mistrust of science. He says it is not just a public relations battle here; science is not immune to erosion of confidence.

4. Stop playing the blame game. Regarding failures in risk communication between experts and the public, as Victor Cohn suggests, there is enough blame to go around. While he acknowledges some journalistic deficiencies, he cites vested interests of some parties, the blind spots of experts, and the myopia of some players — let alone the wide areas of disagreement and uncertainty inherent in risk decision making. The blame game is unhelpful and indeed counterproductive, whether Federal vs. local, industry vs. regulator, media bashing, government bashing, lawyer bashing. After all, in some real sense we are all in this together.

REFERENCES

1. U.S. Environmental Protection Agency, *Reducing Risk: Setting Priorities And Strategies For Environmental Protection,* Science Advisory Board, September 1990, SAB-EC-90-021, Washington, D.C. 20460.
2. T.S. Kuhn, *The Structure of Scientific Revolutions,* 2nd ed., University of Chicago Press, Chicago, 1970.

The
Contributors

THE CONTRIBUTORS

Richard N. L. Andrews is Professor of Environmental Policy in the Department of Environmental Sciences and Engineering, School of Public Health, University of North Carolina at Chapel Hill. Formerly chairman of UNC's Environmental Management and Policy Program and of the Natural Resource Policy and Management Program of the University of Michigan's School of Natural Resources, he is the author of numerous journal articles and book chapters on U.S. and comparative environmental policy, environmental impact and risk assessment, and the uses of science and economics in environmental policy making. He has also served as budget examiner in the U.S. Office of Environmental Studies and Toxicology, and as a member of the EPA's Science Advisory Board Subcommittee on Risk Reduction Strategies.

Jeffrey Arnold is a Ph.D. candidate in the Department of Environmental Sciences and Engineering at the University of North Carolina at Chapel Hill. He received his B.A. and M.A. in History with a focus on medieval history.

Scott R. Baker is the Director of the Health Sciences Group at EA Engineering, Science, and Technology, Inc. He is a toxicologist with broad technical experience in human health and the environment, with 20 years of experience directing and participating in a wide variety of scientific evaluations involving toxicology, health risk assessment, and scientific interpretation of regulatory affairs and risk management issues. In his prior position with the U.S. EPA as Science Advisor to the Assistant Administrator for Research and Development, he earned several citations for his excellent service. Prior to the EPA, he was a Senior Staff Officer at the National Research Council of the National Academy of Sciences. His experience includes scientific evaluations of the effect of chemicals on human health and the environment; assessment of the impacts of legislative initiatives, regulations, and standards on the interests of clients; environmental toxicology investigations; risk assessments; and expert witness testimony. He has related experience in emergency preparedness, indoor air research, pesticide health effects, air toxics, and water quality criteria. He has also chaired and served on a number of committees and task forces related to risk assessment and environmental issues such as chemical safety and the human health effects of chemicals. He received his Doctorate degree in toxicology from Iowa State University in 1978. He has published a number of papers and books and presented a number of papers on topics related to human health and the environment.

Lawrence G. Boyer is a public administration graduate student at The George Washington University. He holds a master's degree in Economics from Rutgers University and a bachelor's degree in Physics from the University of Massachusetts. Mr. Boyer is also a member of the joint GWU-EPA Green University Task Force. His research interests include: intergenerational decision making, risk management, climate change, and non-market valuation techniques.

Donald A. Brown is Director of the Bureau of Hazardous Sites and Superfund Enforcement in the Office of Chief Counsel for the Pennsylvania Department of Environmental Resources. He is interested in, and has written and lectured extensively on, the interface between environmental science, law, economics, and environmental ethics. Mr. Brown represented Pennsylvania at the Earth Summit and was recently the director of a conference held at the United Nations as a follow up to the Earth Summit on the ethical dimensions of the United Nations program on environment and development. He holds a B.S. from Drexel Institute of Technology in Commerce and Engineering Science, a M.A. in Philosophy and Art from Seton Hall University School of Law, and a J.D. from Seton Hall University School of Law. He has also done graduate work toward a Ph.D. in Philosophy at the New School for Social Research. He has worked as an engineer and taught both philosophy and environmental law.

Thomas A. Burke is an Associate Professor of Health Policy and Management at Johns Hopkins School of Hygiene and Public Health. Trained as an epidemiologist, he formerly served as Deputy Commissioner of Health for the State of New Jersey and Director of the Office of Science and Research of the New Jersey Department of Environmental Protection. His experience has shaped his research interest in the interface of science and policy in environmental decision making.

Douglas Crawford-Brown is Professor of Environmental Physics in the Department of Environmental Sciences and Engineering at the University of North Carolina at Chapel Hill. His B.S. and M.S. degrees are in Theoretical Physics and his Ph.D. degree is in Nuclear Science from the Georgia Institute of Technology. Dr. Crawford-Brown teaches and conducts research in risk analysis, philosophy of science, and mathematical modeling of biophysical phenomena. He is Director of Undergraduate Studies in Environmental Science and Policy and of the Institute for Environmental Studies.

Bayard L. Catron is Professor of Public Administration and Policy at George Washington University. His primary research interests are in applied ethics — especially environmental ethics and ethics in government. He was Research Director of a recent study, "Deciding for the Future: Balancing Risks and Benefits Fairly Across Generations", conducted by the National Academy of Public Administration and sponsored by the U.S. Department of Energy. Professor Catron earned his Ph.D. in Social Policies Planning and also a Master of City Planning degree from the University of California at Berkeley. His earlier educational background includes two degrees in Philosophy, a B.A. from Grinnell College (Iowa) and a M.A. from the University of Chicago.

Victor Cohn is one of the nation's leading science reporters and former Science Editor of the *Washington Post*. He began writing about science and medicine for the *Minneapolis Tribune,* joined the *Washington Post* as Science Editor in 1968, and did some of the Post's first environmental reporting.

From 1985 to 1993 he was senior writer and columnist, originating the column "The Patient's Advocate", in the Post's weekly *Health* magazine, writing about the problems of patients and how to get good medical care, as well as the nation's health problems and politics as health care changes. In October 1993, he left the Post to become a research fellow at Georgetown University and work on a book on medical care. In 1986, Georgetown University awarded him the honorary degree of doctor of science for "insightful reporting ... fairness and effectiveness". He is currently a research fellow at the American Statistical Association. Among his publications is a book, *News & Numbers: A Guide to Reporting Statistical Claims and Controversies in Health and Other Fields.*

William Cooper is Professor of Zoology at Michigan State University. He became a full professor in 1972. He became co-director of the Design and Management of Environmental Systems Project, sponsored by the Research Associated with Nations Needs (RANN) section of NSF in 1970. From 1975 until 1988, he was Chairman of the Michigan Environmental Review Board. He was Chairman of the Zoology Department at Michigan State University from 1981–1987. He is presently Senior Consultant for Environmental Science for Public Sector Consultants, Inc. He has been a member of the Science Advisory Board of the Great Lakes Center, the Environmental Cabinet (State of Michigan), Environmental Quality Council (State of Michigan), and consultant for the Michigan Aeronautics Commission. Dr. Cooper holds membership in five professional societies, two editorial boards, and is a lecturer at the Brookings Institute in Washington, DC. He has had two terms on the National Research Council Board on Environmental Studies and Toxicology. He is also the recipient of the Braun InterTec/Dow Chair in Civil Engineering at the University of Minnesota. Dr. Cooper is presently a driving force in the formation of the Hazardous Waste Management Consortium that is directed through the Institute for Environmental Toxicology at Michigan State University. He has an appointment on the U.S. EPA Science Advisory Board and has directed the Relative Risk Assessment Program for Michigan, and currently chairs the Michigan Environmental and Natural Resources Code Commission.

William Freudenburg is a Professor of Rural Sociology and Environmental Studies at the University of Wisconsin-Madison. He has devoted some two decades to the study of technological controversies and the social impacts of environmental and technological change, with a special emphasis on the social-science aspects of risk assessment and risk management. His articles have been published in interdisciplinary journals such as *Science, Risk, Risk Analysis* and *Technological Forecasting and Social Change,* and in numerous sociological journals, including *American Journal of Sociology, American Sociological Review, Annual Review of Sociology, Rural Sociology,* and *Social Forces and Social Problems.* His books include *Public Reactions to Nuclear Power: Are there Critical Masses?* and *Paradoxes of Western Energy Development,* both of which were published by the American Association for the Advancement of Science. His latest book, written with Dr. Robert Gramling and entitled *Oil on Troubled Waters,* compares the risk perceptions related to offshore oil development in Louisiana vs. California. He received a B.A. from the University of Nebraska; his M.A., M.Phil., and Ph.D. are from Yale University.

Jennifer Grund is a Policy Analyst for PRC Environmental Management, Inc. of McLean, Virginia. She is also pursuing a Master of Public Administration degree at the George Washington University. During the 1993–1994 academic year, as a graduate assistant to Professor Bayard Catron, she participated in the Department of Energy's project entitled, "Deciding for the Future". She graduated from University of Hartford with a degree in History and Politics and Government in 1993.

P.J. (Bert) Hakkinen earned a B.A. degree in Biochemistry and Molecular Biology from the University of California at Santa Barbara in 1974, and a Ph.D. in Comparative Pharmacology and Toxicology from the University of California at San Francisco in 1979. From 1979–1982, he was a postdoctoral investigator at the Oak Ridge National Laboratory in toxicology and in exposure and risk assessment. Dr. Hakkinen joined the Procter and Gamble Company (Cincinnati, Ohio) in 1982, and is currently a "Senior Scientist — Toxicology and Risk Assessment". His Procter and Gamble work experience includes human exposure and risk assessment support for numerous types of consumer products. He has chaired the Exposure Assessment Task Group of the Chemical Manufacturers Association (Washington, DC) since 1991. Dr. Hakkinen is a member of the Society of Toxicology, and a charter member of the Society For Risk Analysis (SRA) and International Society of Exposure Analysis (ISEA). He has been an invited expert at several U.S. EPA- and OECD-sponsored workshops held to develop or revise human exposure assessment guidance and resource documents, and has lectured since 1988 on exposure and risk assessment at the University of Cincinnati. Dr. Hakkinen was on the Editorial Board of *Toxicology,* the journal, from 1986 to 1994 and is currently on the Editorial Board of the U.S. EPA, SRA, and an ISEA cooperative agreement effort to develop a book entitled *Residential Exposure: A Source Book.* He has authored and co-authored numerous publications, including ones on consumer product exposure and risk assessments, consumer risk perceptions, toxicological interactions, respiratory tract toxicology, and computer software and databases.

John Hartung is a Special Research Affiliate in the Office of Policy Development of the U.S. Department of Housing and Urban Development and a Research Fellow at the George Washington University Center for Washington Area Studies. He is also pursuing a Masters of Public Administration at the George Washington University. During the 1992–1993 academic year, as a graduate assistant to Professor Bayard Catron, he participated in the efforts leading to the June 1994 conference at which principles referenced in his chapter were articulated. Before coming to Washington, he served as judicial clerk to the Illinois Supreme Court and practiced law with the St. Louis based firm of Brown and James, P.C.

Rachelle D. Hollander has been at the National Science Foundation since 1976. She is Program Director for the Ethics and Values Studies (EVS) program. EVS supports research, educational, and other projects examining ethical and value issues in the interactions between science, technology and society. In 1990–1991, Dr. Hollander was a Visiting Professor in the Department of Science and Technology Studies at Rensselear Polytechnic Institute. She received her doctorate in philosophy in 1979 from the University of Maryland, College Park; she has written articles on applied ethics in numerous fields and on science policy and citizen participation. With Dr. Deborah Mayo, she edited the volume, *Acceptable Evidence: Science and Values in Risk Management.* She is a fellow of the American Association for the Advancement of Science and a AAAS Council Delegate and member of the Committee on Council Affairs. She is on the editorial board of *Risk: Health, Safety and Environment* and the new journal *Science and Engineering Ethics.*

Carolyn J. Leep earned a B.S. degree in Chemistry from Valparaiso University in 1985, and a M.S. degree in Organic Chemistry from Stanford University in 1988. Ms. Leep is currently Associate Director, Risk Issues for the Chemical Manufacturers Association (CMA) in Washington, DC. She serves as staff executive for several CMA task groups involved in risk assessment activities. She also coordinates risk assessment issues across all CMA regulatory affairs programs, involving environmental, product, and occupational risk assessments. Prior to joining CMA, Ms. Leep was a Senior Associate with ICF Incorporated in Fairfax, Virginia, where she was involved in a number of risk-related projects.

Douglas MacLean is Professor of Philosophy at the University of Maryland, Baltimore County. His primary research interests are in moral and philosophical issues in risk analysis and the foundations of policy science. He has published many articles and books on these topics, including *Energy and the Future, The Security Gamble,* and *Values at Risk.* He has been involved in policy making as a consultant or advisor to many government agencies.

Hon. Mike McCormack is a former Member of Congress, former Washington State Legislator, and former research scientist at the Atomic Energy Commission (now Department of Energy) Hanford facility. He is currently Director of the Institute for Science and Society, dedicated to "...enhancing the level of Science Literacy throughout society...", and a member of the Washington State Higher Education Coordinating Board. While a Member of Congress (1970–1980) McCormack was for 8 years chair of a subcommittee on energy of the House Committee on Science and Technology. His experiences on that committee and in shepherding science- and energy-related legislation through the Congress led him to the conviction that members of the scientific community should become active in political affairs and that they should insist on rational consideration of societal issues involving science and technology.

James A. Nash is Executive Director of the Churches' Center for Theology and Public Policy. He is a Lecturer in Social and Ecological Ethics at the Wesley Theological Seminary. His research and writing are focused now on ecology and ethics. His latest book is *Loving Nature: Ecological Integrity and Christian Responsibility.* He is an ordained United Methodist minister who received his S.T.B. degree from the Boston University School of Theology, attended the London School of Economics and Political Science, and received his Ph.D. in Social Ethics from Boston University where he was a Rockefeller Doctoral Fellow in Religion.

Bryan G. Norton received his Ph.D. in Philosophy, specializing in conceptual change in scientific disciplines. Currently Professor of Philosophy of Science and Technology in the School of Public Policy, Georgia Institute of Technology, he writes on intergenerational equity and sustainability theory, with special emphasis on biodiversity and ecosystem characteristics. His current research includes work on intergenerational impacts of global climate change (sponsored by the U.S. Forest Service) and on valuation of long-term impacts of policy choices (sponsored by the U.S. EPA). He is author of *Why Preserve Natural Variety?* (Princeton University Press, 1987), *Toward Unity Among Environmentalists* (Oxford University Press, 1991), and co-editor of *Ecosystem Health: New Goals for Environmental Management* (Island Press, 1992) and *Ethics on the Ark* (Smithsonian Press, 1995). He is especially interested in developing physical measures of ecosystem-level health/illness and relating these to human welfare. Norton serves on numerous panels, including the Ecosystem Valuation Forum, the Risk Assessment Forum (U.S. EPA), and the Environmental Economics Advisory Committee of the EPA Science Advisory Board.

Christopher J. Paterson is a Policy Associate at the Northeast Center for Comparative Risk at the Vermont Law School. He is a Ph.D. candidate in the Environmental Management and Policy Program in the Department of Environmental Sciences and Engineering at the University of North Carolina at Chapel Hill. He received a M.A. in History from the University of North Carolina at Chapel Hill and a B.A. in Cell Biology from the University of California at San Diego.

Van Rensselaer Potter (II), Ph.D., was born on a farm in Northwestern South Dakota in 1911. He and his wife were married in Madison, Wisconsin in 1935. They have three children and six grandchildren. He received a B.S. degree in Chemistry and Biology at the South Dakota State University in Brookings in 1933. He received the Ph.D. in Biochemistry and Medical Physiology in 1938 at the University of Wisconsin in Madison under Professor Conrad Elvehjem. He did post-doctoral research under two Nobel Laureates in Europe in 1938 and 1939: Professor Hans von Euler in Stockholm and Professor Hans Krebs in England, respectively. He and his wife returned to the U.S. in October of 1939, when he continued his fellowship with Professor Thorfin Hogness at the University of Chicago. He received an appointment in the new McArdle Laboratory for Cancer Research at the University of Wisconsin in February 1940 and advanced to Professor in 1947. He received numerous awards for his research on cancer and served as President of the American Association for Cancer Research in 1974–1975 and President of the American Society for Cell Biology in 1964–1965. He is a member of the National Academy of Science and of the American Academy of Arts and Sciences. He coined the word "bioethics" in 1970 and published two books and numerous articles on the subject.

Resha M. Putzrath, Ph.D., DABT, is a Principal of Georgetown Risk Group, where she evaluates the toxicological properties of chemicals and estimates their quantitative risk for hazardous waste sites, occupational exposures, and consumer products. Her focus is on designing innovative methods for improving the accuracy of analysis by combining information including: evaluation of complex chemical mixtures; appropriate application of mechanism of action, biomarker, and genotoxicity data in the assessment of carcinogenicity; and estimation of attendant uncertainties. Prior to establishing her firm, she worked at the National Academy of Sciences, U.S. EPA's Hazardous Waste Enforcement Task Force, and other consulting firms. Dr. Putzrath earned her Ph.D. in Biophysics at the School of Medicine and Dentistry, University of Rochester; she is a Diplomate of the American Board of Toxicology.

Paul Rebers, before his retirement in 1988, worked as a Research Chemist for 28 years on the chemistry of animal disease at the National Animal Disease Center in Ames, Iowa. He has a B.S. and M.S. in Chemical Engineering, and a Ph.D. in Biochemistry from the University of Minnesota. After retirement he served 4 years as a Member of the City Council where he became convinced of the necessity for improving the management of municipal solid waste. He organized a symposium for the American Chemical Society in 1994 entitled, "Municipal Solid Waste: Problems and Solutions." In 1990 he organized a symposium for the American Chemical Society entitled "Ethical Dilemmas of Chemists". His faith in God is responsible for his belief that sound ethical principles are essential to the chemical profession and in the promotion of stewardship on the environment. He is a past Secretary of the Division of Professional Relations of the American Chemical Society, and a member of the

American Society of Microbiology. He was elected to the Graduate Faculty of Iowa State University, and to the American Association of Immunologists. His major research accomplishments are in the field of carbohydrate chemistry where he developed simple and reliable methods for the determination of the total content of carbohydrate in complex biological mixtures. His work on the chemical structure of polysaccharides contributed to the better understanding of their role in the specificity of carbohydrate containing antigens. These studies contributed to the development of a new procedure for the serotyping of *Pasteurella multocida*, the causative agent in fowl cholera, an important disease of poultry. He is a member of the Methodist Church and a member of the Masonic Lodge, AF&FM, and a Past Master of the Lodge.

David W. Schnare is a Senior Policy Analyst in the newly organized Office of Enforcement and Compliance Assurance of the U.S. Environmental Protection Agency, where his duties encompass strategic planning program analysis. His collateral duties include international technical assistance on sustainable environments and free-market environmental economics. He was awarded a baccalaureate degree in Chemistry from Cornell College, as well as a M.S. in Public Health and a Ph.D. in Environmental Management from the School of Public Health at the University of North Carolina at Chapel Hill. His most recent book is *Chemical Contamination and Its Victims,* Quorum Books, New York, 1989.

Virginia A. Sharpe is a faculty member in the Departments of Medicine and Philosophy at Georgetown University and a Charles E. Culpeper Foundation Scholar in Medical Humanities. She teaches medical ethics and environmental philosophy.

Kristin Shrader-Frechette is currently Distinguished Research Professor at the University of South Florida in the Program in Environmental Sciences and Policy and the Department of Philosophy. She has held Professorships at the University of Florida and the University of California and earned undergraduate degrees in mathematics and physics from Xavier University and a doctorate in philosophy of science from the University of Notre Dame. She also did NSF post-doctorates in ecology, economics, and hydrogeology. Author of 185 articles and 12 books that have been translated into 9 languages, her most recent volumes are *Method in Ecology* (Cambridge University, 1993), *Burying Uncertainty: Risk and the Case Against Geological Disposal of Nuclear Waste* (University of California, 1993) and *Risk and Rationality* (University of California, 1991). On the editorial boards of 17 journals, Shrader-Frechette is a member of the U.S. National Academy of Sciences/National Research Council Board on Environmental Studies and Toxicology, its Committee on Risk Characterization, and its EMAP Committee.

Index

Index

A

Accounting system, analytic
framework, 171, 173
Adaptive management, analytic
framework, 169
Aesthetic risk, 341
Algorithms, validation, risk assessment
models, 303
Amenity risks, 341
Amoco Corporation, perceptions, in
risk decision making, 80
Analytic framework
accounting system, 171, 173
anthropocentric view, ethics, human
values and, 181
ignorance, 159
lexicographic, 155
multiple scales, 164
risk
analytic community, 169
decision square, 155, 164, 165
welfare, 157
Automobile industry, global bioethics,
364
Autonomy, validation, risk assessment
models, respect for, 284

B

Baby boom, conflict, environmental
risk and, 18
Balancing, health, ecological, value
judgments in, 8
Behaviors, 341
Benchmarking, validation, risk
assessment models, 304
Beneficence
ethical basis, risk analysis, 259
validation, risk assessment models, 284
Benefit-cost analysis, analytic
framework, 170

"Better Living Through Chemistry,"
conflict, environmental risk and,
20
Bhopal, conflict, environmental risk
and, 19
Bias
availability, conflict, environmental
risk and, 19
overview, environmental risk
decision making, 42
validation, risk assessment models,
285
Bigotry
chemists, citizenship responsibility
of, 350
Biocentrism, ethics, human values and,
182
Biodiversity, analytic framework,
167
Biotechnology, federal research and
development establishment, 345
Biotechnology Research Subcommittee,
federal research and development
establishment, 345
Biotic community, validation, risk
assessment models, 275
Biotic justice
ethics, human values and, 182
moral values, 205, 207
Blind spots, conflict, environmental
risk and, 12, 15, 16, 27
Boundaries, federal research and
development establishment, 338
Burden of proof, validation, risk
assessment models, 293

C

Calibration, conflict, environmental risk
and, 17
California, conflict, environmental risk
and, 29

Caricatures
  chemists, citizenship responsibility
    of, 351
  conflict, environmental risk and, 351
Catastrophic risk, equity,
    intergenerational, 141
Challenger, conflict, environmental risk
    and, 19, 27
Character, 342, 343
Charity, validation, risk assessment
    models, 281, 284
Chemical Manufacturers Association,
    perceptions, in risk decision
    making, 75, 79
Chemist, citizenship responsibility of
  asbestos, 350
  Delaney Clause, 350
  ethos, 350
  nuclear power, 356
  superstition, 350
  zero population growth, 355
Chlorine
  chemists, citizenship responsibility
    of, 355
  conflict, environmental risk and, 355
Civic involvement, chemists,
    citizenship responsibility of, 349
Civil society, global bioethics, 360
Civilization, chemists, citizenship
    responsibility of, hallmarks of,
    357
Classification scheme, 340
Cognitive dissonance, conflict,
    environmental risk and, 18
Coherence, validation, risk assessment
    models, 287
Combined data, data, on value
    judgments, combining, 252
Commission for Racial Justice, conflict,
    environmental risk and, 31
Committee on Environmental and
    Natural Resources, federal
    research and development
    establishment, 346
Committee on Fundamental Science,
    federal research and
    development establishment, 344

Common metric, data, on value
    judgments, combining, 247,
    248, 250
Common sense environmentalism,
    chemists, citizenship
    responsibility of, 355
Community Advisory Panels,
    perceptions, in risk decision
    making, 77
Companies, perceptions, in risk
    decision making, 73, 74
Comparative risk
  assessment, 213–226
  data, on value judgments, combining,
    247
Compassion, global bioethics, 360
Compensable, analytic framework, 165
Complex systems, analytic framework,
    161
Computer
  codes, validation, risk assessment
    models, 305
  models, validation, risk assessment
    models, 292
  verification, validation, risk
    assessment models, 302
Conceptual clarity
  ethical basis, risk analysis, 263
  validation, risk assessment models,
    282
Conduct
  federal research and development
    establishment, 347
Confidence, conflict, environmental
    risk and, public, 30
Conflict, environmental risk and
  antiscientific, 20
  Commission for Racial Justice, 31
  employment trends, 14
  equal rights, 16
  interdependence, 24
  Organizational Amplification of
    Risks, 32
  responsibilities, 29
  social construction, 13
  trust and trustworthiness, 24
  weather forecasters, 18

Consensus Position, validation, risk assessment models, 300
Constitutional convention, analytic framework, 162
Constitutive values, validation, risk assessment models, 285
Constraints
  federal research and development establishment, 339, 347
Consumer
  products, perceptions, in risk decision making, 74, 75
  risk perception, perceptions, in risk decision making, 75
Contest, federal research and development establishment, 338
Contested domain, 341
Contextual values, validation, risk assessment models, 286
Contingent valuation, analytic framework, 158
Correspondence, validation, risk assessment models, 287
Cost-benefit analysis, 207
  moral values, 199, 202
  overview, environmental risk decision making, 56
Cost-effectiveness, 73, 74, 78, 81
Credibility
  conflict, environmental risk and, of science, 26
  validation, risk assessment models, 305
Critics, conflict, environmental risk and, 28
  of technology, 29
Culture
  overview, environmental risk decision making, 41
  validation, risk assessment models, 279
Cumulative impacts, analytic framework, 171
Cyclamates
  chemists, citizenship responsibility of, 350
  conflict, environmental risk and, 350

D

Data, on value judgments, combining
  combined data, 252
  comparative risk, 247
  incremental value, 248
  nonlinerarity, 248–249
  priority rank, 252
  rank order, 250
  relative ranking, 252
  risk aversion, 250
De Chardin, Teilhard, global bioethics, 369
Declaration, global bioethics, 370
Declaration of Independence, global bioethics, 362
Deconstruction, global bioethics, 362
Defined, global bioethics, 369
Delaney Clause
  chemists, citizenship responsibility of, 350
De-legitimation, conflict, environmental risk and, 21
Demographers, conflict, environmental risk and, 18
Denial, overview, environmental risk decision making, 48
Deontology
  ethical basis, risk analysis, 261
  moral theory, sustainability, 270
Department of Energy, validation, risk assessment models, 292
Dimensional analysis, analytic framework, 170
Disciplinary blinders, conflict, environmental risk and, 17
Discipline, validation, risk assessment models, 286
Discounting, equity, intergenerational, 136
Distributive justice, moral values, 201, 204
Diversionary reframing, conflict, environmental risk and, 29
Dow Chemical Company, perceptions, in risk decision making, 79
Driving variables, analytic framework, 170

E

Earth Day
  chemists, citizenship responsibility
    of, 357
  global bioethics, 368
Ecocentric, ethics, human values and, 184
Economic criteria, analytic framework, 169
Economic development, analytic
    framework, 168
Economic impacts, analytic framework,
    173
Economic trade-offs
  chemists, citizenship responsibility
    of, 355
  conflict, environmental risk and, 355
Economists, analytic framework, 166
Ecosystems, analytic framework, 168
Efficiency
  conflict, environmental risk and, 25
  criterion, analytic framework, 159
Einstein, Albert, validation, risk
    assessment models, 303
Emotional, overview, environmental
    risk decision making, 45
Emotionalism
  chemists, citizenship responsibility
    of, 354
  conflict, environmental risk and, 354
Employment trends, conflict,
    environmental risk and, 14
Encoding, validation, risk assessment
    models, 303
Engineers, conflict, environmental risk
    and, 11
Environmental concerns
  chemists, citizenship responsibility
    of, 353
  conflict, environmental risk and, 353
Environmental management
  analytic framework, 161
  perceptions, in risk decision making,
    73
Environmental risk, analytic
    framework, 173
Environmental stewardship,
  perceptions, in risk decision
    making, 73

Epidemiology, conflict, environmental
    risk and, 21
Epistemology
  ethical basis, risk analysis, 256, 258
  rigor, ethical basis, risk analysis, 263
  validation, risk assessment models,
    280, 283, 288
Equal opportunity, equity, justice, 133
Equal rights, conflict, environmental
    risk and, 16
Equity
  across generations, analytic
    framework, 168
  catastrophic risk, 141
  heuristically, 136
  hypothetical risk, 141
  models, 136
  principles for, 139
  rights, of future generations, 133
  trustee, 140
  uncertainty, 134
  validation, risk assessment models, 281
  value judgments, pervasiveness of, 145
Ethical basis, risk analysis
  beneficence, 259
  consequence, 258
  deontological, 261
  epistemology, 256, 257, 258
  ethics, 256, 260
  intentions, 257, 259
  nonmalevolence, 259
  ontology, 255, 257, 258
  psychologistic, 258
  risk, 257, 259
  value, 257, 259, 260, 262
  variability, 258
  virtue, 256, 262, 264
Ethical citizen
  chemists, citizenship responsibility
    of, 358
  conflict, environmental risk and, 358
Ethical discourse, 340
  integration of, into risk assessment
    procedures, 115–130
Ethical obligation
  chemists, citizenship responsibility
    of, 351
  conflict, environmental risk and, 351

Ethical standing, ethical basis, risk
analysis, 261
Ethicists, moral values, 196
Ethics
anthropocentric view, 181
biotic justice, 182
chemists, citizenship responsibility
of, 350
conflict, environmental risk and, 26,
27, 350
ecocentric, 184
ethical basis, risk analysis, 256
federal research and development
establishment, 337, 345, 346
human values and, 180, 183, 191
land ethic, 184
moral values, 196, 197, 201
overview, environmental risk
decision making, 49
philosophy, 178
preferences, 177
public participation, 179
rationality, 180
risk management, 178
skeptic, 179
values, 178
Ethos
chemists, citizenship responsibility
of, 350
conflict, environmental risk and, 350
Expert judgment, perceptions, in risk
decision making, 74, 75
Exponential growth, of human race,
conflict, environmental risk and,
355
Exponential growth of human race,
chemists, citizenship
responsibility of, 355
Exxon Valdez, conflict, environmental
risk and, 19

F

Facts, conflict, environmental risk and,
13
Fairness
conflict, environmental risk and, 15
of outcomes, 341

validation, risk assessment models,
281
Fear
chemists, citizenship responsibility
of, 350
conflict, environmental risk and, 350
mongering, chemists, citizenship
responsibility of, 354
overview, environmental risk
decision making, 46
Federal research and development
establishment
Biotechnology Research
Subcommittee, 345
boundaries, 338
Committee on Environmental and
Natural Resources, 346
domain, contested, 338
ethics, 337, 344, 345, 346, 347
justice, 347
knowledge workers, 345
normative views, 339
outcomes, 347
priority, point of view, 339
risk management, 338, 347
science, technology studies, 337, 338
social science, 337, 344
trade-offs, 339, 347
Feelings, overview, environmental risk
decision making, 45
First-order criteria, analytic framework,
160
Fluorides, chemists, citizenship
responsibility of, 355
Food labeling policy, 341
Fortitude, validation, risk assessment
models, 281
Fossil fuels, chemists, citizenship
responsibility of, 355
Foundationalism, validation, risk
assessment models, 287
Frame, validation, risk assessment
models, 299
Freedom, overview, environmental risk
decision making, 46
Fruitfulness, moral values, 198
Future generations, moral values, 199,
204

G

Genetic engineering, moral values, 206
Geotherapy, global bioethics, 370
Global bioethics
  acceptable survival, 360, 361
  civil society, 360, 361
  intercultural competence, 360
  medical bioethics, 359
  National Academy of Science, 370
  postmodernism, 316, 362, 363
  reason, 361
  sanctity, 360
  survival, 359, 366
  university, 367
Global-scale physical systems, analytic
    framework, 171
Goals, analytic framework, of
    environmental protection, 161
Good, overview, environmental risk
    decision making, 50
Grated pluralism, analytic framework,
    156
Gulf, of mutual ignorance, conflict,
    environmental risk and, 350

H

Hallmarks, of civilization, conflict,
    environmental risk and, 357
Hanford
  validation, risk assessment models,
    301
  Washington, validation, risk
    assessment models, 294
Hatred
  chemists, citizenship responsibility
    of, 350
  conflict, environmental risk and,
    350
Havel, Vaclav, global bioethics, 361
Hazard index, data, on value
    judgments, combining, 250
Hazardous chemicals
  chemists, citizenship responsibility
    of, 350
  conflict, environmental risk and, 350
Health
  analytic framework, 171

  ecological flexibility, 10
  ecologist, 3
  judgment, 3, 8
  perception, 4
  professionalism, 8
  protection, 8
  redundancy, 10
  safety, 8
  social values, 5
  survive, 4
  trade-offs, 9
  value, 3, 8
Helleger, Andre, global bioethics,
    369
"Heuristics," conflict, environmental
    risk and, 33
Hierarchical, analytic framework, 169
High-level waste, validation, risk
    assessment models, 298
Holistic view, federal research and
    development establishment, 339
Honesty, validation, risk assessment
    models, 286
Hope, validation, risk assessment
    models, 281
Horse betting, conflict, environmental
    risk and, 18
Human health risks, analytic
    framework, 157
Human life, overview, environmental
    risk decision making, 43
Human rights, moral values, 203
Human valuation, analytic framework, 161
Human welfare, analytic framework, 166
Humility
  global bioethics, 360
  moral values, 209
  validation, risk assessment models,
    283
Hypothesis, validation, risk assessment
    models, 299, 304
Hypothetical risk, 141

I

Idaho National Laboratory, validation,
    risk assessment models, 302
Ignorance
  analytic framework, 159

chemists, citizenship responsibility
    of, 350
conflict, environmental risk and, 350
gulf of, conflict, environmental risk
    and, 350
validation, risk assessment models,
    292, 293
Inconsistency, validation, risk
    assessment models, 297
Incremental value, data, on value
    judgments, combining, 248
Individual welfare, analytic framework,
    156
Induction, validation, risk assessment
    models, 304
Industry, perceptions, in risk decision
    making, 74
Informing of public, of risks, 103–112
Institute for Environmental Studies,
    global bioethics, 368
Instrumental values, moral values, 206
Integration, analytic framework, 163
Integrity
    analytic framework, 156, 172
    overview, environmental risk
        decision making, 43
    validation, risk assessment models,
        286
Intellectual obligation
    ethical basis, risk analysis, 258
    validation, risk assessment models,
        288
Intentions, ethical basis, risk analysis,
    257
Interconectedness, federal research and
    development establishment, 339
Intercultural competence, global bioethics,
    360
Interdependence
    conflict, environmental risk and,
        24
    validation, risk assessment models,
        274
Intergenerational, analytic framework,
    163
Intergenerational justice, 140
    moral values, 204
International borders, global bioethics,
    359

Interpersonal chemistry, conflict,
    environmental risk and, 23
Intertemporal, analytic framework, 159
Intrinsic value
    analytic framework, 163
    moral values, 202
Intuition, global bioethics, 364
Ironizing radiation
    chemists, citizenship responsibility
        of, 356
    conflict, environmental risk and, 356
Irrationality, conflict, environmental
    risk and, 21
Irreversibility, analytic framework, 169

J

Judgment, health, ecological, value
    judgments in, 3, 8
Justice
    federal research and development
        establishment, 347
    overview, environmental risk
        decision making, 46
    principles of, equity, 133
    validation, risk assessment models,
        281, 285, 286

K

Kant, validation, risk assessment
    models, 280, 281
Keystone resource, analytic framework,
    158
Knowledge, federal research and
    development establishment, 338
    workers, 338

L

Land Grant College, global bioethics,
    365
Landscape-level ecosystems, analytic
    framework, 167
Legitimation, federal research and
    development establishment, 338
Leopold, Aldo
    analytic framework, 163
    global bioethics, 369

validation, risk assessment models,
    274
Level of concern, data, on value
    judgments, combining, 249
Lexicographic, analytic framework, 155
Life
    global bioethics, 360
    overview, environmental risk
        decision making, value of, 51
Lincoln, Abraham, global bioethics,
    365
Local Emergency Planning
    Committees, perceptions, in risk
        decision making, 77
Logicality, validation, risk assessment
    models, 286
Long-lasting hazardous wastes, analytic
    framework, 159
Long-range wisdom, global bioethics,
    366

Moral values
    arbitrariness, 200
    distributive justice, 201, 202, 204,
        206
    future generations, 199, 203, 204
    humility, 209
    neutral ground, 209
    objectivity, 198
    public ethics, 209
    risk assessment, 195, 196, 198, 210
    sustainability, 204, 205
    as utility values, 207
Motivation problem, equity, 134
Multiple generations, analytic
    framework, 162
Multiple scales, analytic framework,
    164
Multiscalar, analytic framework, 161
Multi-tiered models, analytic
    framework, 156

M

Magnitude
    analytic framework, 169
    of impacts, analytic framework, 165
Mainstream economist, analytic
    framework, 164, 165, 167
Materials policy, analytic framework,
    158, 159
Maxey Flats, validation, risk
    assessment models, 296
Mead, Margaret, global bioethics, 365
Media coverage, conflict,
    environmental risk and, 23
Medical bioethics, global bioethics, 359
Methodological rigor
    ethical basis, risk analysis, 262
    validation, risk assessment models,
        282
Millennialism, global bioethics, 371
Miracles
    chemists, citizenship responsibility
        of, 355
    conflict, environmental risk and, 355
Modeling, global bioethics, 370
Models, validation, risk assessment
    models, 302
Monetary values, 342

N

National Academy of Science, global
    bioethics, 370
National Science and Technology
    Council, 344
Natural cycles, analytic framework, 157
Neutral ground, moral values, 209
News, entertainment media, chemists,
    citizenship responsibility of,
    353
Nonanthropocentric, analytic
    framework, 156
Nonlinearity, data, on value judgments,
    combining, 248
Nonmalevolence
    ethical basis, risk analysis, 259
    validation, risk assessment models,
        284
Nonmoral values, moral values, 195
Nonscalar, analytic framework, 156,
    173
Nonviable values, moral values, 196
Nonwelfare, analytic framework,
    173
Normative views, federal research
    and development establishment,
    339

Norms
    federal research and development
        establishment, 347
    and structures, 341
Nuclear power
    chemists, citizenship responsibility
        of, 356
    conflict, environmental risk and, 356
Nuclear technologies, conflict,
        environmental risk and, 16
Nuclear waste
    conflict, environmental risk and, 11,
        23
    validation, risk assessment models,
        291

O

Obedience, validation, risk assessment
        models, 285
Objectivity
    moral values, 198
    validation, risk assessment models,
        291
Occupation, conflict, environmental
        risk and, risky, 18
Offshore oil, conflict, environmental
        risk and, 11, 28
Oil embargo, conflict, environmental
        risk and, 14
Ontology, ethical basis, risk analysis,
        255, 258
Opponents, conflict, environmental risk
        and, 27
Opportunities, analytic framework, 158
Organizational Amplification of Risks,
        conflict, environmental risk and,
        32
Overconfidence, conflict,
        environmental risk and, 17, 18,
        28
Overview, environmental risk decision
        making
    biased, 42
    cultures, 41
    denial, 48
    good, 50
    integrity, 43
    quality of life, 46

    right, 50
    theology, 52
    value judgments, 40
    worldview, 64

P

Pacific Northwest, analytic framework,
    168
Page, Talbot, analytic framework, 156,
    158
Paradigm, analytic framework, 164
Participation, public, 381
Perceptions
    Amoco Corporation, 80
    Chemical Manufacturers Association,
        75, 76, 79
    companies, 73, 74
    consumer products, 74, 75
    cost-effectiveness, 73, 74, 78, 81
    environmental stewardship, 73
    expert judgment, 74, 75
    Local Emergency Planning
        Committees, 77
    proactive risk assessment, 74
    public perceptions, 73, 75
    tiered approach, 75
    Toxics Release Inventory, 79
Philosophy, ethics, human values and, 178
Physical systems, analytic framework, 171
Planting programs, analytic framework,
    168
Platt, John R., global bioethics, 366
Plausibility, moral values, 209
Pluralism, analytic framework, 168
Pneumonia, conflict, environmental risk
    and, 17
Policy
    analysis, analytic framework, 155
    conflict, environmental risk and,
        29
    discourse, 340
    federal research and development
        establishment, 347
    questions, conflict, environmental
        risk and, 15
Political cartoonists
    chemists, citizenship responsibility
        of, 351

Political institutions, analytic
 framework, 162
Political involvement, chemists,
 citizenship responsibility of, 349
Political pressure, conflict,
 environmental risk and, 26
Pollution
 chemists, citizenship responsibility
  of, 355
 conflict, environmental risk and, 355
Positivism, moral values, 198
Possibility, ethical basis, risk analysis,
 260
Postmodernism, global bioethics, 316,
 363
Practical ethics, overview,
 environmental risk decision
 making, 66
Practicality, validation, risk assessment
 models, 282
Precautionary principle
 analytic framework, 170
 equity, 141
Prejudices, conflict, environmental risk
 and, 25
Priority
 data, on value judgments, combining,
  250
 federal research and development
  establishment, point of view,
  339
 rank, data, on value judgments,
  combining, 252
 setting, data, on value judgments,
  combining, 251
Proactive risk assessment, perceptions,
 in risk decision making, 74
Probability, validation, risk assessment
 models, 296
Procter & Gamble Company,
 perceptions, in risk decision
 making, 74
Programs, validation, risk assessment
 models, 303
Progress, global bioethics, 366
Protection, health, ecological, value
 judgments in, 8
Prudence, validation, risk assessment
 models, 281, 284

Public, informing, of risks, 103–112
Public education, conflict,
 environmental risk and, 20
Public ethics, moral values, 209
Public health
 officers, global bioethics, 359
 role of, 95, 96, 100
  community, 95
   values, 93, 95
Public relations, conflict, environmental
 risk and, 28, 29
Public service, conflict, environmental
 risk and, 350
Public support, conflict, environmental
 risk and, 23
Public trust, conflict, environmental
 risk and, 30

Q

Quality of life, overview,
 environmental risk decision
 making, 46
Quantitative issues, 151–153

R

Radiation hazard, validation, risk
 assessment models, 296
Radioactive contamination, validation,
 risk assessment models, 302
Radioactive liquids, validation, risk
 assessment models, 299
Radon
 chemists, citizenship responsibility
  of, 350
 conflict, environmental risk and, 350
Rank order, data, on value judgments,
 combining, 250
Rationality, ethical basis, risk analysis,
 256
Rawlsian contractual obligation,
 analytic framework, 159
Reason, global bioethics, 361
Recreancy, conflict, environmental risk
 and, 29
Reduction, analytic framework, 171
Redundancy, health, ecological, value
 judgments in, 10

Reintroduction of predators, analytic
framework, 164
Relationships, sustainability, 271
Relative ranking, data, on value
judgments, combining, 252
Relative risk, data, on value judgments,
combining, 252
Relevance strategies, validation, risk
assessment models, 287
Reliability
analysis, validation, risk assessment
models, 197
validation, risk assessment models,
291, 305
Religion, global bioethics, 361
Repository, validation, risk assessment
models, 292, 298
Repression
chemists, citizenship responsibility
of, 350
Research, federal research and
development establishment,
339
Resolution, analytic framework, 169
Responsibility
conflict, environmental risk and,
25, 27, 29
global bioethics, 360
overview, environmental risk
decision making, 48
Revelation, global bioethics, 361
Reversibility, analytic framework, 165,
170
Right
of future generations, equity, 133
overview, environmental risk
decision making, 50
Rights-based theory, sustainability, 271
Risk
acceptance, 340
analysis
analytic framework, 156, 171
validation, risk assessment models,
280
analytic framework, 155, 157, 173
assessment
analytic framework, 155, 159, 173
moral values, 195
value laden, 377

communication
conflict, environmental risk and,
20, 21, 28
overview, environmental risk
decision making, 40
decision models, overview,
environmental risk decision
making, 55
decision space, analytic framework,
167
decision square, analytic framework,
155, 165, 167
ethical basis, risk analysis, 257
judgment, 381
management, 341
analytic framework, 173
ethics, human values and, 178
federal research and development
establishment, 338
moral values, 195, 210
overview, environmental risk
decision making, 52
model, analytic framework, 156
moral values, 198
policy, 340
validation, risk assessment models,
279
Risk-based, perceptions, in risk
decision making, 81
Risk-based priority setting
overview, environmental risk
decision making, 73
perceptions, in risk decision making,
78, 79
Risk-decision criteria, analytic
framework, 156
Risk-management team, analytic
framework, 171
Rolling present, equity, 145

S

Safety, health, ecological, value
judgments in, 8
Sale, analytic framework, 166
Sanctity, global bioethics, 360
Savannah River, validation, risk
assessment models, 302
Scalar aspects, analytic framework, 161

Scale, analytic framework, 161
Scale-sensitive, 171
Scapegoats, conflict, environmental risk
    and, 21
Science
    conflict, environmental risk and, 12, 19
        technology and, 26
    federal research and development
        establishment, 338
    technology studies, 338
Scientific truths, overview,
    environmental risk decision
    making, 42
Scientifically illiterate public
    chemists, citizenship responsibility
        of, 350
    conflict, environmental risk and,
        350
Scientist, conflict, environmental risk
    and, 12, 15, 16, 24, 25
Second-order criterion, analytic
    framework, 160
Secular program, global bioethics, 369
Simulation, validation, risk assessment
    models, 297
Single-tiered, analytic framework, 156
Situational rationality, validation, risk
    assessment models, 286
Skeptic, ethics, human values and, 179
Social construction
    conflict, environmental risk and, 13
    of reality, 378
    of risk, 343
Social learning, analytic framework,
    169
Social science, 340, 343
    conflict, environmental risk and, 28
    federal research and development
        establishment, 337, 345, 347
Social values
    analytic framework, 169, 170, 171
    health, ecological, value judgments
        in, 5
Spatial scale, analytic framework, 165
Statistics, conflict, environmental risk
    and, 13
Stewardship
    equity, 140
    ethic, 311–332

overview, environmental risk
    decision making, 46
Strategic research priorities, 344
Subjectivity, impact on objectivity
    belief-laden, 89
    belief-neutral, 89
    bias, 90
    certainty, 90
    consensus, 91
    cost-benefit analysis, 89
    management, 85
    objectivity, 85
    professional scientific judgment, 86
    risk assessment-risk management
        paradigm, 85
Substitutability, analytic framework,
    160, 166, 170
Substitutes, analytic framework, 166
Superstition, chemists, citizenship
    responsibility of, 350
Survival
    global bioethics, 359
    health, ecological, value judgments
        in, 4
Sustainability
    analytic framework, 168
    deontological moral theory, 270
    ecology, 267, 271
    interdependencies, 271
    moral values, 205
    preference utilitarianism, 273
    relationships, 271
    rights-based theory, 271
    sustainability, 268
    utilitarianism, 272
Systems theory, analytic framework, 161

T

Technical specialists, conflict,
    environmental risk and, 19
Technological controversies, conflict,
    environmental risk and, 11, 12
Temperance, validation, risk assessment
    models, 281
Temporal scale, analytic framework,
    156
Testability, validation, risk assessment
    models, 291

Texas, conflict, environmental risk and, 28

Theology, overview, environmental risk decision making, 52

Thresholds, analytic framework, 166

Tiered approach, perceptions, in risk decision making, 75

Time magazine, global bioethics, 368

Times Beach, conflict, environmental risk and, 19

Titanic, global bioethics, 364

Top priority, federal research and development establishment, 347

Toxics Release Inventory, perceptions, in risk decision making, 79

Trade-offs
equity, 136, 139
federal research and development establishment, 339, 347
health, ecological, value judgments in, 9

True environmentalist, chemists, citizenship responsibility of, 356

Trust
conflict, environmental risk and, 24
federal research and development establishment, 347

Trustee, equity, 140

Truth, global bioethics, 363

Truthfulness, validation, risk assessment models, 283

TV script writers
chemists, citizenship responsibility of, 351
conflict, environmental risk and, 351

Two-tiered system, analytic framework, 170

U

Uncertainty
conflict, environmental risk and, 17
ethical basis, risk analysis, 256
moral values, 208
overview, environmental risk decision making, 65
validation, risk assessment models, 293

Unethical values, moral values, 201

United Press International, conflict, environmental risk and, 33

University, global bioethics, 367

Unsustainability, moral values, 204

U.S. Atomic Energy Commission, conflict, environmental risk and, 28

U.S. Environmental Protection Agency, validation, risk assessment models, 301

Utilitarianism
analytic framework, 137
sustainability, 272
validation, risk assessment models, 280

Utopias, global bioethics, 365

V

Validation
affirming consequent, 303, 304, 305
burden of proof, 293, 301
contextual values, 286
epistemology, 280, 282, 288
Idaho National Laboratory, 302
Kant, 280, 281
ontological reflection, 282
rationality, 279, 281, 282, 283, 285
simulation, 297
U.S. Environmental Protection Agency (EPA), 301

Value. See also Moral values
analytic framework, 161
conflict, environmental risk and, 16, 25, 267
ethical basis, risk analysis, 257, 262
ethics, human values and, 178
expert (scientific), 380
health, ecological, value judgments in, 3, 8
moral values, 198
overview, environmental risk decision making, 40, 41
public, 380
type of, 378
validation, risk assessment models, 279

Value judgments
moral values, 198

overview, environmental risk
    decision making, 40
Verification, validation, risk assessment
    models, 303, 305
Vigilance, conflict, environmental risk
    and, 26
Virtue
    ethical basis, risk analysis, 262
    global bioethics, 360
    validation, risk assessment models,
        279, 280, 282, 284

W

Waste repositories, 305
Water supplies, validation, risk
    assessment models, 295
Weaknesses, conflict, environmental
    risk and, 25
Weather forecasters, conflict,
    environmental risk and, 18

Welfare
    economics, analytic framework, 157
    model, analytic framework, 160
    states of future people, analytic
        framework, 158
Worldview, overview, environmental
    risk decision making, 64

Y

Yucca Mountain, validation, risk
    assessment models, 291, 293,
    295, 296, 297, 298, 299, 300,
    301, 302, 303, 305

Z

Zero population growth
    chemists, citizenship responsibility
        of, 355
    conflict, environmental risk and, 355